Springer Tracts in Natural Philosophy

Volume 28

Daniel D. Joseph

Stability of Fluid Motions II

With 39 Figures

Springer-Verlag
Berlin Heidelberg New York 1976

DANIEL D. JOSEPH
University of Minnesota, Department of Aerospace Engineering and Mechanics, Minneapolis, Minnesota 55455/U.S.A.

AMS Subject Classification (1970): 34 Dxx, 35 A 20, 35 B 10, 35 B 15, 35 B 30, 35 B 35, 35 Cxx, 35 P 15, 35 P 20, 35 Q 10, 49 F 25, 49 G 05, 49 Gxx, 49 H 05, 73 Bxx, 76 Axx, 76 Dxx, 76 S 05, 85 A 30

ISBN 3-540-07516-X Springer-Verlag Berlin Heidelberg New York
ISBN 0-387-07516-X Springer-Verlag New York Heidelberg Berlin

Library of Congress Cataloging in Publication Data. Joseph, Daniel D. Stability of fluid motions. (Springer tracts in natural philosophy; v. 27–28) Includes bibliographical references and indexes. 1. Fluid dynamics. 2. Stability. I. Title. QA911.J67. 532′.053. 76-4887.

Typesetting and printing: Zechnersche Buchdruckerei, Speyer. Bookbinding: Konrad Triltsch, Würzburg.

Preface

The study of stability aims at understanding the abrupt changes which are observed in fluid motions as the external parameters are varied. It is a demanding study, far from full grown, whose most interesting conclusions are recent. I have written a detailed account of those parts of the recent theory which I regard as established.

Acknowledgements

I started writing this book in 1967 at the invitation of Clifford Truesdell. It was to be a short work on the energy theory of stability and if I had stuck to that I would have finished the writing many years ago. The theory of stability has developed so rapidly since 1967 that the book I might then have written would now have a much too limited scope. I am grateful to Truesdell, not so much for the invitation to spend endless hours of writing and erasing, but for the generous way he has supported my efforts and encouraged me to higher standards of good work. I have tried to follow Truesdell's advice to write this work in a clear and uncomplicated style. This is not easy advice for a former sociologist to follow; if I have failed it is not due to a lack of urging by him or trying by me.

My research during the years 1969–1970 was supported in part by a grant from the Guggenheim foundation to study in London. I enjoyed the year in London and I am grateful to Trevor Stuart, Derek Moore and my other friends at Imperial College for their warm hospitality. I welcome the opportunity to acknowledge the unselfish assistance of the world's best Maths librarian, Miss J. Pindelska of Imperial College. In the summer and fall of 1973 a grant from the British Science Research Council made it possible for me to work in England again, this time with L. A. Peletier, D. Edmunds and other mathematicians in Sussex. In the summer of 1974 I taught a short course on stability at L'École d'Eté in Bréau Sans Nappe, France. The French students were able and well-trained engineers with graduate degrees and a good background in mathematics. It was a stimulating group and some of the better results which are set down in § 15, Chapters XI and XIV and in the Addendum to Chapter X stem directly from questions raised at the summer school.

My research in stability theory has been funded from the beginning by the fluid mechanics branch of the National Science Foundation. The contribution which their funding has made to this book can scarcely be overestimated. My work was also made easier by the good conditions which prevail in the mechanics department of the University of Minnesota and by the presence there of good friends and fine colleagues, by an understanding chairman, P. Sethna and by the two cheerful and efficient secretaries, Jean Jindra and Susan Peterson, who typed the various drafts of this manuscript.

Many persons have contributed to this book in different ways. Fritz Busse, Ta Shen Chen, Stephen Davis, Daniel Jankowski, Klaus Kirchgässner, Simon Rosenblat, Robert Sani, William Warner and Hans Weinberger read and criticized various parts of the text. The contributions of W. Hung to Chapters V and VI, of Bruce Munson to Chapter VII, of Ved Gupta to Chapter XII and Appendix C, and of E. Dussan V to Chapter XIV deserve special mention. Chapter XI is based on joint work with D. A. Nield. Nield also worked through the whole book in the final stages of preparation for printing. I do not have words sufficient to convey the depth of gratitude I feel for Nield's careful and efficient assistance with the onerous task of reading proofs.

As a graduate student I was strongly influenced by James Serrin's beautiful article on fluid mechanics in the Handbuch der Physik. I later had the good fortune to find a position at the University of Minnesota where, through contact with Serrin, I became interested in the energy theory of stability. I am really indebted to Serrin, for his support in the early days, and for the continued inspiration which I derive from seeing good mathematics applied to problems at the foundations of mechanics.

Finally I want to acknowledge all that I have learned from my colleagues, F. H. Busse, T. S. Chen, E. Dussan V, V. Gupta, W. Hung, B. Munson, D. A. Nield and D. Sattinger, with whom I have collaborated in stability studies. My view of stability theory as a branch of mathematics has been particularly influenced by David Sattinger, and as a branch of physics, by Fritz Busse.

"Like most philosophers, I am much indebted to conversations I have had with others over the years... I will not indulge in the conventional fatuity of remarking that they are not responsible for the errors this book may contain. Obviously, only I can be *held* responsible for these: but, if I could recognize the errors, I should have removed them, and, since I cannot, I am not in a position to know whether any of them can be traced back to the opinions of those who have influenced me."[1]

My wife, Ellen, read the entire manuscript with me and together in our effort to achieve good writing we studied Truesdell's letters and Gowers' splendid little book[2] "The Complete Plain Words". Ellen has agreed to take responsibility for lack of clarity and precision in the writing and for all mathematical and conceptual mistakes in the presentation. It is a pleasure for me to dedicate my first book to this lovely lady.

---

[1] Michael Dummett "Frege: Philosophy of Language" Duckworth: London, 1973.

[2] Sir Ernest Gowers "The Complete Plain Words" Penguin Books Ltd.: Middlesex, England, 1973.

Plan of the Work

The plan of this work is given in the table of contents. The book is divided into two parts. Part I gives the general theory of stability, instability, bifurcation and some discussion of the problem of repeated bifurcation and turbulence. The general theory is developed in Chapters I and II and is applied to flows between concentric cylinders (Chapters III—VI) and between concentric spheres (Chapter VII). Part I is self-contained. The six chapters (VIII—XIV) of Part II take up special topics of general interest. The topics are selected to develop extensions of the general theory, to introduce new techniques of analysis, to extend the scope of application of the theory and to demonstrate how stability theory is essential in understanding the mechanics of the motion of fluids. The purpose of each chapter in both parts is set out in the introduction to that chapter. Attributions are given in the text, where a new result first appears, or in bibliographical notes at the end of each chapter. Many results are given here for the first time. Other results, not known to most readers, are not new but are not well known or are not available in a form which can be understood by interested persons with training only in classical analysis. All results new and old, have been reworked to fit the plan of this work.

Remarks for Students

I expect readers of this book to know calculus and parts of the theory of differential equations. If you know more, so much the better. Students at the required level of preparation will greatly improve their knowledge of useful techniques of analysis which are required in the study of different aspects of the theory of hydrodynamic stability. Among these, bifurcation theory and variational methods for linear and nonlinear eigenvalue problems defined on solenoidal fields have received an especially thorough treatment. My explanation of these topics has been guided by my desire to make the theory accessible to a wide audience of potentially interested persons. The appendices are to help beginners who wish to learn the details of the mathematical procedures considered by me to be important for stability studies but inappropriate for the main line of development in the text. I have formulated about 230 exercises to help students learn and to elaborate and extend results which are developed in the text.

Introductory Remarks for Part II

This is the second of a two part work on the stability of fluid motions. The first four of the six chapters of Part II are about stability problems in convecting fluids in which motion is induced by density differences associated with gradients of temperature and chemical composition. The equations governing such motion, motionless solutions of these equations, and criteria for their stability, are given in Chapter VIII. In Chapter IX, we consider the stability of some motionless solution in a heterogeneous fluid like salt water. The large difference in the values of the diffusion constants for heat and salt allow for a new mechanism of instability and the equations may be studied by an interesting type of generalized energy analysis. A similar generalization applies to magnetohydrodynamic flows. Chapters X, XI and XII treat the problem of convection in porous materials. In Chapter X, the problem is posed in an impermeable container with insulated side walls and admits an elementary separation of variables which leads to an equally elementary analysis of bifurcation and stability at eigenvalues of higher multiplicity. The results in Chapter X are all new. Many of the results in Chapter XI, which treats the problem of wave number selection through stability, are new results based on joint work with D.A. Nield. Chapter XII gives a fairly complete discussion of the variational theory of turbulence applied to convection in porous materials. Chapter XIII gives new methods of analysis for studying the flow and stability of flow of viscoelastic fluids. In Chapter XIV we compare the static theory of interfacial stability which involves minimizing the free energy with the new nonlinear dynamic theory of E. Dussan V.

Though occasional reference is made to results given in Part I, Part II is essentially self-contained and may be studied independently.

Table of Contents

Contents of Part I

Chapter VIII

The Oberbeck-Boussinesq Equations. The Stability of Constant Gradient Solutions of the Oberbeck-Boussinesq Equations

The presence of density gradients in a fluid means that gravitational potential energy can be converted into motion through the action of buoyant forces. Density differences can be induced by heating the fluid and by forcing concentration differences in mixtures like salt water. In the Oberbeck-Boussinesq (OB) equations, the fluid is assumed to have a uniform density; density differences are recognized only in those terms which drive the motion. The OB equations are given in § 54. In § 55 we discuss boundary conditions of various kinds and complete the statement of the initial-boundary-value problem, IBVP, for the OB equations.

In §§ 56—58, we formulate an energy stability theory for basic solutions of the IBVP of the OB equations. This problem is very much like the one considered in Chapter I but now there are more equations and the boundary conditions are more complicated. In this context the energy functional is typically a linear combination of the kinetic energy and other energy-like integrals. To study the stability problem, we first fix the coupling constants in the linear combination and find a critical stability number for global stability. This critical value of energy theory is again defined by a problem of variational calculus. Each choice of coupling constants gives a different "energy" and leads to a different critical stability number. The coupling constants which lead to the largest stability number are called "optimal".

In the usual circumstance, temperature differences which are imposed at the boundary of a region of space occupied by a fluid will induce convection currents. The importance of such currents can scarcely be overestimated; we need only to recall that the circulation of the earth's atmosphere could not be explained without reference to convective motions induced by solar heating. Motionless solutions of the OB equations which support heat conduction without convection are also possible only under certain stringent conditions. The motionless solutions are easiest to treat theoretically and they allow one to consider the mechanisms which lead to convective instability without complications due to motion.

Motionless solutions of the OB equations are given in § 59. In § 60 we give a physical description of mechanisms which induce instability. Motionless states may become top heavy and lose their stability to motion if they are heated too much from below[1]. Necessary and sufficient conditions for the stability of certain

[1] Motion may also begin when a fluid layer which is open to the atmosphere loses its ability to resist tangential tractions due to surface-tension gradients associated with temperature fluctuations.

motionless solutions of a heterogeneous fluid are given in § 61. (Some interesting problems in heterogeneous fluids are considered in Chapter IX.) A detailed study of the stability of a motionless homogeneous fluid heated from below (the Bénard problem) is given in § 62.

The basic mechanisms which produce instability of motionless states are frequently dominant even when there is a basic motion. Examples of basic motions which also have constant temperature gradients are considered in §§ 63 and 64.

§ 54. The Oberbeck-Boussinesq Equations for the Basic Flow

If the boundary of a container of water which is otherwise motionless is heated, say, on one side, it will be noticed that motion ensues. That motion is driven by buoyant forces which exist when one fluid element in a gravity field is less dense (hotter) than its neighbour. The motion can even be turbulent under the influence of such heating. Large motions can be driven by density variations of the order of one per cent.

There are very important situations in which the density variations are produced by temperature and concentration differences and not by pressure differences. For example, it is easier to change the density of water in the container by heating than by "squeezing". At 25 °C (77 °F) and 1.02 atmospheres of pressure, we have for water

$$\rho=\rho_0\left[1+\left(\frac{1}{\rho}\frac{\partial\rho}{\partial T}\right)_0(T-T_0)+\left(\frac{1}{\rho}\frac{\partial\rho}{\partial P}\right)_0(P-P_0)\right],$$

where

$$\alpha_T\equiv-\left(\frac{1}{\rho}\frac{\partial\rho}{\partial T}\right)_0=2.64\times10^{-4}\,{}^\circ\mathrm{C}^{-1},$$

and

$$\left(\frac{1}{\rho}\frac{\partial\rho}{\partial P}\right)_0=5.11\times10^{-5}\text{ bars.}^{-1}\,.$$

To produce the same change of density as a temperature difference of 1 °C, one needs to change the pressure by roughly five atmospheres. It follows that even vigorous motions of the water will not introduce important buoyant forces other than those associated with the temperature variations. The temperature-induced density changes in such motions can be very small, but they are motive power for the convective part of the motion.

Gases are very compressible and if we continuously deformed the container walls we could cause very large changes of density with pressure. But in isochoric motions, say plane Couette flow or no flow at all, the pressure changes do not enter. In gases (Jeffreys (1930), Spiegel and Veronis (1960)) one can induce *essentially* isochoric motions in which small changes of density, induced by temperature but not pressure variations, drive convection. The low Mach number motions of the earth's atmosphere are essentially isochoric.

To account for convective motions of an *essentially* isochoric kind, one can avoid the intractable problem associated with the exact, compressible Navier-Stokes equations by taking advantage of simplifying features which characterize the motion. These are:

(1) The motion is as if incompressible except that density changes are not ignored in the body-force terms of the momentum equations;

(2) The density changes are induced by changes of temperature (T) and concentration (C) but not by pressure (P);

(3) The velocity gradients are sufficiently small so that the effect on the temperature of conversion of work to heat can be ignored.

The equations which result from invoking these simplifications are to be called the Oberbeck-Boussinesq equations.

We shall call the OB equations "simplified" if, in addition to assumptions (1), (2) and (3), it is also assumed that:

(4) The dynamic viscosity μ, the thermal conductivity k and the specific heats are parametric constants;

(5) The equation of state $\rho(T, C)$ is linearized (see (54.1)).

There are many situations in which all five assumptions strongly characterize the flow. For instance, they are satisfied in the (Bénard) problem of the breakup of the heat conduction regime in a thin layer of air or water heated from below.

More typically, assumptions (4) and (5) do not hold. For example, a nonlinear equation of state $\rho(T)$ cannot be avoided when treating convection near the critical point (4 °C) at which the density of water has a relative maximum (Veronis (1963)). In very large scale systems (typical in geo-astrophysical applications) the variations of material properties cannot be neglected.

For flows in which the Mach number is sensibly different from zero, the solenoidality of the velocity field and the independence of density on pressure are lost. Then the OB equations would not be relevant to the physical problem.

When all of the simplifying features are present, the Navier-Stokes equations for compressible, heat conducting and diffusive flow of a viscous, nonhomogeneous fluid can be approximated by the following set of (OB) equations:

(i) The equation of state of the OB fluid is

$$\rho=\rho_0(1-\alpha_T(T-T_0)+\alpha_C(C-C_0)) \tag{54.1}$$

where ρ_0, T_0 and C_0 are reference density, temperature and concentration, respectively, and α_T and α_C are chosen for convenience in matching the OB equation of state (54.1) with the equation of state $\rho=\rho(T, C, P)$.

(ii) The momentum balance equation is

$$\rho_0\left(\frac{\partial \mathbf{U}}{\partial t}+(\mathbf{U}\cdot\nabla)\mathbf{U}\right)=-\nabla P+\rho_0\{1-\alpha_T(T-T_0)+\alpha_C(C-C_0)\}\mathbf{g}+\nabla\cdot\mathbf{S} \tag{54.2}$$

where $\mathbf{T}=-P\mathbf{1}+\mathbf{S}$ is the stress, $\mathbf{S}=2\mu\mathbf{D}[\mathbf{U}]$ is the extra stress, $\mathbf{U}$ is the solenoidal velocity and $\mathbf{g}$ is a body-force field (typically gravity).

(iii) The evolution equation for the temperature is

$$\frac{\partial T}{\partial t} + \mathbf{U} \cdot \nabla T = \kappa_T \nabla^2 T + Q_T(\mathbf{x}, t) \tag{54.3}$$

where $\kappa_T = k_T/\rho_0 C_P$ is the thermal diffusivity, k_T is the thermal conductivity, C_P is the specific heat at constant pressure and Q_T is a prescribed heat source field.

(iv) The evolution equation for the solute concentration is

$$\frac{\partial C}{\partial t} + \mathbf{U} \cdot \nabla C = \kappa_C \nabla^2 C + Q_C(\mathbf{x}, t) \tag{54.4}$$

where κ_C is the solute diffusivity and $Q_C(\mathbf{x}, t)$ is a prescribed field specifying the distribution of solute sources.

By OB equations we have in mind the simplified Eqs. (54.1, 2, 3 and 4) which are generally attributed to Boussinesq (1903) who derives them by decree from the list of assertions (1—5).

"Can be approximated" means that solutions of the exact compressible Navier-Stokes equations are continuous in some parameters, say $(\varepsilon_1, \varepsilon_2)$ in the neighbourhood of (0,0), and the limiting solutions $\mathbf{U}_0, C_0, T_0, P_0$ (where, for example, $T(\varepsilon_1, \varepsilon_2) \to T(0,0) \equiv T_0$) satisfy the limiting equations (54.2, 3, 4) and $\operatorname{div} \mathbf{U} = 0$. Two steps are required in such a demonstration:

(1) the parameters must be identified and some formal perturbation scheme defined, (2) it must be shown that the perturbation can be carried out and converges, in some sense, to the nonlinear solution ($T(\varepsilon_1, \varepsilon_2)$, etc.).

Step 1 was first carried out by Oberbeck (1879), and more recently and completely by Mihaljan (1962).

The approximations generally attributed to Boussinesq were actually of earlier origin and were used by Oberbeck (1891) in meteorological studies of the Hadley regime. But Oberbeck's first use of these equations, in 1879, is more substantial than his later 1891 application, and the equations which he sets out in the earlier study are just exactly the ones generally attributed to Boussinesq. In some respects, Oberbeck's 1879 treatment of these equations is superior to Boussinesq's; for example, Boussinesq obtains the simplified equations as a consequence of a list of assertions, but in Oberbeck's work the equations arise as the lowest-order terms of a power series development in the expansion coefficient α. And included in Oberbeck's fundamental paper is an application of both the convection equations and the series ordering by which they are derived to the problem of convection induced by differential heating of stationary concentric spheres.

It cannot be said that the Oberbeck study went unnoticed, for in 1881 L. Lorenz published a celebrated study of free convection along a heated flat plate in which he used Oberbeck's equations to derive a formula relating, for the first time, the Nusselt number to given flow data. Jakob [1949, p. 443] calls this formula "... a triumph of classical theory, having revealed for the first time the complex nature of the coefficient of heat transfer by free convection ... in a form which has been proved valid with good approximation through more than half a century". The "Rayleigh" number also appears for the first time in the Lorenz study; it is called α there and $\mathscr{R}^2$ here.

The attribution of the equations to Boussinesq is probably due to Rayleigh (1916) who may have been unaware of the earlier papers, though it cannot be said that these earlier papers are in any sense obscure. Such was the prestige of Rayleigh, especially in England, that this attribution stuck, despite the fact that the engineering heat transfer literature makes use of the "Boussinesq" equations as if they were created at $t = 0$, without reference to Boussinesq but with a properly humble deference to the important results of Lorenz.

The OB equations have not yet been completely justified. Convergence of the perturbation scheme in which the OB equations appear at lowest order has yet to be established. However, Fife (1970) has shown convergence for steady solutions of a not necessarily Newtonian fluid with density $\rho = \rho_0(1 - \alpha(T - T_0)) + \rho_*(T)$ where $\rho_*(T)$ is nonlinear and "small enough". Convergence of iterates is proved in Fife's paper but only for sufficiently small, but not zero, norms for the nonlinear motion.

These results all carry over to the fluid mixture if (54.1) and (54.4) are postulated. The derivation of the OB equations from the exact equations for a two-component fluid has not yet been taken quite up to step (1), though some partial justification is available in the paper of Vertgeim (1955).

It is perhaps pertinent to remark that there is no special reason, besides our lack of proofs, to doubt the validity of the nonlinear OB equations in some limit of small parameters. It would, however, be good to have a demonstration of the convergence of *Oberbeck-Boussinesq* solutions with large norms to solutions of the compressible Navier-Stokes equations in some limit of small parameters.

Exercise 54.1 (Mihaljan, 1962): (a) The density of a homogeneous fluid depends on the pressure P and the temperature T. Show that the enthalpy h, internal energy e and the specific heat at constant pressure of all special fluids whose density $\rho(T)$ depends on T alone are given by $h=\hbar(T)+\frac{P}{\rho}\left[1+\frac{T}{\rho}\rho'\right]$, $e=h-\frac{P}{\rho}$, $C_P=C(T)-PT(1/\rho)''$ where primes denote differentiation and $\hbar(T)$ and $C(T)$ are functions of T alone. *Hint:* Show that

$$\left(\frac{\partial \mathscr{S}}{\partial P}\right)_T=-\left(\frac{\partial 1/\rho}{\partial T}\right)_P, \quad \left(\frac{\partial C_P}{\partial p}\right)_T=-T\left(\frac{\partial^2 1/\rho}{\partial T^2}\right)_P \quad \text{and}$$

$$\left(\frac{\partial h}{\partial P}\right)_T=\frac{1}{\rho}+\frac{T}{\rho^2}\left(\frac{\partial \rho}{\partial T}\right)_P$$

where $\mathscr{S}$ is the specific entropy.

(b) Show that all special fluids whose density depends on T alone obey the equation

$$\rho C_P\frac{dT}{dt}+T\frac{\rho'}{\rho}\frac{dP}{dt}=\nabla\cdot(k\nabla T)+2\mathbf{D}:\mathbf{S} \tag{54.5}$$

where $2\mathbf{D}:\mathbf{S}$ is the stress power.

Construct a rough argument based on the presumed smallness of the velocity gradients and the density variation ρ'/ρ to deduce that

$$\rho_0 C(T)\frac{dT}{dt}=\nabla\cdot(k\nabla T) \tag{54.6}$$

where $\rho_0=\rho(T_0)$. If $k(T)$ and $C(T)$ are regarded as constants k_0, C_0 equal to their values at the reference temperature T_0, (54.6) may be written in classic form as $dT/dt=\kappa_0\nabla^2 T$ where $\kappa_0=k_0/\rho_0 C_0$.

Remark: Mihaljan introduces four basic dimensionless parameters

$$[\varepsilon_1=\alpha\Delta T,\ \varepsilon_2=\kappa_0^2/C_0 l^2\Delta T,\ \mathscr{P},\ \mathscr{P}\mathscr{R}^2]$$

where ΔT is the temperature difference across a horizontal layer of fluid of height l, $\mathscr{P}$ is the Prandtl number, $\mathscr{R}^2$ is the Rayleigh number and all material parameters are evaluated at the reference temperature T_0. He shows that the simplified OB equations arise in the limit $\varepsilon_1\to 0$, $\varepsilon_2\to 0$. He estimates that in a layer of water with $l=1$ cm and $\Delta T=1\,^\circ$C

$$\varepsilon_1\sim 10^{-4} \quad \text{and} \quad \varepsilon_2\sim 10^{-11}.$$

Mihaljan notes that the principal difference between (54.5) and (54.6) is proportional to ε_2. The difference between the momentum and continuity equations for a compressible fluid and the OB approximation to these equations is proportional to ε_1. (For a similar analysis which makes use of different scales see Cordon and Velarde, 1975).

§ 55. Boundary Conditions

We wish here to consider the situation which holds on an interface S separating the region occupied by fluid from the exterior region (see Fig. 55.1). Let $\mathbf{x} \in S$ designate a point of the boundary surface. Surface and volume symbols in the region occupied by fluid are denoted with a subscript f, surface and volume symbols and field variables on the other side of S are in a region exterior to the fluid and are designated with a subscript e. Field variables in the fluid are written without subscripts.

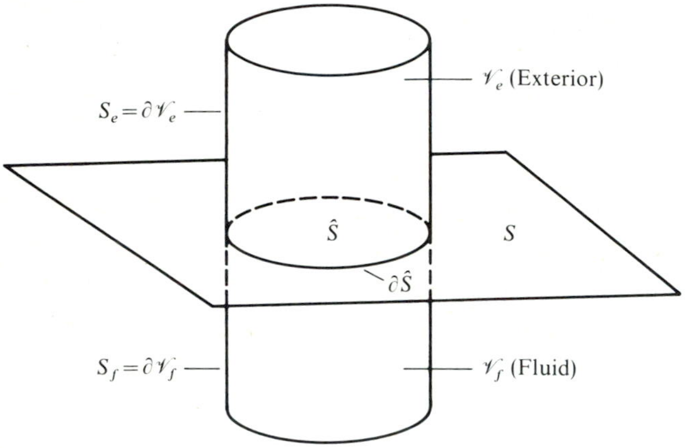

Fig. 55.1: The pillbox limit at a material surface S. The surface $\hat{S} = \mathscr{V} \cap S$ is fixed while $\mathscr{V} = \mathscr{V}_e \cup \mathscr{V}_f \to 0$

The boundary of the fluid is a material surface across which the temperature, velocity, the normal component of the heat flux vector and the mechanical stress are continuous. These coupling conditions are approximated with prescribed conditions.

In reality, boundary conditions are *coupling* conditions between adjacent physical systems. The usual procedure of mathematical physics is to approximate *coupling* conditions with *a priori* conditions which allow one to avoid problems involving simultaneous solution of coupled sets of field equations. This program is carried out below.

(a) Temperature Conditions

Temperature coupling conditions can be approximated by prescribed boundary conditions. We assume the continuity of temperature across the interface separating two material bodies. The continuity of the normal components of the heat flux vector $\mathbf{q}_T$ then follows from an energy balance. For the fluid and exterior material the balance of energy takes form, respectively, as

$$\int_{\mathscr{V}_f} \circ + \int_{\partial \mathscr{V}_f} \circ \mathbf{q}_T \cdot \mathbf{n} = 0\,,$$

$$\int_{\mathscr{V}_e} \circ + \int_{\partial \mathscr{V}_e} \circ \mathbf{q}_T \cdot \mathbf{n} = 0\,.$$

The volume integrals have continuous integrands and these integrals vanish as $\mathscr{V}=\mathscr{V}_e\cup\mathscr{V}_f\to 0$. Subtracting, and passing to the limit $\mathscr{V}\to 0$ with $\hat{S}=\mathscr{V}\cap S$ fixed, we find that

$$\int_{\hat{S}}[\mathbf{q}_T]\cdot\mathbf{n}=0 \tag{55.1}$$

where $\mathbf{n}$ is the normal to the arbitrary subelement $\hat{S}$ of S which points away from the fluid. Since $\hat{S}$ is arbitrary (55.1) implies that

$$\mathbf{n}\cdot[\mathbf{q}_T]=\mathbf{n}\cdot\mathbf{q}_T-\mathbf{n}\cdot(\mathbf{q}_T)_e=0\,.$$

Already in the derivation of the field equations (54.3), we have used a linear relation for the heat flux vector

$$-\mathbf{q}_T=k_T\nabla T \qquad \text{(Fourier law)}\,.$$

From this relation and the continuity requirements, one finds coupling conditions

$$T=T_e \quad\text{and}\quad k_T\frac{\partial T}{\partial n}=(k_T)_e\frac{\partial T_e}{\partial n}\,. \tag{55.2a}$$

To obtain prescribed conditions from these coupling conditions, one assumes that the solution in exterior regions can be expressed at the boundary by empirical functions $\hbar_T(\mathbf{x},t)$ through the equation

$$(k_T)_e\frac{\partial T_e}{\partial n}=-\hbar_T(T_e-T_R) \tag{55.2b}$$

where $T_R(x,t)$, $\mathbf{x}\in S$, is a reference temperature, which, with unbounded exterior domains $\mathscr{V}_e$, is taken as the value of T_e at infinity[2].

Direct combination of (55.2a, b) leads to prescribed conditions on S:

$$k_T\frac{\partial T}{\partial n}+\hbar_T T=\text{prescribed function}\,. \tag{55.3}$$

We allow $\hbar_T$ to be piecewise continuous so that different conditions can be prescribed over different parts of the boundary. The limit $\hbar_T\to\infty$ corresponds to a prescribed temperature and $\hbar_T=0$ to a prescribed heat flux. The symbol S_T is used to designate the part of the boundary S on which the temperature T is prescribed.

---

[2] The heat-transfer coefficients $\hbar_T$ are sometimes called Biot numbers. The determination of these coefficients is an important topic in heat-transfer research. A reasonably complete discussion of this topic and further references can be found in Bird, Stewart and Lightfoot (1960, pp. 396—405). For many common geometries and flow situations, the functions $\hbar_T$ are considered to be "known". But, of course, $\hbar_T$ and T_R are substitutes for the true solution in $\mathscr{V}_e$ evaluated at the boundary S. This specious procedure for solving the exterior problem is clearly a concession to the intractable character of the coupled problem.

(b) Concentration Boundary Conditions

We shall consider systems for which the diffusion of solute can be approximated as the direct analogue of the diffusion of heat. Then, on S,

$$k_C \frac{\partial C}{\partial n} + \mathscr{h}_C C = \text{prescribed function}. \tag{55.4}$$

Here $\mathscr{h}_C(\mathbf{x}, t)$ is a mass-transfer coefficient (Sherwood number). The symbol S_C is like S_T.

(c) Velocity Boundary Conditions

Coupling conditions for velocities are formed from a mass balance, the adherence condition[3] for viscous fluids and the stress principle of Cauchy. Adherence implies the continuity of velocity across S

$$\mathbf{U}(\mathbf{x}) = \mathbf{U}_e(\mathbf{x}). \tag{55.5}$$

Given (55.5), the mass balance equation $\mathbf{U}\cdot\mathbf{n} = \mathbf{U}_e\cdot\mathbf{n}$ is automatically satisfied. When the bounding surface is rigid and its future motion is known in advance, the RHS of (55.5) can be prescribed. One cannot accurately prescribe the RHS in all situations. It is less restrictive, but not completely satisfactory, to prescribe the normal component $\mathbf{U}_e\cdot\mathbf{n}$ of the RHS of (55.5), so that

$$\mathbf{U}\cdot\mathbf{n}|_S = \text{prescribed function}. \tag{55.6}$$

Even with $\mathbf{U}_e\cdot\mathbf{n}$ prescribed, it remains to find two other prescribed conditions for the other components of $\mathbf{U}$. There are many possible prescriptions and all of them are, to a degree, approximations; a common one is to prescribe the tangential component of the stress vector $\mathbf{t} = \mathbf{T}\cdot\mathbf{n} = -P\mathbf{n} + \mathbf{S}\cdot\mathbf{n}$. This leads to an equation of the form

$$\mathbf{S}\cdot\mathbf{n} - \mathbf{n}(\mathbf{n}\cdot\mathbf{S}\cdot\mathbf{n}) = \text{prescribed tangential vector} \tag{55.7}$$

on $S - S_U$, where S_U is the set of boundary points where the tangential component of velocity is prescribed.

If the surfaces can sustain forces, then the stress need not be continuous across S. In addition, the tension $\sigma(T, C)$ per unit length may vary and, as a consequence, drive the motion of the fluid in $\mathscr{V}_f$. Assuming that surface tension

[3] Viscous fluids satisfy *no-slip* conditions but they need not *adhere*. One fluid may replace another along a line of contact on a solid wall. The fluid on the wall is displaced without slipping (see text following (96.14) for further discussion).

forces exert only membrane stresses[4], we would be obliged to replace (55.7) with the condition

$$\mathbf{S}\cdot\mathbf{n}-\mathbf{n}(\mathbf{n}\cdot\mathbf{S}\cdot\mathbf{n})-\nabla_{\text{II}}\sigma=\text{prescribed tangential vector} \tag{55.8}$$

where

$$\nabla_{\text{II}}=\nabla-\mathbf{n}(\mathbf{n}\cdot\nabla)=(\mathbf{x}_\xi\wedge\mathbf{n}\partial_\zeta-\mathbf{x}_\zeta\wedge\mathbf{n}\partial_\xi)/[\mathbf{n}\cdot(\mathbf{x}_\xi\wedge\mathbf{x}_\zeta)]$$

is the projected gradient (see Exercise 55.2).

To prove (55.8) we consider a balance of forces in the volume $\mathscr{V}=\mathscr{V}_e\cup\mathscr{V}_f$. Here

$$\int_{\mathscr{V}}\tilde{\mathbf{A}}+\int_{S_e}\mathbf{T}\cdot\mathbf{n}+\int_{S_f}\mathbf{T}\cdot\mathbf{n}+\oint_{\partial\hat{S}}\hat{\boldsymbol{\tau}}\sigma=0 \tag{55.9}$$

where $\tilde{\mathbf{A}}$ is a uniformly-bounded, body-force field (per unit volume) and could include a contribution from acceleration, $S_e=\partial\mathscr{V}_e-\hat{S}$ and $S_f=\partial\mathscr{V}_f-\hat{S}$, $\hat{\boldsymbol{\tau}}$ is the outward normal to the curve $\partial\hat{S}$ in the surface S, $\mathbf{T}\cdot\mathbf{n}$ is a stress vector, $\mathbf{T}$ is the stress tensor and $\mathbf{n}$ the outward normal on S_e or S_f. In the limit $\mathscr{V}\to 0$, $\partial\hat{S}$ fixed, we find that

$$\int_{\hat{S}}(\mathbf{T}_e-\mathbf{T})\cdot\mathbf{n}+\oint_{\partial\hat{S}}\hat{\boldsymbol{\tau}}\sigma=0\,.$$

This last relation can be transformed, using the divergence theorem for surfaces [Weatherburn, 1955, p. 240],

$$\oint_{\partial\hat{S}}\hat{\boldsymbol{\tau}}\sigma=\int_{\hat{S}}[\nabla_{\text{II}}\sigma+2H\sigma\mathbf{n}]$$

where $2H$ is the sum of the principal curvatures (H is the mean curvature), into an integral over $\hat{S}$ alone

$$\int_{\hat{S}}[(\mathbf{T}_e-\mathbf{T})\cdot\mathbf{n}+\nabla_{\text{II}}\sigma+2H\sigma\mathbf{n}]=0\,.$$

Since $\hat{S}\subset S$ is arbitrary,

$$(\mathbf{T}_e-\mathbf{T})\cdot\mathbf{n}+\nabla_{\text{II}}\sigma+2H\sigma\mathbf{n}=0 \tag{55.10}$$

on S. Eq. (55.8) is the tangential projection of (55.10).

Exercises 55.2, 3, 4 below give equations of geometry which will be used to study interfacial stability in Chapter XIV.

Exercise 55.1: The ratio of the dynamic viscosity of water to the dynamic viscosity of air is about one hundred. Describe some flow configurations in which the shear *strain* in the water at an air-water interface is small. How should the right side of (55.7) be prescribed for these flow configura-

---

[4] Membrane stresses in a surface are like pressure forces in a fluid and the membrane stress theory can be considered to be an "inviscid" theory of surfaces. The membrane stresses are tensile, they do not resist shear or dilatation of the surface. A "Newtonian" theory of surfaces due to L.E. Scriven is discussed by Aris (1962).

tions? Write (55.7) in component form. Obtain (55.8) from (55.10). Show that, when air friction is neglected.

$$-(P_e-P)-\mathbf{n}\cdot\mathbf{S}\cdot\mathbf{n}+2\sigma H=0 \tag{55.11}$$

and

$$\mathbf{S}\cdot\mathbf{n}-\mathbf{n}(\mathbf{n}\cdot\mathbf{S}\cdot\mathbf{n})-\nabla_{\text{II}}\sigma=0 \tag{55.12}$$

on S. Formulate the problem of stability and bifurcation of the state of rest in a fluid heated from below when the top surface is free to deflect and to move under the action of tangential gradients of the surface tension (see § 94).

Exercise 55.2: Let $J_{\hat{S}}=\mathbf{n}\cdot(\mathbf{x}_\xi\wedge\mathbf{x}_\zeta)$ where ξ and ζ are time-independent coordinates embedded in $\hat{S}$,[5] $\mathbf{x}(\xi,\zeta,t)$ is the position vector for points on $\hat{S}$ and $\mathbf{n}J_{\hat{S}}=\mathbf{x}_\xi\wedge\mathbf{x}_\zeta$. For any scalar $\phi(\mathbf{x}(\xi,\zeta,t),t)$ defined on $\hat{S}$

$$d\phi=d\mathbf{x}(\xi,\zeta,t)\cdot\nabla\phi=\phi_\xi d\xi+\phi_\zeta d\zeta$$

where

$$\phi_\xi=\mathbf{x}_\xi\cdot\nabla\phi\,,\qquad \phi_\zeta=\mathbf{x}_\zeta\cdot\nabla\phi\,.$$

Show that

$$\phi_\xi=\mathbf{x}_\xi\cdot\nabla_{\text{II}}\phi\,,\qquad \phi_\zeta=\mathbf{x}_\zeta\cdot\nabla_{\text{II}}\phi$$

where

$$\nabla_{\text{II}}=J_{\hat{S}}^{-1}[\mathbf{n}\wedge\mathbf{x}_\xi\partial_\zeta-\mathbf{n}\wedge\mathbf{x}_\zeta\partial_\xi]=\nabla-\mathbf{n}(\mathbf{n}\cdot\nabla)$$

is the projected gradient. Show that

$$\left.\frac{\partial J_{\hat{S}}}{\partial t}\right|_{\xi,\zeta}\equiv\frac{d}{dt}J_{\hat{S}}=\frac{\partial}{\partial t}J_{\hat{S}}\big|_{\mathbf{x}}+\mathbf{V}\cdot\nabla_{\text{II}}J_{\hat{S}}=J_{\hat{S}}\nabla_{\text{II}}\cdot\mathbf{V} \tag{55.13}$$

where $\mathbf{V}=d\mathbf{x}(\xi,\zeta,t)/dt$ is the velocity of the point (ξ,ζ) which moves with $\hat{S}$.

Exercise 55.3: Show that

$$\frac{d}{dt}\int_{\hat{S}(t)}d\hat{S}=\iint\frac{d}{dt}J_{\hat{S}}d\xi d\zeta=\int_{\hat{S}(t)}\nabla_{\text{II}}\cdot\mathbf{V}d\hat{S}=\oint_{\partial\hat{S}}\boldsymbol{\tau}\cdot\mathbf{V}dl-2\int_{\hat{S}(t)}H\mathbf{V}\cdot\mathbf{n}d\hat{S} \tag{55.14}$$

Let (ξ,ζ) and (η,μ) be surface coordinates on $\hat{S}$ which are related to one another by an invertible transformation. Let

$$\mathbf{x}(\xi,\zeta,t)=\mathbf{x}(\xi(\eta,\mu,t),\,\zeta(\eta,\mu,t),\,t)=\hat{\mathbf{x}}(\eta,\mu,t)$$

be the position vector to the surface $\hat{S}$ and

$$\mathbf{V}=\frac{\partial\mathbf{x}(\xi,\zeta,t)}{\partial t},\qquad \hat{\mathbf{V}}=\frac{\partial\hat{\mathbf{x}}(\eta,\mu,t)}{\partial t}.$$

Suppose further that

$$\hat{\phi}(\eta,\mu,t)=\phi(\xi(\eta,\mu,t),\,\zeta(\eta,\mu,t),\,t)=\phi(\xi,\zeta,t)\,.$$

Show that

$$\frac{\partial\phi(\xi,\zeta,t)}{\partial t}-\mathbf{V}\cdot\nabla_{\text{II}}\,\phi=\frac{\partial\hat{\phi}(\eta,\mu,t)}{\partial t}-\hat{\mathbf{V}}\cdot\nabla_{\text{II}}\,\hat{\phi}$$

[5] $J_{\hat{S}}=\sqrt{g}$ where $g=\det g_{\alpha\beta}$ is the first fundamental form of a surface.

and that

$$\frac{d}{dt}\int_{\hat{S}}\phi d\hat{S}=\int_{\hat{S}}\left[\frac{\partial\hat{\phi}(\eta,\mu,t)}{\partial t}-\hat{\mathbf{V}}\cdot\nabla_{\text{II}}\hat{\phi}-2H\hat{\phi}\mathbf{V}\cdot\mathbf{n}\right]d\hat{S}+\oint_{\partial\hat{S}}\hat{\phi}\boldsymbol{\tau}\cdot\mathbf{V}dl \tag{55.15}$$

Exercise 55.4: Two surfaces $S_1(t)$ and $S_2(t)$ coincide within and on a closed curve $\partial S(t)$. The normal velocities of the two surfaces are the same at the points where they contact but their tangential velocity need not be the same. Show that

$$\frac{d}{dt}\int_{S_1(t)}\phi dS_1=\frac{d}{dt}\int_{S_2(t)}\phi dS_2+\oint_{\partial S}\phi\boldsymbol{\tau}\cdot(\mathbf{V}_1-\mathbf{V}_2)dl \tag{55.16}$$

Exercise 55.5: The ratio of the thermal conductivity of water to the thermal conductivity of air is about thirty. Construct an argument to support the assertion that a disturbance θ of the temperature T at an air-water interface can be assumed to satisfy an equation of the form

$$\frac{\partial\theta}{\partial n}+h\theta=0$$

where the constant h is a numerically small value which in the first approximation can be taken as zero.

§ 56. Equations Governing Disturbances of Solutions of the OB Equations

Our main interest is in the stability of motionless solutions of the OB equations. We start the analysis more generally, however, and single out some solutions of equations of (54.2, 3, 4) and the boundary conditions (55.3, 4, 5, 6, 7). We call this solution the basic motion. The dimensionless description of the basic motion

$$[\mathbf{U},\ \mathbf{D}[\mathbf{U}],\ \boldsymbol{\eta}_T,\ \boldsymbol{\eta}_C,\ \boldsymbol{\eta},\ \mathbf{x},\ t,\ h_T,\ h_C] \tag{56.1}$$

is obtained from the dimensional description

$$[\mathbf{U},\ \mathbf{D}[\mathbf{U}]),\ \nabla T,\ \nabla C,\ \mathbf{g},\ \mathbf{x},\ t,\ \hbar_T,\ \hbar_C]$$

by dividing the dimensional variables by the scale factors

$$\left[U',\ \frac{U'}{l},\ \frac{T'}{l},\ \frac{C'}{l},\ g',\ l,\ \frac{l^2}{\nu},\ \frac{k_T}{l},\ \frac{k_C}{l}\right].$$

The scale factors T', C', U' and g' are typical values (to be chosen later) for temperature, concentration, velocity and gravity for the basic motion.

A disturbance of the basic motion is the difference between the basic motion and another initially different motion satisfying the same equations and prescribed conditions. The dimensionless velocity, strain-rate, temperature, concentration and pressure are designated as

$$[\mathbf{u},\ \mathbf{D}[\mathbf{u}],\ \theta,\ \gamma,\ p]. \tag{56.2}$$

The dimensional disturbances may be formed by multiplying the dimensionless variables by scale factors

$$\left[\frac{\nu}{l}, \frac{\nu}{l^2}, \sqrt{\frac{\nu^3 T'}{\kappa_T \alpha_T g' l^3}}, \sqrt{\frac{\nu^3 C'}{\kappa_C \alpha_C g' l^3}}, \frac{\rho_0 \nu^2}{l^2}\right].$$

The equations which govern the evolution of a disturbance of the basic motion are

$$(\partial_t + R\mathbf{U}\cdot\nabla + \mathbf{u}\cdot\nabla)\begin{bmatrix}\mathbf{u}\\ \mathscr{P}_T\theta\\ \mathscr{P}_C\gamma\end{bmatrix} + \mathbf{u}\cdot\begin{bmatrix}R\nabla\mathbf{U}\\ \mathscr{R}\boldsymbol{\eta}_T\\ \mathscr{C}\boldsymbol{\eta}_C\end{bmatrix} = \begin{bmatrix}-\nabla p - (\mathscr{R}\theta - \mathscr{C}\gamma)\boldsymbol{\eta} + 2\nabla\cdot\mathbf{D}\\ \nabla^2\theta\\ \nabla^2\gamma\end{bmatrix} \tag{56.3a}$$

$$\operatorname{div}\mathbf{u} = 0, \quad \mathbf{u}\cdot\mathbf{n} = 0 \text{ on } S \tag{56.3b}$$

$$\mathbf{u} = 0 \text{ on } S_U, \quad \mathbf{D}[\mathbf{u}]\cdot\mathbf{n} - \mathbf{n}(\mathbf{n}\cdot\mathbf{D}[\mathbf{u}]\cdot\mathbf{n}) = 0 \quad \text{on} \quad S - S_U \tag{56.3c}$$

$$\theta = 0 \text{ on } S_T, \quad \frac{\partial\theta}{\partial n} + h_T\theta = 0 \quad \text{on} \quad S - S_T \tag{56.3d}$$

$$\gamma = 0 \text{ on } S_C, \quad \frac{\partial\gamma}{\partial n} + h_C\gamma = 0 \quad \text{on} \quad S - S_C. \tag{56.3e}$$

The stability parameters in (56.3a) are the Rayleigh numbers for heat and solute ($\mathscr{R}^2$ and $\mathscr{C}^2$) and a Reynolds number R given by

$$\mathscr{R}^2 = \frac{\alpha_T T' g' l^3}{\nu\kappa_T}, \quad \mathscr{C}^2 = \frac{\alpha_C C' g' l^3}{\nu\kappa_C} \quad \text{and} \quad R = \frac{U' l}{\nu}.$$

The Prandtl number $\mathscr{P}_T = \nu/\kappa_T$ and Schmidt number $\mathscr{P}_C = \nu/\kappa_C$ enter our considerations, but not as critical values for stability problems.

In dimensionless variables, the energy identities are

$$\tfrac{1}{2}\frac{d}{dt}\langle|\mathbf{u}|^2\rangle = -R\langle\mathbf{u}\cdot\nabla\mathbf{U}\cdot\mathbf{u}\rangle - \mathscr{R}\langle\theta\boldsymbol{\eta}\cdot\mathbf{u}\rangle + \mathscr{C}\langle\gamma\boldsymbol{\eta}\cdot\mathbf{u}\rangle - 2\langle\mathbf{D}[\mathbf{u}]:\mathbf{D}[\mathbf{u}]\rangle, \tag{56.4}$$

$$\tfrac{1}{2}\mathscr{P}_T\frac{d}{dt}\langle\theta^2\rangle = -\mathscr{R}\langle\theta\mathbf{u}\cdot\boldsymbol{\eta}_T\rangle - \langle|\nabla\theta|^2\rangle - \langle h_T\theta^2\rangle_S, \tag{56.5}$$

$$\tfrac{1}{2}\mathscr{P}_C\frac{d}{dt}\langle\gamma^2\rangle = -\mathscr{C}\langle\gamma\mathbf{u}\cdot\boldsymbol{\eta}_C\rangle - \langle|\nabla\gamma|^2\rangle - \langle h_C\gamma^2\rangle_S, \tag{56.6}$$

and

$$\frac{d}{dt}\langle\theta\gamma\rangle = -\frac{\mathscr{R}}{\mathscr{P}_T}\langle\gamma\mathbf{u}\cdot\boldsymbol{\eta}_T\rangle - \frac{\mathscr{C}}{\mathscr{P}_C}\langle\theta\mathbf{u}\cdot\boldsymbol{\eta}_C\rangle - \left(\frac{1}{\mathscr{P}_C} + \frac{1}{\mathscr{P}_T}\right)\langle\nabla\theta\cdot\nabla\gamma\rangle - \left\langle\left(\frac{h_T}{\mathscr{P}_T} + \frac{h_C}{\mathscr{P}_C}\right)\theta\gamma\right\rangle_S. \tag{56.7}$$

Here, as before

$$\langle f\rangle = \frac{1}{\mathcal{M}(\mathcal{V})}\int_{\mathcal{V}} f$$

and, in addition

$$\langle f\rangle_S = \frac{1}{\mathcal{M}(\mathcal{V})}\int_S f\,.$$

Eq. (56.4, 5, 6, 7) are called, loosely, energy identities. Though $\langle|\mathbf{u}|^2\rangle/2$ is proportional to the kinetic energy, the other quadratic integrals $\langle\theta^2\rangle$ and $\langle\gamma^2\rangle$ cannot be called energies in any strict sense. Certainly the coupling integral $\langle\theta\gamma\rangle$ is not easy to identify as an energy. In these equations R, $\mathcal{R}$ and $\mathcal{C}$ are stability parameters and the integrals which they multiply are the production integrals. Production integrals can have the right sign for the *production* of energy of the disturbance from the energy of the basic flow. The other integrals lead to a dissipation of the energy.

Exercise 56.1: Suppose $\mathbf{u}$ is a solenoidal field and $\mathbf{u}\cdot\mathbf{n}=0$ on S. Show that

$$2\langle\mathbf{D}[\mathbf{u}]:D[\mathbf{u}]\rangle = \langle|\nabla\mathbf{u}|^2\rangle + \langle(\mathbf{u}\cdot\nabla\mathbf{u})\cdot\mathbf{n}\rangle_S$$

where the boundary integral vanishes on free planar elements of S. Derive (56.4, 5, 6, 7).

§ 57. The λ Family of Energy Equations

A linear combination of Eqs. (56.4, 5, 6) with coupling parameters λ_T and λ_C can be put into the form

$$\frac{d\mathcal{E}}{dt} = \mathbb{R}\mathcal{I} - \mathcal{D}\,, \tag{57.1}$$

where

$$\mathbb{R} = \sqrt{R^2 + \mathcal{R}^2 + \mathcal{C}^2}$$

is the stability parameter,

$$\mathcal{E}[\mathbf{u}, \sqrt{\lambda_T}\theta, \sqrt{\lambda_C}\gamma] = \tfrac{1}{2}\langle|\mathbf{u}|^2 + \lambda_T\mathcal{P}_T\theta^2 + \lambda_C\mathcal{P}_C\gamma^2\rangle \tag{57.2}$$

is the "energy" of the disturbance,

$$\begin{aligned}\mathcal{D}[\mathbf{u}, \sqrt{\lambda_T}\theta, \sqrt{\lambda_C}\gamma] = 2\langle\mathbf{D}[\mathbf{u}]:\mathbf{D}[\mathbf{u}]\rangle + \lambda_T\langle|\nabla\theta|^2\rangle + \lambda_T\langle h_T\theta^2\rangle_S \\ + \lambda_C\langle|\nabla\gamma|^2\rangle + \lambda_C\langle h_C\gamma^2\rangle_S\end{aligned} \tag{57.3}$$

is the "dissipation" and

$$-\mathscr{I}[\mathbf{u}, \theta, \gamma; \lambda_T, \lambda_C] = \mathscr{A}_R\langle \mathbf{u}\cdot\nabla\mathbf{U}\cdot\mathbf{u}\rangle + 2\mathscr{A}_{\mathscr{R}}\langle \mathbf{u}\cdot\mathbf{P}_T\theta\rangle + 2\mathscr{A}_{\mathscr{C}}\langle \mathbf{u}\cdot\mathbf{P}_C\gamma\rangle \tag{57.4}$$

is the production. Here

$$\mathscr{A}_R = R/\mathrm{I\!R}, \quad \mathscr{A}_{\mathscr{R}} = \mathscr{R}/\mathrm{I\!R}, \quad \mathscr{A}_{\mathscr{C}} = \mathscr{C}/\mathrm{I\!R},$$
$$\mathbf{P}_T = \tfrac{1}{2}(\lambda_T\boldsymbol{\eta}_T + \boldsymbol{\eta})$$

and

$$\mathbf{P}_C = \tfrac{1}{2}(\lambda_C\boldsymbol{\eta}_C - \boldsymbol{\eta}).$$

When (57.2, 3, 4) are written in the variables

$$\theta' = \sqrt{\lambda_T}\,\theta, \quad \gamma' = \sqrt{\lambda_C}\,\gamma, \tag{57.5}$$

we have

$$\begin{aligned} &\mathscr{E}[\mathbf{u}, \theta', \gamma'] = \tfrac{1}{2}\langle |\mathbf{u}|^2 + \mathscr{P}_T\theta'^2 + \mathscr{P}_C\gamma'^2\rangle \\ &\mathscr{D}[\mathbf{u}, \theta', \gamma'] = 2\langle \mathbf{D}[\mathbf{u}] : \mathbf{D}[\mathbf{u}]\rangle \\ &\qquad + \langle |\nabla\theta'|^2\rangle + \langle h_T\theta'^2\rangle_S + \langle |\nabla\gamma'|^2\rangle + \langle h_C\gamma'^2\rangle_S \end{aligned} \tag{57.6}$$

and

$$\begin{aligned} -\mathscr{I}[\mathbf{u}, \theta'/\sqrt{\lambda_T}, \gamma'/\sqrt{\lambda_C}; \lambda_T, \lambda_C] &= \mathscr{A}_R\langle \mathbf{u}\cdot\nabla\mathbf{U}\cdot\mathbf{u}\rangle + 2\mathscr{A}_R\lambda_T^{-1/2}\langle \mathbf{u}\cdot\mathbf{P}_T\theta'\rangle \\ &\quad + 2\mathscr{A}_C\lambda_C^{-1/2}\langle \mathbf{u}\cdot\mathbf{P}_C\gamma'\rangle. \end{aligned} \tag{57.7}$$

In these variables only the functional $\mathscr{I}$ depends explicitly on λ_T and λ_C.

As in our earlier work, we may use the evolution equation to find sufficient conditions for stability. We will again need to define a class of admissible initial conditions.

§ 58. Kinematic Admissibility, Sufficient Conditions for Stability

The reader's attention is directed to the boundary conditions (58.1 b, c), below. Note that at each point of the boundary, there is specified some condition which must be satisfied by the solution. For example, (58.1 b) specifies the part of the boundary where the disturbances must vanish. On the other parts of the boundary where (58.1 c) holds, the values of the transfer coefficient h_T relate the solution to the value of its normal derivative.

The conditions (58.1 b) are called "zero boundary conditions" and the conditions (58.1 c) are called "natural boundary conditions". There is a difference in the way these two kinds of boundary conditions enter into our extended definition of kinematically admissible vectors. To describe the difference it helps to think about the eigenvalue problem

$$\nabla^2\theta+\Lambda\theta=0\,; \qquad \theta|_{S_T}=0\,, \qquad \frac{\partial\theta}{\partial n}+h_T\theta|_{S-S_T}=0\,. \tag{58.1 a, b, c}$$

It is well known that the eigenvalues of this problem can be characterized by extremum problems[6] of the form

$$\Lambda^{-1}=\max_H \frac{\langle\theta^2\rangle}{\langle|\nabla\theta|^2\rangle+h_T\langle\theta^2\rangle_{S_T}} \tag{58.2}$$

where, for convenience, $h_T=\text{const}$. Here we take H as the space of functions which satisfy (58.1 b) but do not necessarily satisfy (58.1 c). The boundary condition (58.1 c) is, in a sense, already expressed as a part of the problem (58.2). The reason is that the prescription of boundary values on $S-S_T$ is essentially a prescription of the value h_T. This news from the boundary is already incorporated in the form of the functional and it need not be stated as a separate condition of admissibility. In fact boundary condition (58.1 c) must necessarily be satisfied by the functions that win in the competition for the maximum of (58.2). In this sense the boundary conditions (58.1 c) are generated "naturally" by the maximum problem and they need not be posed, like (58.1 b), as a separate condition (see B.3).

We shall call a function kinematically admissible if it satisfies the zero boundary conditions (but it need not satisfy the natural boundary conditions). Vector fields are kinematically admissible if they satisfy the zero boundary conditions and are solenoidal.

In convection problems we need to consider the union of kinematically admissible vectors **H** and functions H. We think of $(\mathbf{u},\theta,\gamma)$ as a five-component vector and also call the combined space **H**. Kinematically admissible functions are such that

$$\mathscr{D}^{1/2}=(\langle 2\mathbf{D}[\mathbf{u}]:\mathbf{D}[\mathbf{u}]+|\nabla\theta'|^2+|\nabla\gamma'|^2\rangle+\langle h_C|\gamma'|^2+h_T|\theta'|^2\rangle_S)^{1/2}$$

is bounded. Then,

$$\mathbf{H}=\{(\mathbf{u},\theta',\gamma'):\operatorname{div}\mathbf{u}=0,\ \mathbf{u}|_{S_U}=0,\ \theta'|_{S_T}=0,\ \gamma'|_{S_C}=0\}\,. \tag{58.3}$$

It is convenient and should be sufficiently general to say that *the set of admissible initial conditions for the disturbance coincides with the set* **H** *of kinematically admissible vectors.*

We now seek the form of the disturbance which makes the energy $\mathscr{E}(t,\lambda_T,\lambda_C)$, satisfying equation (57.1), increase initially at the smallest value of $\mathbb{R}$ (Shir and Joseph, 1968)[7].

Suppose that $\lambda_T>0$ *and* $\lambda_C>0$ *are given arbitrarily and*

$$\mathbb{R}=\sqrt{R^2+\mathscr{R}^2+\mathscr{C}^2}<\mathbb{R}_{\mathscr{E}}$$

[6] These problems are discussed in detail in Section B.3 of Appendix B.

[7] A related result has been given by Rionero (1971A).

where

$$\mathbb{R}_{\mathscr{E}}^{-1}=\max_{\mathbf{H}}\mathscr{I}\left[\mathbf{u},\frac{\theta'}{\sqrt{\lambda_T}},\frac{\gamma'}{\sqrt{\lambda_C}};\lambda_T,\lambda_C\right]/\mathscr{D}[\mathbf{u},\theta',\gamma'] . \tag{58.4}$$

Then there exists a constant $a^2>0$ *such that*

$$\mathscr{E}(t;\lambda_T,\lambda_C)\leqslant\mathscr{E}(0;\lambda_T,\lambda_C)\exp\{-a^2t(1-\mathbb{R}_{\mathscr{E}}/\mathbb{R})\} , \tag{58.5}$$

The initial condition which maximizes $\mathbb{R}_{\mathscr{E}}^{-1}$ *makes the energy* $\mathscr{E}(0;\lambda_T,\lambda_C)$ *increase initially at the smallest value of* $\mathbb{R}$.

The parameters λ_T and λ_C are free and can be selected to optimize the limit for monotonic, global stability. The value

$$\tilde{\mathbb{R}}_{\mathscr{E}}=\mathbb{R}_{\mathscr{E}}(\tilde{\lambda}_T,\tilde{\lambda}_C)=\sup_{\lambda_T,\lambda_C}\mathbb{R}_{\mathscr{E}}(\lambda_T,\lambda_C) \tag{58.6}$$

is called the optimum stability boundary.

To simplify the study of the problem (58.6) consider homogeneous fluids for which $\gamma'\equiv 0$, $\lambda_C=0$, $\lambda_T=\lambda$ and $\mathbb{R}_{\mathscr{E}}(\lambda_T,0)=\mathbb{R}_\lambda$. Then $\tilde{\mathbb{R}}_{\mathscr{E}}=\max_{\lambda>0}\mathbb{R}_\lambda$. We shall show that $\mathbb{R}_\lambda\to 0$ for large and small values of λ. Hence

$$\tilde{\mathbb{R}}_{\mathscr{E}}=\mathbb{R}_{\tilde{\lambda}} \quad\text{where}\quad \left.\frac{d\mathbb{R}_\lambda}{d\lambda}\right|_{\tilde{\lambda}}=0 . \tag{58.7}$$

To show that $\mathbb{R}_\lambda\to 0$ we rewrite (58.4) as

$$\frac{1}{\mathbb{R}_\lambda}=\max_{\mathbf{H}}(\mathscr{I}_1+\sqrt{\lambda}\mathscr{I}_2+\mathscr{I}_3/\sqrt{\lambda})/\mathscr{D} \tag{58.8}$$

where

$$\mathscr{I}_1=\mathscr{A}_R\langle\mathbf{u}\cdot\mathbf{D}\cdot\mathbf{u}\rangle , \qquad \mathscr{I}_2=\tfrac{1}{2}\langle\boldsymbol{\eta}_T\cdot\mathbf{u}\theta'\rangle , \qquad \mathscr{I}_3=\tfrac{1}{2}\langle\boldsymbol{\eta}\cdot\mathbf{u}\theta'\rangle .$$

Since $\mathscr{D}=2\langle\mathbf{D}[\mathbf{u}]:\mathbf{D}[\mathbf{u}]+|\nabla\theta'|^2\rangle+\langle h_T\theta'^2\rangle_S$ is quadratic in θ' it follows that the sign, but not the magnitude, of $\mathscr{I}_2/\mathscr{D}$ and $\mathscr{I}_3/\mathscr{D}$ changes under the transformation $\theta'\to-\theta'$.

Since $\mathscr{I}_1/\mathscr{D}$, $\mathscr{I}_2/\mathscr{D}$ and $\mathscr{I}_3/\mathscr{D}$ are bounded functionals in $\mathbf{H}$, we have

$$\frac{1}{\sqrt{\lambda}}\frac{\mathscr{I}_1}{\mathscr{D}}\to 0 , \qquad \frac{1}{\lambda}\frac{\mathscr{I}_3}{\mathscr{D}}\to 0$$

as $\lambda\to\infty$. Hence, as $\lambda\to\infty$,

$$\frac{1}{\mathbb{R}_\lambda}=\sqrt{\lambda}\max_{\mathbf{H}}\frac{\mathscr{I}_2}{\mathscr{D}}+0(1)\to\infty.$$

Similarly,

$$\frac{1}{\mathbb{R}_\lambda}=\frac{1}{\sqrt{\lambda}}\max_{\mathbf{H}}\frac{\mathscr{I}_3}{\mathscr{D}}+0(1)\to\infty$$

as $\lambda\to 0$.

Since $\mathbb{R}_\lambda\to 0$ for large and small λ and is positive on interior intervals, it must have a maximum for a finite, nonzero value of λ. At this maximum (58.7) must hold.

To draw the main consequences of (58.7), we differentiate

$$\mathbb{R}_\lambda^{-1}\mathscr{D}=\mathscr{I}_1+\sqrt{\lambda}\mathscr{I}_2+\mathscr{I}_3/\sqrt{\lambda}$$

with respect to $\sqrt{\lambda}$ to find that

$$\frac{d\mathbb{R}_\lambda^{-1}}{d\sqrt{\lambda}}\mathscr{D}=\mathscr{I}_2-\mathscr{I}_3/\lambda\,. \tag{58.9}$$

Here, using (B 4.16), we have noted that a weak form of Euler's equation for (58.8) may be written as

$$\mathbb{R}_\lambda^{-1}\frac{\partial\mathscr{D}}{\partial\sqrt{\lambda}}=\frac{\partial\mathscr{I}_1}{\partial\sqrt{\lambda}}+\sqrt{\lambda}\frac{\partial\mathscr{I}_2}{\partial\sqrt{\lambda}}+\frac{1}{\sqrt{\lambda}}\frac{\partial\mathscr{I}_3}{\partial\sqrt{\lambda}}\,. \tag{58.10}$$

The formula

$$\mathscr{I}_2-\mathscr{I}_3/\lambda=\langle(\boldsymbol{\eta}_T-\boldsymbol{\eta}/\tilde{\lambda})\cdot\tilde{\mathbf{u}}\tilde{\theta}'\rangle=0 \tag{58.11}$$

where variables with the tilde overbar satisfy (58.8), (58.9) and (58.11), is the main consequence of (58.9) and (58.10).

Eq. (58.11) can be satisfied in several ways.

The formula $\tilde{\lambda}\langle\boldsymbol{\eta}_T\cdot\tilde{\mathbf{u}}\tilde{\theta}'\rangle=\langle\boldsymbol{\eta}\cdot\tilde{\mathbf{u}}\tilde{\theta}'\rangle$ is the basis for obtaining approximate values of the coupling parameter $\tilde{\lambda}$. When $\boldsymbol{\eta}(\mathbf{x})=\boldsymbol{\eta}_T(\mathbf{x})$ (as in the Bénard problem), it gives the exact result, $\tilde{\lambda}=1$. There are rare cases, like plane Couette flow heated from above, in which $\boldsymbol{\eta}_T=-\boldsymbol{\eta}$ where (58.11) does not lead to a positive value of λ. In these cases (58.11) is satisfied because $\tilde{\theta}'\equiv 0$.

Exercise 58.1: Prove (58.5). Define a^2. Formulate all the technical conditions under which (58.5) holds.

Exercise 58.2: Prove the energy stability theorem II of Davis and von Kerczek (1975; see § 4) for basic solutions of the OB equations.

Exercise 58.3: Prove that two almost periodic solutions of the OB equations are the same almost everywhere when $\mathbb{R}<\mathbb{R}_{\mathscr{E}}$.

Exercise 58.4: Prove that two solutions of the IBVP for the OB equations which start with the same initial values and satisfy (57.1) are the same almost everywhere.

Exercise 58.5: Suppose that $h_T \geq 0$ and $h_C \geq 0$ on the whole boundary S of $\mathscr{V}$ and that there are parts of S (with positive area) on which the inequality holds. Suppose further that $\mathbf{u}=0$ on a part of S (with positive area). Show that $\mathscr{E}/\mathscr{D}$ is bounded.

Exercise 58.6: Consider the maximum problem (58.8) when $\boldsymbol{\eta}_T = -\boldsymbol{\eta}$. Show that one can always find admissible θ' such that $(\sqrt{\lambda}\mathscr{I}_2 + \mathscr{I}_3/\sqrt{\lambda})/\mathscr{D}$ is not negative. Show that the smallest value of $1/\mathbb{R}_\lambda$ is the one for which $\lambda=1$. Prove that $\theta'=0$ when $\lambda=1$.

Exercise 58.7 (Shir and Joseph, 1968): Show that $\mathbb{R}_\lambda$ is monotone and convex in $\hbar$.

Exercise 58.8: Show that the best value of λ_c for the maximum value of $\mathscr{I}/\mathscr{D}$ satisfies the equation

$$\langle (\boldsymbol{\eta}_C - \boldsymbol{\eta}/\lambda_C)\cdot \tilde{\mathbf{u}}\tilde{\gamma}\rangle = 0\,. \tag{58.12}$$

§ 59. Motionless Solutions of the Oberbeck-Boussinesq Equations

Motionless states have $\mathbf{U}=0$ and support distributions of temperature and mass which are determined by the requirements of mass diffusion and heat conduction alone. In nature, there are configurations for which the domain and boundary conditions are of such a character that the motionless solution is possible and is observed when the parameters have some values and not others. For example, a container of fluid heated from below can theoretically support a linear temperature decrease for any temperature contrast, but convective motion actually starts when the temperature contrast exceeds a certain value.

Suppose that the gravity vector $\mathbf{g}(x)$ *is a field which is derivable from a potential. Then a motionless solution of the Oberbeck-Boussinesq equations exists only if the density gradient and gravity are parallel vector fields*[8].

---

[8] The criterion ($\mathbf{g}\wedge\nabla\rho=0$) for the existence of a motionless state was clearly stated and perhaps first stated by Euler in his (1764) essay *On the motion of fluid arising from different degrees of heat.* (Excerpt from C. Truesdell's introduction to Vol. II.12 of Euler's work.) In his discussion of the physical origins of convection, Euler writes that

"The subject ... is not only entirely new, since no one hitherto has tried to determine geometrically the motion of fluids arising from heat, but it belongs also to that part of analysis which up to now has not begun to be treated ... Some time ago the author showed ... that for conserving the air in equilibrium it is requisite that at equal altitudes there be equal density and also an equal degree of heat. Such being true, when equilibrium is destroyed by different degrees of heat a motion of the fluid will follow, and the author, with that new analysis more abundantly cultivated by him, now has begun to investigate (the matter): and in the present dissertation he most clearly shows whence the forces generating such a motion are to be sought and what will be that motion arising from them, and by easy reasoning he concludes that if in a vessel sufficiently full the water of one portion is rendered hotter ..., then the water in the lower portion will continually rise from the cold part to the hot, and contrariwise in the higher portion will be carried down from the hot part toward the cold ..."

"When formerly I examined more accurately the theory of equilibrium of fluids, I showed plainly that neither the atmosphere nor any other fluid can be in equilibrium unless the degree of heat is the same everywhere at equal altitudes." At that time I did not venture to specify the nature of the motion resulting from different degrees of heat, since such an investigation seemed to require a more copious knowledge. But looking at the matter afresh, I realized that this remarkable property always holds: if the fluid is warmer in one place than in another at the same altitude, then in the lower region the fluid will flow from the colder to the warmer; in the upper, from the warmer to the colder. "Since this conclusion is marvelously confirmed by experience, as we observe such motions in chimneys and in heated sweating chambers, it seems altogether worthwhile that I should work it out more accurately and that I should explain clearly the generation of this motion, in so far as possible."

To prove this we note that when $U \equiv 0$, Eqs. (54.2a), (54.3) and (54.4) become

$$\nabla \frac{P}{\rho_0} = (1 - \alpha_T(T - T_0) + \alpha_C(C - C_0))\mathbf{g}, \tag{59.1}$$

$$\frac{\partial T}{\partial t} = \kappa_T \nabla^2 T + Q_T, \tag{59.2}$$

$$\frac{\partial C}{\partial t} = \kappa_C \nabla^2 C + Q_C. \tag{59.3}$$

Taking the curl of Eq. (59.1) we find that

$$\operatorname{curl} \mathbf{g}[\alpha_T(T - T_0) - \alpha_C(C - C_0)] = -\mathbf{g} \wedge (\alpha_T \nabla T - \alpha_C \nabla C) = \mathbf{g} \wedge \nabla \frac{\rho}{\rho_0} = 0. \tag{59.4}$$

Given the necessary condition (59.4), every motionless solution must also satisfy (59.2) and (59.3). We shall not consider (59.2, 3) in full generality, but instead treat steady solutions with $Q_C = 0$. We shall require that ∇T, ∇C, $\nabla \rho$ and $\mathbf{g}$ be parallel vectors. For the most part, it will suffice to take $\mathbf{g}$ as a constant gravity field pointing "down". Here, "up" is the direction of x_3 increasing, and $\mathbf{e}_z$ is a unit vector in this direction. By integration, (59.2, 3) determines functions $T(x_3)$ and $C(x_3)$ and through these, $P(x_3)$. It is customary to consider the stability of motionless states in fluid layers of the kind just mentioned. But the existence of motionless solutions $T(x_3)$ and $C(x_3)$ does not require that one consider special domains like layers; the domain can be arbitrary.

To express the temperature and concentration fields in the dimensionless form in which they appear in the energy identities (56.4—7) we need to choose the scaling parameters T', C' and g' which appear in the definition of the stability parameters $\mathscr{R}$ and $\mathscr{C}$ and the basic-state vectors $\boldsymbol{\eta}_T$, $\boldsymbol{\eta}_C$, and $\boldsymbol{\eta}$. From now on, we shall consider situations for which $g' = |\mathbf{g}|$ is the constant gravitational acceleration and $\boldsymbol{\eta} = -\mathbf{e}_z$.

The steady distributions of heat and salt without sources satisfy (59.2, 3, 4) and must have the form

$$T(x_3) = A_T x_3 + T(0)$$

and (59.5)

$$C(x_3) = A_C x_3 + C(0).$$

The scale factors for Eqs. (56.4—7) are now selected as

$$\frac{T'}{l} = |A_T| \quad \text{and} \quad \frac{C'}{l} = |A_C|.$$

Then

$$\eta_T\begin{cases}=-\mathbf{e}_z & \text{(heated below)},\\ =\mathbf{e}_z & \text{(heated above)},\end{cases}$$

and

$$\eta_C\begin{cases}=-\mathbf{e}_z & \text{(salty below)},\\ =\mathbf{e}_z & \text{(salty above)}.\end{cases}$$

If the thermal conductivity $k_T(T)$ depends on the temperature, the conduction solution satisfies the equation $k_T(T)dT/dx_3=\text{const.}$

In our study of the stability of conduction with distributed sources we consider only chemically homogeneous fluids and set $Q_T(x_3)=Q(x_3)$ and $\kappa_T=\kappa$. The heat conduction solution then has the form

$$\frac{dT(x_3)}{dx_3}-\frac{dT(0)}{dx_3}+\frac{1}{\kappa}\int_0^{x_3} Q(y)dy=0\,.$$

This can be written as

$$\frac{d}{dz}(T/T')+(1+\xi\hat{g}(z))=0 \tag{59.6}$$

where $z=x_3/l$, $T'/l=-dT(0)/dx_3$ (>0 for our problems)

$$\xi\hat{g}(z)=\frac{l^2}{\kappa T'}\int_0^z Q(ly)dy\,,$$

and

$$\eta_T=\mathbf{e}_z(d(T/T')/dz)=-\mathbf{e}_z(1+\xi\hat{g}(z))\,. \tag{59.7}$$

Here ξ is a heat source parameter.

Eq. (59.7) also describes the instantaneous temperature gradient field for a fluid without heat sources, but such that

$$\frac{\partial T(x_3,t)}{\partial t}=-Q(x_3)\,. \tag{59.8}$$

In certain circumstances slow time changes in $T(x_3, t)$ can be represented approximately by (59.7) and a suitable choice of $Q(x_3)$.

Exercise 59.1 (An OB solution with motion): Consider a fluid layer confined by parallel rigid conducting plates extending to infinity in two directions. The lower plate is hotter than the upper plate. Suppose that the plates are not horizontal. Let the normal to the plates be $\mathbf{e}_{x_3}$. The direction of motion will be $\mathbf{U}=\mathbf{e}_{x_1}U(x_3)$ where $\mathbf{e}_{x_1}$ lies in the line defining the intersection of a plane parallel to the plates and a vertical plane perpendicular to the plates. Find the velocity field $U(x_3)$ and a coupled temperature field in the form $T(x_3)=C_1+C_2x_3$ where C_1 and C_2 are constants.

Exercise 59.2: Derive an expression giving the most general form which a motionless solution of the OB equations can take.

Exercise 59.3 (Chandrasekhar, 1961): Consider the problem of an equilibrium temperature distribution in a self-gravitating sphere or spherical annulus. Let the gravitational potential $\Phi(r)$ $(\mathbf{g}=-\nabla\Phi)$ and heat source distribution $Q(r)$ be arbitrary functions of r alone where r is the radius in spherical polar coordinates. Express the temperature distribution in the spherical annulus in terms of the given heat source distribution and the given temperatures at the boundaries.

§ 60. Physical Mechanisms of Instability of the Motionless State

When the fluid is heavier above, it tends to be gravitationally unstable. It may be more dense above if it is saltier or cooler there. A lighter fluid element displaced upward will experience a lift (Archimedes principle) which will tend to move the displaced fluid element further up. In the course of its upward travel, the density difference between the fluid element and its surroundings is diminished by diffusion and conduction. If the upward velocity is fast relative to the rate of conduction and diffusion, the driving buoyant force (density difference) will not be erased and the motion will be unstable.

The following argument shows how the Rayleigh number is the natural stability parameter. For simplicity, consider a fluid element of radius a in a temperature field of gradient $\Delta T/a$. Let the element center be ΔT degrees hotter than its environment in the horizontal plane containing its center. Assume that the element moves up fast enough so that in the distance a, it loses no more than ΔT degrees. This will be so if in time t, $\Delta T/t \simeq \kappa\Delta T/a^2$; that is, if it moves up with velocity of order κ/a. The motion will persist if the buoyant lift $(\frac{4}{3}\rho\pi a^3)\,\alpha g\Delta T$ on the particle exceeds its viscous drag. This drag, for slow convection velocities, can be approximated by Stokes' formula

$$6\pi\mu a U = 6\pi\mu a\,\kappa/a\,.$$

Hence, for instability, we should expect that

$$\alpha g\Delta T a^3/\nu\kappa > 1\,. \tag{60.1}$$

This combination of parameters defines a Rayleigh number, but is not a quantitative estimate of the instability criterion because there is no way to estimate the value of a. The best known example of buoyancy-induced instability occurs in the problem of a fluid layer heated from below which was studied by Bénard (1900, 1901).

Cellular convective motion of the Bénard type can be induced by surface-tension forces as well as by buoyancy forces. When a layer of liquid is heated from below (or above), small variations in the surface temperature can induce surface tractions, which cause the liquid below to flow, and under certain circumstances the induced motion will persist leading to heat transport by convection as well as conduction. Such convection could not, of course, be initiated in a confined layer without a free surface.

In shallow layers of liquids, the surface-tension instability can be produced at temperature gradients which are much smaller than those required for buoyancy-driven convection. In fact, Bénard's original observations of ordered hexagonal motion (see Fig. 1 in Chandrasekhar) in open layers of spermacetti (0.5—1.0 mm depth) are inconsistent with the buoyancy mechanism; the temperature gradients in Bénard's experiments are too small to produce buoyancy-driven convection, and the observed surface deflection is not consistent with such convection. The agreement between Rayleigh's (1916) predictions of the ratio of cell spacing to layer depth (unitless wave number = 3.26) and Bénard's (1900, 1901) observations so impressed Bénard and other early workers in the theory of convective instability, that the applicability of Rayleigh's buoyancy theory to the Bénard observations was assumed without question for a half-century. The observations of Cousens (cited in Pearson, 1958, and Block, 1956) and the analyses of Pearson (1958), Sternling and Scriven (1959), Scriven and Sternling (1964), Nield (1964) and Smith (1966) show conclusively that Bénard's hexagons were driven by surface tension.

A basic mechanism by which the surface-tension instability may be induced is in evidence in the force balance leading to (55.12)[9]. This relation, when applied to a disturbance $\mathbf{S}'$, σ' of $\mathbf{S}, \sigma$, leads to the expression

$$\mathbf{S}'\cdot\mathbf{n}-\mathbf{n}(\mathbf{n}\cdot\mathbf{S}'\cdot\mathbf{n})-\nabla\sigma'+\mathbf{n}(\mathbf{n}\cdot\nabla)\sigma'=0\,.$$

In liquids the interfacial tension (force/length) σ generally decreases with increasing temperature. Then $\sigma=\sigma_0-s(T-T_R)$, where T_R is a reference temperature and $s>0$, gives the linear approximation to the equation of state for the surface tension; and $\sigma'=-s\theta'$ where θ' is a disturbance of $T-T_R$.

Literature on surface tension instability is reviewed in the paper of Berg, Acrivos and Boudart (1966).

§ 61. Necessary and Sufficient Conditions for Stability

When linear and energy stability limits coincide, all stable disturbances, whatever their size, will decay exponentially from the initial instant. In this situation, subcritical instabilities cannot occur. The energy and linear limits coincide for

[9] Surface-tension problems are among those for which the position of the interface cannot be prescribed in advance. Some authors (see Smith, 1966) prescribe a flat surface but allow surface-tension gradients. Using this approximation, Davis (1969 B) has given an energy stability analysis for the stability of the rest state and Gumerman and Homsy (1975) have given a linear and energy stability analysis for plane Couette flow of superposed fluids heated from below. Gumerman and Homsy (1975) have developed an energy theory for problems in which the boundary temperatures or velocities are changed impulsively. They apply their theory to the problem of an impulsively cooled liquid layer susceptible to instabilities driven by surface-tension gradients.

a class of problems which include the constant gradient case (the Bénard problem). It is easy to give *a priori* conditions which are sufficient to guarantee this coincidence.

If the basic state is motionless, then $\mathbf{U}=0$. The spectral problem of linear theory can be obtained from (56.3) by linearization and the use of the exponential time factor $e^{-\sigma t}$. One finds that

$$-\sigma\mathbf{u} = -\nabla p - \mathbb{R}\boldsymbol{\eta}(\theta\mathscr{A}_{\mathscr{R}} - \gamma\mathscr{A}_{\mathscr{C}}) + \nabla^2 u\,, \tag{61.1 a}$$

$$-\sigma\mathscr{P}_T\theta = -\mathbb{R}\mathscr{A}_{\mathscr{R}}\mathbf{u}\cdot\boldsymbol{\eta}_T + \nabla^2\theta\,, \tag{61.1 b}$$

and

$$-\sigma\mathscr{P}_C\gamma = -\mathbb{R}\mathscr{A}_{\mathscr{C}}\mathbf{u}\cdot\boldsymbol{\eta}_C + \nabla^2\gamma \tag{61.1 c}$$

where

$$\mathbb{R} = (\mathscr{R}^2 + \mathscr{C}^2)^{1/2}\,.$$

The energy equations for the functional (58.4) are obtained as in § 56 and have the form

$$\frac{\mathbb{R}}{2}\begin{Bmatrix} \mathscr{A}_{\mathscr{R}}(\sqrt{\lambda_T}\boldsymbol{\eta}_T + \boldsymbol{\eta}/\sqrt{\lambda_T})\theta + \mathscr{A}_{\mathscr{C}}(\sqrt{\lambda_C}\boldsymbol{\eta}_C - \boldsymbol{\eta}/\sqrt{\lambda_C})\gamma \\ \mathscr{A}_{\mathscr{R}}(\sqrt{\lambda_T}\boldsymbol{\eta}_T + \boldsymbol{\eta}/\sqrt{\lambda_T})\cdot\mathbf{u} \\ \mathscr{A}_{\mathscr{C}}(\sqrt{\lambda_C}\boldsymbol{\eta}_C - \boldsymbol{\eta}/\sqrt{\lambda_C})\cdot\mathbf{u} \end{Bmatrix} = \nabla^2\begin{Bmatrix}\mathbf{u}\\ \theta \\ \gamma\end{Bmatrix} - \begin{Bmatrix}\nabla p\\ 0\\ 0\end{Bmatrix}. \tag{61.2 a, b, c}$$

Here, the values of λ_T and λ_C satisfy the conditions

$$\lambda_T = \frac{\langle\boldsymbol{\eta}\cdot\mathbf{u}\theta\rangle}{\langle\boldsymbol{\eta}_T\cdot\mathbf{u}\theta\rangle}, \qquad \lambda_C = \frac{-\langle\boldsymbol{\eta}\cdot\mathbf{u}\gamma\rangle}{\langle\boldsymbol{\eta}_C\cdot\mathbf{u}\gamma\rangle} \tag{61.3}$$

required on the optimum energy stability boundary. The boundary conditions are as in (56.3).

Now we are ready to state necessary and sufficient conditions for the global stability of the motionless solution. *Let the spatial variations of the gravity-vector field and temperature and concentration gradient fields of the motionless state be proportional so that* $\boldsymbol{\eta} = \boldsymbol{\eta}_T = -\boldsymbol{\eta}_C$. *Then,* $\mathbb{R}_L = \mathbb{R}_{\mathscr{E}}$. *Moreover, every solution of* (61.1) *has* $\operatorname{im}(\sigma)=0$ *(exchange of stability holds).*

Subcritical instabilities of this motionless state cannot exist when

$$(\mathscr{R}^2 + \mathscr{C}^2)^{1/2} < \mathbb{R}_L\,.$$

This theorem does not mean that exchange of stability and the nonexistence of subcritical instabilities are equivalent. Proof: Let $\bar{f}$ designate the complex conjugate of f. Since σ is a complex number, Eqs. (61.1) can have complex values.

One finds from (61.1) that

$$-\sigma\langle|\mathbf{u}|^2+\mathscr{P}_T|\theta|^2+\mathscr{P}_C|\gamma|^2\rangle=-\mathbb{R}_L\mathscr{A}_{\mathscr{R}}\langle\overline{\mathbf{u}}\cdot\boldsymbol{\eta}\theta+\mathbf{u}\cdot\boldsymbol{\eta}\overline{\theta}\rangle \tag{61.4}$$
$$+\mathbb{R}_L\mathscr{A}_{\mathscr{C}}\langle\overline{\mathbf{u}}\cdot\boldsymbol{\eta}\gamma+\mathbf{u}\cdot\boldsymbol{\eta}\overline{\gamma}\rangle+\langle\overline{\mathbf{u}}\cdot\nabla^2\mathbf{u}+\overline{\theta}\nabla^2\theta+\overline{\gamma}\nabla^2\gamma\rangle\,.$$

The right side of (61.4) is real for any of the boundary conditions of (56.3). Hence, $\mathrm{im}(\sigma)=0$.

Given $\boldsymbol{\eta}=\boldsymbol{\eta}=-\boldsymbol{\eta}_C$, one finds from (61.3) that $\lambda_T=\lambda_C=1$. Then, comparing (61.1) with $\sigma=0$ and (61.2) with $\lambda_T=\lambda_C=1$, one sees that $\mathbb{R}_{\mathscr{E}}=\mathbb{R}_L$, proving the theorem.

§ 62. The Bénard Problem

The best known and best understood instance of thermally-induced instability is the Bénard problem. In this problem $\boldsymbol{\eta}_T=\boldsymbol{\eta}=-\mathbf{e}_z$ (heated from below) and the fluid is homogeneous. Under these conditions the stability theorem of § 61 holds[10].

Since subcritical solutions of the Bénard problem are not possible we may conclude that the first critical Rayleigh number $\mathscr{R}=\mathscr{R}_B$ (that is, the smallest $\mathscr{R}>0$ for which $\sigma=0$) of the spectral problem (62.1) for Bénard convection is the global stability limit $\mathscr{R}_B=\mathscr{R}_G$. We have global stability when $\mathscr{R}<\mathscr{R}_B$ and instability when $\mathscr{R}>\mathscr{R}_B$.

The spectral problem for Bénard convection is

$$-\sigma\mathbf{u}=-\nabla p+\mathscr{R}\theta\mathbf{e}_z+\nabla^2\mathbf{u}\,, \tag{62.1 a}$$

$$-\sigma\theta=\mathscr{R}w+\nabla^2\theta\,,\qquad \operatorname{div}\mathbf{u}=0 \tag{62.1 b, c}$$

$$\mathbf{u}\cdot\mathbf{n}|_S=0\,,\qquad \theta|_{S_T}=0\,,\qquad \frac{\partial\theta}{\partial n}+h\theta|_{S-S_T}=0\,, \tag{62.1 d, e, f}$$

$$\mathbf{u}|_{S_U}=0\quad\text{and}\quad \mathbf{D}[\mathbf{u}]\cdot\mathbf{n}-\mathbf{n}(\mathbf{n}\cdot\mathbf{D}[\mathbf{u}]\cdot\mathbf{n})|_{S-S_U}=0\,. \tag{62.1 g, h}$$

Though the Bénard problem cannot be "integrated in general", one can always "integrate" the equation $\operatorname{div}\mathbf{u}=0$ by introducing the poloidal potential χ and the toroidal potential Ψ. Then, instead of three unknown components of $\mathbf{u}$

---

[10] The first proof of the principle of exchange of stability in a bounded domain is due to Sorokin (1953, 1954). Sorokin was also the first to show that $\mathscr{R}_B=\mathscr{R}_G$. For steady flows, Sorokin's result was rediscovered by Ukhovski and Yudovich (1963) and by Sani (1964). This same result was reproduced for statistically stationary motions in fluid layers by Howard (1963). Joseph (1965, 1966) proved the $\mathscr{R}_B=\mathscr{R}_G$ result independently under slightly more general boundary conditions and for variable $\boldsymbol{\eta}$ and $\boldsymbol{\eta}_T$.

we need only to find two fields, χ and Ψ. Referring to the decomposition discussed in Appendix B.6 we note that any solenoidal fluid may be written as

$$\mathbf{u} = \operatorname{curl}^2 \mathbf{e}_z \chi + \operatorname{curl} \mathbf{e}_z \Psi = \mathbf{u}_1 + \mathbf{u}_2$$

where $\mathbf{u}_1$ has no vertical vorticity and $\mathbf{u}_2$ no vertical velocity.

The Bénard problem may be considerably simplified when the vertical vorticity $\zeta = \mathbf{e}_z \cdot \operatorname{curl} \mathbf{u}$ vanishes. Then $\mathbf{u} = \mathbf{u}_1$, and the velocity field is purely poloidal. When the velocity field is purely poloidal (62.1 a, b) may be reduced to

$$\sigma \nabla^2 \chi + \nabla^4 \chi - \mathscr{R}\theta = 0 , \qquad \sigma\theta - \mathscr{R}\nabla_2^2 \chi + \nabla^2 \theta = 0 . \tag{62.2 a, b}$$

The vertical vorticity of two-dimensional motions in roll cells always vanishes. Only very special three-dimensional motions can have a vanishing vertical vorticity.

The spectral problem for Bénard convection has this special property of vanishing vertical vorticity when the region occupied by the fluid is an infinitely extended horizontal layer. This problem is very thoroughly treated in the papers of Pellew and Southwell (1940) and of Reid and Harris (1958). The results of these authors are summarized in the book by Chandrasekhar (1961)[11].

Consider problem (62.1) when the domain is a fluid layer and $\mathbf{u}$ and p are assumed to be almost periodic functions of x and y. The boundary conditions (62.1 d—h) are assumed to be uniform, but possibly different, on each of the planes $z = 0, 1$. Thus, at $z = 0, 1$

$$\mathbf{u} = 0 \quad \text{or} \quad \left(w, \frac{\partial u}{\partial z}, \frac{\partial v}{\partial z} \right) = 0 \tag{62.1 i, j}$$

and

$$\frac{\partial \theta}{\partial z} + h\theta = 0 \tag{62.1 k}$$

where h is constant on $z = 0$ and $z = 1$.

Every eigensolution of (62.1 a, b, c, i, j, k) with eigenvalue $\sigma \leqslant \min \langle |\nabla \zeta|^2 \rangle / \langle \zeta^2 \rangle$ has

$$\zeta \equiv \mathbf{e}_z \cdot \operatorname{curl} \mathbf{u} = \frac{\partial v}{\partial x} - \frac{\partial u}{\partial y} = 0 .$$

[11] Computational analysis of the Bénard problem in a bounded domain is more difficult. Davis (1967) has given a Galerkin approximate analysis of the linearized Bénard problem in a rectangular box. Charlson and Sani (1970) have given a similar analysis when the fluid is contained in a right circular cylinder. Charlson and Sani (1975) have also given a Galerkin approximate analysis of properties and stability of axisymmetric bifurcating solutions in the cylinder. Stork and Muller (1972, 1975) have reported experiments which show good agreements with the computations of Davis (1967) and Sani (1970). The review of Koschmieder (1974) gives a critical evaluation of theory and experiment for the nonlinear Bénard problem.

To prove this we note that $-\sigma\zeta=\nabla^2\zeta$ and either $\zeta=0$ or $\dfrac{\partial\zeta}{\partial z}=0$ when $z=0,1$. Hence,

$$\sigma\langle\zeta^2\rangle=\langle|\nabla\zeta|^2\rangle\,. \tag{62.3}$$

Since $\zeta=0$ in the fluid layer (62.1 a, b, c, i, j, k) may, without loss of generality, be reduced to (62.2 a, b) where

$$\chi=\frac{\partial\chi}{\partial z}=0 \quad \text{at a rigid surface, and} \tag{62.2 c}$$

$$\chi=\frac{\partial^2\chi}{\partial z^2}=0 \quad \text{at a free surface}\,. \tag{62.2 d}$$

Eq. (62.2 a, b) can be reduced to ordinary differential equations by the separability hypothesis,

$$\begin{Bmatrix}\chi(x,y,z)\\ \theta(x,y,z)\end{Bmatrix}=\begin{Bmatrix}\hat\chi(z)\\ \hat\theta(z)\end{Bmatrix}\phi(x,y) \tag{62.4 a, b}$$

of Pellew and Southwell (1940). Here, $\phi(x,y)$ is called a "plan form", and it satisfies $\nabla_2^2\phi=-a^2\phi$. Every almost periodic, plan-form function is necessarily the uniform limit of polynomials of exponentials with terms of the form

$$\exp\{i(\alpha x+\beta y)\}$$

where $\alpha^2+\beta^2=a^2$. Not every plan form function is almost periodic. For example, the Bessel function $J_0(\lambda r)$ has $\nabla_2^2 J_0+\lambda^2 J_0=0$ but is not almost periodic. In fact, any quadratically integrable function must have a vanishing almost periodic norm (property 8, App. A) and could not be almost periodic.

The assumption that solutions of (62.2 a, b) can be written as a series with terms of the form (62.4) leads through orthogonality $\langle\phi_n\phi_l\rangle=\delta_{nl}$ to the problem

$$\sigma\begin{bmatrix}L\hat\chi\\ \hat\theta\end{bmatrix}+\begin{bmatrix}L^2\hat\chi\\ L\hat\theta\end{bmatrix}-\mathscr{R}\begin{bmatrix}\hat\theta\\ -a^2\hat\chi\end{bmatrix}=0 \tag{62.5 a, b}$$

with $L=D^2-a^2$ where

$$D\hat\theta+h\hat\theta=\hat\chi=0 \quad \text{and} \quad D\hat\chi \text{ or } D^2\hat\chi=0 \quad \text{at} \quad z=0,1\,. \tag{62.5 c, d}$$

The stability limit $\mathscr{R}=\mathscr{R}_B$ for the linearized Bénard problem is now found as the smallest of eigenvalues $\mathscr{R}(a^2)$ when $\sigma=0$, that is

$$\tilde{\mathscr{R}}_B=\min_{a^2}\mathscr{R}(a^2)=\mathscr{R}(\tilde a^2)\,.$$

In our example, we will favor the conducting boundary condition $\hat\theta=0$ at $z=0,1$. This corresponds to prescribed temperature conditions on the boundary; it may be obtained as a limiting case in which $h\to\infty$ (see Exercise 62.1).

Exercise 62.1: Show that

$$\frac{\partial \mathscr{R}(a^2,h)}{\partial h} > 0 \quad \text{and} \quad \frac{\partial^2 \mathscr{R}(a^2,h)}{\partial h^2} \leqslant 0 .$$

(*Hint*: see Appendix B4.3).

Exercise 62.2: (a) Show that the linearized Bénard problem (62.1) is self-adjoint. (b) Prove that σ is real. (c) Find an upper bound for $\sigma(\mathscr{R})$. (d) Find a variational characterization for the critical values (a^2). (e) Are there a finite number of eigenvalues $\sigma(\mathscr{R})$? What can be said about the multiplicity of the eigenvalues $\sigma(\mathscr{R})$ (see Appendix B9)?

Exercise 62.3: Consider the eigenvalues σ at criticality $\sigma(\mathscr{R}(a^2))=0$. Show that for each critical value $\mathscr{R}>0$, there is also an eigenvalue $-\mathscr{R}<0$. Show that there is a smallest critical alue $\mathscr{R}_B$ and there are no critical values of $\mathscr{R}$ in the interval $(-\mathscr{R}_B, \mathscr{R}_B)$.

Exercise 62.4: Consider the linearized Bénard problem in a right vertical cylinder of arbitrary cross section bounded at $z=0,1$ by free, conducting surfaces on which

$$\theta = w = \partial u/\partial z = \partial v/\partial z = 0 . \tag{62.6}$$

Show that the solutions of this problem are necessarily separable:

$$\begin{Bmatrix} \theta(x,y,z) \\ w(x,y,z) \end{Bmatrix} = \sin n\pi z \begin{Bmatrix} \theta_n(x,y) \\ w_n(x,y) \end{Bmatrix} \qquad (n>0)$$

and

$$\begin{Bmatrix} u(x,y,z) \\ v(x,y,z) \\ p(x,y,z) \end{Bmatrix} = \cos n\pi z \begin{Bmatrix} u_n(x,y) \\ v_n(x,y) \\ p_n(x,y) \end{Bmatrix}.$$

Find the boundary value problem satisfied by w_n, u_n, v_n, θ_n and p_n.

Exercise 62.5: Consider the linearized Bénard problem for the fluid layer with boundary condition (62.6). Show that this problem may be reduced to

$$-\sigma\begin{bmatrix} (n^2\pi^2+a^2)\chi_n \\ -\theta_n \end{bmatrix} + \begin{bmatrix} (n^2\pi^2+a^2)^2\chi_n \\ -(n^2\pi^2+a^2)\theta_n \end{bmatrix} - \mathscr{R}\begin{bmatrix} \theta_n \\ -a^2\chi_n \end{bmatrix} = 0$$

where χ_n and θ_n are constants and $n=1,2\ldots$. Show that

$$\mathscr{R}_B = (27\pi^4/4)^{1/2} .$$

Exercise 62.6 (Hales, 1937): Consider the problem posed in Exercise 62.4 when the cylinder has a circular cross section. Expand the functions $w_n(r,\theta)$, θ_n, u_n, etc. into Fourier series in the periodic variable $0 \leqslant \theta \leqslant 2\pi$ and find the ordinary differential equations which govern the Fourier coefficients.

Exercise 62.7 (Rabinowitz, 1968): Suppose that $\sigma=0$ and that $\hat{\theta}=0$ at $z=0,1$. Reduce problem (62.5) to a single integral equation whose kernel is the composition of Green functions for L and L^2. Use the results cited in Appendix D to show that this kernel is an "oscillation kernel". What can than be said about the eigenvalues $\mathscr{R}(a^2)$? Invert the problem (62.5) when $\sigma \neq 0$ and discuss the conditions under which the kernel of the resulting integral equation is an oscillation kernel.

§ 63. Plane Couette Flow Heated from below

An OB fluid is confined between two infinite horizontal planes each of which is translating with a uniform velocity in its own plane. The difference between the plate speeds is U, the distance between the plates is l and the bottom plate is ΔT degrees hotter than the top. The basic flow is a plane Couette flow

$$\boldsymbol{\eta}=\boldsymbol{\eta}_T=-\mathbf{e}_z\,,\qquad \mathbf{U}=2\mathbf{e}_x z\,. \tag{63.1}$$

We consider prescribed boundary conditions: $\mathbf{u}=\theta=0$ at $z=\pm\frac{1}{2}$. The disturbances of (63.1) satisfy the evolution equation (57.1) with $\gamma=h_T=h_C=0$.

The basic flow (63.1) *is globally and monotonically stable when*

$$\mathbb{R}^2\equiv R^2+\mathscr{R}^2<\mathbb{R}_{\mathscr{E}}^2=1708\,. \tag{63.2}$$

The proof of (63.2) follows from the solution of the problem

$$\frac{1}{\mathbb{R}_{\mathscr{E}}}=\max_{\mathbf{H}}\frac{\left\langle w\left[2\mathscr{A}_R u-\mathscr{A}_{\mathscr{R}}\dfrac{1+\lambda}{\sqrt{\lambda}}\theta'\right]\right\rangle}{\langle|\nabla\mathbf{u}|^2+|\nabla\theta'|^2\rangle}\,. \tag{63.3}$$

This problem is left for the reader in Exercise (63.1).

The maximizing field for (63.3) is independent of the direction x of the basic motion. The difference between this problem and the Bénard problem, to which it is closely allied, is that the presence of a shearing flow gives the problem a basic directionality. The x-independent motions do appear to arise in experiments on the stability of shear flows heated from below when the shearing is not too great[12].

The basic stability problem for (63.1) simplifies considerably if the disturbances are assumed from the start to be independent of x. Then we may set $\nabla^2=\partial_{yy}^2+\partial_{zz}^2$

$$\frac{\partial v}{\partial y}+\frac{\partial w}{\partial z}=0\,, \tag{63.4}$$

and

$$(\partial_t+v\partial_y+w\partial_z)\begin{bmatrix}u\\ v\\ w\\ \mathscr{P}_T\theta\end{bmatrix}+\begin{bmatrix}2Rw\\ 0\\ -\mathscr{R}\theta\\ -\mathscr{R}w\end{bmatrix}=\begin{bmatrix}\nabla^2 u\\ \nabla^2 v-\partial_y P\\ \nabla^2 w-\partial_z P\\ \nabla^2\theta\end{bmatrix}\,. \tag{63.5}$$

The disturbance equations for v, w, θ may be identified as those which govern roll cell disturbances of a motionless fluid heated from below (Bénard problem);

[12] Graham (1933) and others have observed convection rolls with axes perpendicular to the shear at small rates of shear. Nield (1975) has argued that the appearance of cross-rolls is an effect of a mean temperature gradient in the direction of the shear.

u may be computed once v and w are known. From this identification, after introducing the energy

$$\mathscr{E}(t)=\tfrac{1}{2}\langle v^2+w^2+\mathscr{P}_T\theta^2\rangle,$$

we can prove that:

Disturbances of plane Couette flow heated from below which are x-independent decay under the inequality

$$\mathscr{E}(t)<\mathscr{E}(0)\exp\{-2\Lambda(1-\mathscr{R}/\sqrt{1708})t\}$$

whenever

$$\mathscr{R}<\sqrt{1708} \tag{63.6}$$

independent of the value of R. Moreover, the criterion (63.6) *is necessary for stability as well as sufficient for stability to x-independent disturbances.*

The decay constant Λ is defined by (58.2) with $h_T=0$.

Gumerman and Homsy (1975) have generalized the analysis just given to two-phase flows separated by a flat interface which can move under the action of surface tension gradients. They also have analyzed the spectral problem under the same assumptions about the interface.

Exercise 63.1 (Joseph, 1966): Prove (63.2) and (63.3). *Hint*: reduce the problem to the one treated in § 48 and show that the maximizing functions are x-independent. Then reduce the problem to the one treated in § 62.

Exercise 63.2: Consider x-independent disturbances of the parallel motion between tilted parallel planes which is given in Exercise 59.1. Find

(i) an energy identity for the temperature disturbance,
(ii) an energy identity for the longitudinal component of the disturbance velocity,
(iii) an energy identity for the other two components,
(iv) a coupling identity for the product of the temperature and longitudinal component of velocity.

The problem of global stability of flow between heated-tilted planes has been considered by Ayyaswamy (1974). The spectral problem for the stability of this flow has been studied by Gershuni and Zhukhovitskii (1969) and by Hart (1971).

§ 64. The Buoyancy Boundary Layer

Consider the motion of an OB fluid in the half-space $x>0$ bounded by an infinite vertical wall at $x=0$. At each height z this wall is ΔT degrees hotter than the fluid at infinity; the vertical temperature stratification at $x=0$ and $x=\infty$ is stabilizing: $\partial T/\partial z=c>0$ where c is a constant. In this configuration buoyant forces drive a boundary layer flow along the wall. Our interest is in the stability of the flow; more particularly, in the aspects of energy stability theory which are introduced by the fact that this unbounded domain cannot be contained in a strip. Our analysis modifies one given by Dudis and Davis (1971 A).

The buoyancy boundary layer is an exact solution of the OB equations (54.2) and (54.3) for a homogeneous fluid ($\alpha_C = 0$) without heat sources ($Q_T = 0$) which satisfies the boundary conditions

$$\mathbf{U} = 0\,, \qquad T = \Delta T + cz|_{x=0}$$

and

$$\mathbf{U} \to 0\,, \qquad T \to cz|_{x\to\infty}\,.$$

The solution (Gill, 1966) is

$$\mathbf{U} = (U, V, W) = (0, 0, W(x))\,, \tag{64.1 a}$$

$$W = W_0 e^{-x/L} \sin(x/L) \tag{64.1 b}$$

and

$$T = (\Delta T) e^{-x/L} \cos(x/L) + cz \tag{64.1 c}$$

where

$$L = (4\nu\kappa/\alpha g c)^{1/4} \quad \text{and} \quad W_0 = (\alpha g \kappa / \nu c)^{1/2} \Delta T\,.$$

Prandtl's (1952) "mountain and valley winds in stratified air" contains this solution as a special case when the surface under consideration is vertical rather than slanted. The specific solution used here was given in Gill (1966) in connection with the problem of thermal convection in a rectangular cavity (see also Batchelor, 1954). It is an approximation to the solution at the vertical midpoint of the cavity, far away from the horizontal walls at the top and bottom of the cavity. Solutions of similar form have been found by Barcilon and Pedlosky (1967) for vertical boundary layers in a rotating, strongly stratified fluid. The buoyancy boundary layer consistently recurs in stratified fluid systems sustaining horizontal temperature gradients.

The OB equations which govern disturbances of (64.1) are the first two of Eqs. (56.3 a) (with $\gamma = 0$). The disturbances must also take on prescribed values at the wall ($x = 0$) and far away from the wall:

$$\mathbf{u} = \theta|_{x=0} = 0 \quad \text{and} \quad \mathbf{u} \to 0\,, \qquad \theta|_{x\to\infty} \to 0\,.$$

The disturbances are assumed to be almost periodic in y and z and absolutely integrable on $0 \leqslant x < \infty$.

The energy identities for the buoyancy boundary layer in dimensionless variables are

$$\frac{1}{2}\frac{d}{dt}\langle|\mathbf{u}|^2\rangle = -[\![RuwdW/dx + 2\boldsymbol{\eta}\cdot\mathbf{u}\theta + \nabla\mathbf{u}:\nabla\mathbf{u}]\!]\,, \tag{64.2a}$$

$$\frac{1}{2}\mathscr{P}\frac{d}{dt}\langle\theta^2\rangle = -[\![\mathscr{P}R\theta\mathbf{u}\cdot\nabla T + \nabla\theta\cdot\nabla\theta]\!]\,, \tag{64.2b}$$

where $\boldsymbol{\eta}=(0, 0, -1)$, the Reynolds number R is defined by $R=W_0L/\nu=\Delta T(4\alpha g\kappa^3/\nu^5c^3)^{1/4}$ and the Prandtl number by $\mathscr{P}=\nu/\kappa$. The physical variables may be obtained by multiplying dimensionless variables by scale factors $[L, W_0, \Delta T, L^2/\nu]=[$length, velocity, temperature, time$]$. The basic state in dimensionless variables is

$$W=e^{-x}\sin x$$

and

$$T=e^{-x}\cos x+2z/\mathscr{P}R$$

defined in

$$\mathscr{V}=[(x, y, z): x\geqslant 0, -\infty<y, z<\infty]\,.$$

The bracket $[\![\ ]\!]$ is defined by

$$[\![\bar{\circ}]\!]=\int_0^\infty \bar{\circ}\,dx$$

where

$$\bar{\circ}=\lim_{L\to\infty}\frac{1}{4L^2}\int_{-L}^{L}\int \circ\,dy\,dz\,.$$

Consider the energy equation obtained by adding (64.2a) and (64.2b)

$$\frac{d\mathscr{E}}{dt}=R\mathscr{I}-\mathscr{D} \tag{64.4}$$

where

$$\mathscr{D}=[\![|\nabla u|^2+|\nabla v|^2+|\nabla w|^2+|\nabla\theta|^2]\!]$$

and

$$\mathscr{I}=-[\![uwdW/dx+\mathscr{P}u\theta\partial T/\partial x]\!]\,.$$

Note that terms involving the product $w\theta$ do not appear in $\mathscr{I}$. Further, we define

$$\frac{1}{R_{\mathscr{E}}}=\sup \mathscr{I}/\mathscr{D} \tag{64.5}$$

where the supremum is taken over functions (u, v, w, θ) which have $\operatorname{div}\mathbf{u}=0$, vanish at $x=0$, are absolutely integrable in $0\leqslant x\leqslant\infty$ and are almost periodic in y and z.

It needs to be shown that a finite value $R_{\mathscr{E}}$ solving (64.5) exists. It suffices to exhibit an upper bound for $\mathscr{I}/\mathscr{D}$; then $\mathscr{I}/\mathscr{D}$ has a least upper bound. We first note using Schwarz's inequality and the inequality $2ab\leqslant a^2+b^2$ that

$$\begin{aligned}\mathscr{I}&\leqslant\tfrac{1}{2}[\![\sqrt{(dW/dx)^2}\,(u^2+w^2)+\mathscr{P}\sqrt{(\partial T/\partial x)^2}\,(u^2+\theta^2)]\!]\\&\leqslant\tfrac{1}{2}[\![\sqrt{(dW/dx)^2+\mathscr{P}^2(\partial T/\partial x)^2}\,(u^2+v^2+w^2+\theta^2)]\!]\,.\end{aligned} \tag{64.6}$$

Since

$$dW/dx\leqslant\sqrt{2}e^{-x}$$

and

$$dT/dx\leqslant\sqrt{2}e^{-x},$$

we have that

$$\mathscr{I}/\mathscr{D}\leqslant\frac{\sqrt{2}}{2}\sqrt{1+\mathscr{P}^2}[\![e^{-x}\mathbf{q}\cdot\mathbf{q}]\!]/\mathscr{D}\equiv\sqrt{\frac{1+\mathscr{P}^2}{2}}\,M(\mathbf{q}) \tag{64.7}$$

where $\mathbf{q}=(u, v, w, \theta)$.

Let M_1 be the supremum of M over admissible functions. These have $\operatorname{div}\mathbf{u}=0$ and vanish at $x=0$ and $x=\infty$. Let M_2 be the supremum of M over a widened class of admissible functions which include $\operatorname{div}\mathbf{u}\neq 0$ and $\mathbf{u}<\infty$ and $\theta<\infty$ at $x=\infty$ but which retain the conditions $\mathbf{u}=\theta=0$ at $x=0$. Clearly,

$$M_1\leqslant M_2=[\![e^{-x}|\mathbf{q}|^2]\!]/\mathscr{D}\leqslant[\![e^{-x}|\mathbf{q}|^2]\!]/[\![|\partial\mathbf{q}/\partial x|^2]\!]\,. \tag{64.8}$$

Then, with $\phi^2=\overline{|\mathbf{q}|^2}$ we have

$$\phi\frac{d\phi}{dx}=\overline{\mathbf{q}\cdot\frac{\partial\mathbf{q}}{\partial x}}\leqslant\sqrt{\overline{|\mathbf{q}|^2}}\sqrt{\overline{\left|\frac{\partial\mathbf{q}}{\partial x}\right|^2}}$$

and, upon squaring,

$$\left|\frac{d\phi}{dx}\right|^2\leqslant\overline{\left|\frac{\partial\mathbf{q}}{\partial x}\right|^2}\,.$$

Hence,

$$M_2 \leqslant \int_0^\infty \phi^2 e^{-x} dx \Big/ \int_0^\infty \left|\frac{d\phi}{dx}\right|^2 dx. \tag{64.9}$$

On the other hand the vector $\mathbf{q}_1 = (\phi, \phi, \phi, \phi)$ has $\mathbf{q}_1 = 0$ at $x=0$ and $|\mathbf{q}_1| < \infty$ at $x = \infty$ since these are properties of $\phi = \phi(x)$. Hence, $\mathbf{q}_1$ is a vector admissible in the competition for the supremum of M_2; we find that

$$[\![e^{-x}\mathbf{q}_1 \cdot \mathbf{q}_1]\!] / \mathscr{D} = \int_0^\infty e^{-x}\phi^2 dx \Big/ \int_0^\infty \left|\frac{d\phi}{dx}\right|^2 dx \leqslant M_2. \tag{64.10}$$

Comparison of (64.9) and (64.10) shows that the supremum for M_2 can be found among functions $\phi(x)$ of x alone which vanish at $x=0$ and are bounded at $x=\infty$.

To find M_2 it is convenient to map the semi-infinite interval $0 \leqslant x \leqslant \infty$ into the unit strip $1 \geqslant \hat{x} \geqslant 0$:

$$\hat{x} = e^{-x/2}.$$

This leads to the variational problem

$$M_2 = \sup\left\{\int_0^\infty e^{-x}\phi^2 dx \Big/ \int_0^\infty \left|\frac{d\phi}{dx}\right|^2 dx\right\} = 4 \sup \int_0^1 \hat{x}\phi^2 d\hat{x} \Big/ \int_0^1 \hat{x} \left|\frac{d\phi}{d\hat{x}}\right|^2 d\hat{x}$$

where the supremum is over functions $\phi(\hat{x})$ which have $\phi(0) < \infty$ and $\phi(1) = 0$. This supremum is attained by an admissible function which satisfies the Euler equation

$$\frac{d^2\phi}{d\hat{x}^2} + \frac{1}{\hat{x}}\frac{d\phi}{d\hat{x}} + \frac{4}{M_2}\phi = 0, \qquad \phi(0) < \infty, \qquad \phi(1) = 0.$$

Hence, $\phi = J_0(2\hat{x}/\sqrt{M_2})$ and $2/\sqrt{M_2}$ is the first positive zero of $J_0(x)$; that is, $2/\sqrt{M_2} = 2.405$. Returning now to (64.5) and (64.7) we find

$$\frac{1}{R_{\mathscr{E}}} \leqslant \sqrt{\frac{1+\mathscr{P}^2}{2}}\, \frac{4}{(2.405)^2}.$$

Hence, there is stability for the buoyancy boundary layer when

$$R \leqslant (2.405)^2/2\sqrt{2(1+\mathscr{P}^2)} \leqslant R_{\mathscr{E}}.$$

The values $R_{\mathscr{E}}(\mathscr{P})$ for the problem (64.5) have been calculated numerically by Dudis and Davis (1971A) and compared with values R_L computed by Gill and Davey (1969) in their study of the stability of the buoyancy boundary layer from the point of view of linear stability theory. In a comparison paper Dudis and Davis (1971B) also give an energy stability analysis for the Ekman boundary layer.

Chapter IX

Global Stability of Constant Temperature-Gradient and Concentration-Gradient States of a Motionless Heterogeneous Fluid

Motionless states are more complicated in heterogeneous fluids than in homogeneous fluids. Instability can be associated with a difference in the ability of the fluid to diffuse different fields. For example, in the ocean the diffusion of heat can be one hundred times faster than the diffusion of salt. Unlike the motionless homogeneous fluid, the motionless heterogeneous fluid can lose its stability to a time-periodic motion which can exist even when the heavier fluid is on the bottom.

The stability problem for motionless states of a heterogeneous fluid has especially interesting and unexpected physical properties. In addition, this problem occupies an important place in our global theory of stability because we are able to carry a generalized energy analysis of it through to completion.

§ 65. Mechanics of the Instability of the Conduction-Diffusion Solutions in a Motionless Heterogeneous Fluid

It is frequently true that the density differences which drive buoyant motions stem from differences in mass concentration (salty water, dissimilar gases, etc.) as well as from differences in temperature. The conduction-diffusion solutions are here regarded as the totality of motionless solutions of the Boussinesq equations when the density field depends on both the temperature and concentration fields of a two-component mixture.

In § 59 we showed that the conditions for the existence of motionless solutions are stringent. If the external conditions are steady and there are no sources of heat or mass, then the temperature and concentration fields have constant gradients parallel to gravity. There remain four possibilities corresponding to various combinations of heating and salting below. (Now and henceforth we speak of salt but mean solute.) We shall examine the global stability problem for these constant gradient solutions.

Heat and salt fields are called stabilizing if they promote stabilizing density gradients. Heating from above and salting from below are stabilizing. Heated above and salted below is easily shown (Exercise 65.1) to be absolutely, monotonically and globally stable. We will show in § 66 that for heated below and salty above, like the Bénard problem, the critical parameters of energy and linear theory coincide. The resulting criterion is both necessary and sufficient for global stability.

The two solutions in which the heat and salt gradients compete for the stability of the motionless solution are the interesting ones in that they allow for effects without parallel in the Bénard problem. In the competing fields problem, unlike the Bénard problem, convection can occur when the motionless density field is gravitationally stable; the convection can occur as a sustained oscillation rather than as a monotonic growth (PES does not hold) and the whole configuration can be subcritically unstable.

We shall here consider the problems associated with heating and salting from below. The other interesting problem, heating and salting from above (the ocean), can be obtained by merely interchanging the heat and salt parameters in the final result.

Heating and salting from below is perhaps best described in physical terms in the context of the "salt pond" (Tabor, 1963; Tabor and Matz, 1965; Weinberger, 1964).

The "pond" is a contained fluid layer which is both heated and salted below, so that the upper fluid layers thermally insulate the lower. Like the Dead Sea, the pond is washed by fresh water at its free surface and salted at its bottom, ensuring the existence of a stabilizing salt gradient in the vertical. The dark bottom of the pond is an effective absorber of radiative energy of the sun, which has the effect of heating the pond from below. Without the stable salt gradient, the limit of heating that could be achieved in this way is determined by the stability condition for the onset of convective motions. The stabilizing salt gradient enables significantly larger temperature differences to develop before the fluid turns over. This configuration and the insulating property induced by the stable salt gradient also seem to explain features of the observed temperature distributions in the Red Sea (Swallow and Crease, 1965) and in certain Hungarian (Weinberger, 1964) and Antarctic (Shirtcliffe, 1964; Hoare, 1966) salt lakes.

The ocean may, assuming generosity of the reader, be likened to a salt lake with the "up" side down. There is no doubt that in the typical configuration at sea, the fresh cool water is found in the depths with salty hot water above[1]. Stommel, Aarons and Blanchard (1956) have described an "oceanographical curiosity" by noting that if a long vertical tube was lowered into the ocean in such a manner that its bottom was exposed to cold fresh water and its top to warm saline water, a continuous motion could be maintained therein after priming the fountain. Their explanation is that the ascending (or descending) water in the tube would exchange heat but not salinity with the ambient ocean and would be accelerated by its deficit in salt and density relative to fluid at the same level outside the tube. This fountain is an example of convection in a gravitationally-stable field (the mean density of salt water outside the tube decreasing upward).

The phenomenon known as "salt fingering" is nature's way of simulating the "salt fountain" through a mechanism which is associated with the great differences of molecular diffusivities of salt and heat:

[1] An exception is the north Atlantic west of Gibraltar where fresher and cooler Atlantic water overlays hot and salty water from the Mediterranean.

$$\tau = \frac{\mathscr{P}_C}{\mathscr{P}_T} = \frac{\kappa_T}{\kappa_C} = \frac{1.5 \times 10^{-3}\,\mathrm{cm}^2\mathrm{sec}^{-1}}{1.3 \times 10^{-5}\,\mathrm{cm}^2\mathrm{sec}^{-1}} \simeq 100$$

for salinity of 3.5 % at 20° C. This mechanism was first described by Stern (1960). Consider the situation in which hot salty water overlies the fresh cooler water, but let the mean density decrease upward (gravitationally stable). Now displace downward a particle of hot salt water of small radius. The particle will lose both heat and salt to its fresher cooler environment, but the loss of heat will be roughly 100 times faster than the loss of salt. In this way, depending on the speed of the displacement, the displaced particle can become heavier than its neighbors and a self-excited motion can ensue[2].

The analogue mechanism for "heat fingering" which would be appropriate in the salt pond problem (if one could find a brand of salt for which $\tau \simeq \frac{1}{100}$) requires no special explanation. In the analysis we allow the parameters to take on all possible values. We shall treat the problem in which the fluid is heated and salted below. To have the result for heated and salted above, one needs only to interchange $\mathscr{R}$ and $\mathscr{C}$, $\mathscr{P}_T$ and $\mathscr{P}_C$ and h_T and h_C.

The bases for the stability analyses are the disturbance Eqs. (56.3) and the energy identities (56.4—7) with $R=0$.

Exercise 65.1: Consider a layer of salty fluid heated from above and salted from below ($\boldsymbol{\eta} = -\boldsymbol{\eta}_T = \boldsymbol{\eta}_C = -\mathbf{e}_z$). Find a positive definite functional of the disturbance which decays monotonically independent of the values $\mathscr{R}, \mathscr{C}$ or the size of the initial disturbance.

§ 66. Energy Stability of Heated below and Salted above

For this configuration $\boldsymbol{\eta} = \boldsymbol{\eta}_T = -\mathbf{e}_z = -\boldsymbol{\eta}_C$. Consider the energy equation (57.1). We can determine the values of λ_T and λ_C which appear in (57.1) directly, without solutions, from (58.11) (with $\tilde{\lambda} = \lambda_T$) and (58.12). The result is that $\lambda_T = \lambda_C = 1$. Then using (57.1) (with $\mathbf{U}=0$), one finds that

$$d\mathscr{E}/dt \leqslant -\mathscr{D}[1 - \mathscr{R}/\mathscr{R}_{\mathscr{E}}] \tag{66.1}$$

with

$$\frac{1}{\mathscr{R}_{\mathscr{E}}(\alpha)} = \max_H \frac{2\langle w\theta\rangle - 2\alpha\langle w\gamma\rangle}{\mathscr{D}[\mathbf{u}, \theta, \gamma]}, \tag{66.2}$$

where $\alpha = \mathscr{C}/\mathscr{R}$ and $\mathscr{D}$ is defined by (57.3). The Euler equations and boundary conditions for (66.2) are exactly (56.3) when in (56.3a), $R = \partial/\partial t = 0$, $\mathscr{R} = \tilde{\mathscr{R}}_{\mathscr{E}}$ and the equations are linearized. Since the energy limit $\tilde{\mathscr{R}}_{\mathscr{E}}(\alpha)$ coincides with the linear limit for neutral disturbances, $\mathscr{R}_L(\alpha)$, their common locus is the global

---

[2] An extensive list of references on salt fingering and other problems of convection in heterogeneous fluids can be found in the book of Turner (1973) and in the review paper of Schechter, Velarde and Platten (1974). Turner's book also contains a large number of photographs.

stability-instability boundary. Following the proof of the energy stability theorem (§ 4) we may prove the following theorem:

Theorem IX.1: *The constant-gradient conduction-diffusion solution for heated below and salted above is unconditionally and monotonically stable when*

$$\mathscr{R} < \mathscr{R}_{\mathscr{E}}(\alpha), \qquad \alpha = \mathscr{C}/\mathscr{R}$$

and is unstable when

$$\mathscr{R} > \mathscr{R}_{\mathscr{E}}(\alpha).$$

There can exist no steady subcritical convection when

$$\mathscr{R} < \tilde{\mathscr{R}}_{\mathscr{E}}(\alpha).$$

The strong stability here is accompanied by a decay estimate of the form (4.7).

§ 67. Heated and Salted from below: Linear Theory

Here $\boldsymbol{\eta} = \boldsymbol{\eta}_T = \boldsymbol{\eta}_C = -\mathbf{e}_z$. It will be convenient to distinguish three cases:

(i) Problem A: The domain is arbitrary and any appropriate combination of conditions (56.3 c, d, e) holds.

(ii) Problem B: The domain is arbitrary and $S = S_T = S_C$ ($\theta = \gamma = 0$ on the whole boundary). The boundary can be composed of both stress-free and rigid elements.

(iii) Problem C: The domain is a layer bounded by stress-free surfaces on which θ and γ are required to vanish. The disturbances are two-dimensional.

The spectral equations for each of these problems are:

$$-\sigma \mathbf{u} = -\nabla p + \mathbf{e}_z \mathscr{R}(\theta - \alpha\gamma) + \nabla^2 \mathbf{u}, \qquad \operatorname{div} \mathbf{u} = 0, \tag{67.1 a}$$

$$-\sigma \mathscr{P}_T \theta = \mathscr{R} w + \nabla^2 \theta, \tag{67.1 b}$$

and

$$-\sigma \mathscr{P}_C \gamma = \mathscr{R} \alpha w + \nabla^2 \gamma. \tag{67.1 c}$$

Here we have chosen $\alpha = \mathscr{C}/\mathscr{R}$ and seek the values $\mathscr{R} = \mathscr{R}_L(\alpha)$ corresponding to $\operatorname{re}(\sigma) = 0$ as critical values of linear stability theory.

We note that the value

$$\mathscr{R}_B^{-1} = \max_H \frac{2\langle w\theta \rangle}{\mathscr{D}[\mathbf{u}, \theta]} \tag{67.2}$$

is the smallest critical value for the Bénard problem.

Linear stability results which are to be used in subsequent analysis are summarized below.

Problem A: *In the limit* $\mathscr{P}_C \to \infty$,

$$\mathscr{R}_L(\alpha) \to \mathscr{R}_L(0) = \mathscr{R}_B . \tag{67.3}$$

Problem B: *The constant-gradient solution for heated and salted from below is unstable whenever*

$$\mathscr{R}^2 - \mathscr{C}^2 > \mathscr{R}_B^2 . \tag{67.4}$$

Problem C: *The constant-gradient solution for heated and salted from below is unstable whenever*

$$\mathscr{R}^2 > \begin{cases} \mathscr{R}_B^2 + \mathscr{C}^2 , & (\tau < 1) \\ \mathscr{R}_B^2 + \mathscr{C}^2 , & (\tau > 1, \mathscr{C}/\mathscr{C}^* < 1) . \\ \dfrac{\mathscr{R}_B^2(\tau+1)(1+\tau\mathscr{P}_T)}{\tau^2 \mathscr{P}_T} + \dfrac{(1+\tau\mathscr{P}_T)\mathscr{C}^2}{(1+\mathscr{P}_T)\tau^2} , & (\tau > 1, \mathscr{C}/\mathscr{C}^* > 1) \end{cases} \tag{67.5}$$

where $\mathscr{R}_B^2 = 657$ and $\mathscr{C}^{*2} = \mathscr{R}_B^2(\mathscr{P}_T + 1)/\mathscr{P}_T(\tau - 1)$.

The criterion (67.5) follows from explicit calculations of Sani (1963) and Veronis (1964). A relatively complete linear theory for problem A when the domain is a layer has been formulated by Nield (1967). Eqs. (67.3, 4) are taken from the work of Shir and Joseph (1968).

Proof of 67.3): Let an overbar designate the complex conjugate. Then from (67.1) one finds that

$$\begin{aligned} |\text{im}(\sigma)| &= \mathscr{R}\alpha \frac{|\text{im}\langle \gamma\bar{w} - \bar{\gamma}w \rangle|}{\langle |\mathbf{u}|^2 + \mathscr{P}_T|\theta|^2 + \mathscr{P}_C|\gamma|^2 \rangle} \\ &\leqslant \mathscr{R}\alpha \frac{2\langle |\gamma w| \rangle}{\langle |w|^2 + \mathscr{P}_C|\gamma|^2 \rangle} \leqslant \frac{\mathscr{R}\alpha}{\sqrt{\mathscr{P}_C}} \end{aligned}$$

where we have used the inequality $2\sqrt{\mathscr{P}_C}\langle |\gamma w| \rangle \leqslant \langle |w|^2 + \mathscr{P}_C|\gamma|^2 \rangle$. Suppose that at criticality $\sigma(\mathscr{R}) = i\,\text{im}\,\sigma(\mathscr{R})$ where

$$\text{im}(\sigma) = \frac{C_1 \mathscr{R}\alpha}{\sqrt{\mathscr{P}_C}} , \qquad -1 \leqslant C_1 \leqslant 1 , \qquad C_1 \neq 0 , \tag{67.6}$$

and since $\text{re}(\sigma) = 0$, (67.1) may be rewritten as

$$\nabla^2 \mathbf{u} + \mathscr{R}\theta \mathbf{e}_z - \nabla p = \mathscr{R}\alpha\gamma \mathbf{e}_z + \frac{C_1 \mathscr{R}\alpha \mathbf{u}}{\sqrt{\mathscr{P}_C}} i ,$$

$$\nabla^2 \theta + \mathscr{R} w = \frac{C_1 \mathscr{R}\alpha \mathscr{P}_T \theta}{\sqrt{\mathscr{P}_C}} i$$

and

$$C_1 \mathscr{R}\alpha\gamma i = \frac{1}{\sqrt{\mathscr{P}_C}}\{\mathscr{R}\alpha w + \nabla^2\gamma\}$$

in the limit $\mathscr{P}_C \to \infty$

$$\gamma \to 0\,, \quad \nabla^2\mathbf{u} + \mathscr{R}\theta\mathbf{e}_z - \nabla p \to 0\,, \quad \nabla^2\theta + \mathscr{R}w \to 0\,.$$

These equations hold relative to any domain and for any of the natural boundary conditions and define the critical value of linear stability theory for the Bénard problem $\mathscr{R} = \mathscr{R}_B$.

Proof of (67.4): Consider Problem B. We seek a solution of the problem with $\sigma = 0$. From (67.1 b, c), we see that $\theta/\mathscr{R} - \gamma/\alpha\mathscr{R}$ is harmonic and since this sum vanishes on the boundary, it must be that

$$\alpha\theta = \gamma\,. \tag{67.7}$$

Replacing γ in (67.1 a, b) with $\alpha\theta$ we find that

$$0 = -\nabla p + \mathbf{e}_z\theta\mathscr{R}(1-\alpha^2) + \nabla^2\mathbf{u}\,,$$

and

$$0 = \mathscr{R}w + \nabla^2\theta\,.$$

These equations do have eigensolutions for the boundary conditions of problem B. The smallest eigenvalue is

$$\mathscr{R}_B^2 = \mathscr{R}^2(1-\alpha^2) = \mathscr{R}^2 - \mathscr{C}^2\,.$$

Proof of (67.5): Consider Problem C. Here, $\theta = \gamma = w = \partial^2_{zz}w|_{z=\pm 1/2} = 0$. The analysis starts from (67.1 b, c) and the 4th-order equation for w which arises as the vertical component of the curl² of (67.1 a). These equations show that all the even-order derivatives of θ, γ, w wanish at $z = \pm\frac{1}{2}$, so that admissible solutions are series with terms of the form

$$\exp\{i(\tilde{\alpha}x + \tilde{\beta}y)\}\cos n\pi z\,, \quad \tilde{\alpha}^2 + \tilde{\beta}^2 = a^2$$

The equations for $\mathbf{e}_z \cdot \mathrm{curl}^2\mathbf{u}$, θ and γ may then be written as:

$$(n^2\pi^2 + a^2)(n^2\pi^2 + a^2 - \sigma)w_0 - a^2\mathscr{R}\theta_0 + a^2\mathscr{R}\alpha\gamma_0 = 0\,, \tag{67.8 a}$$

$$\mathscr{R}w_0 - (n^2\pi^2 + a^2 - \sigma\mathscr{P}_T)\theta_0 = 0\,, \tag{67.8 b}$$

and

$$\mathscr{R}\alpha w_0 - (n^2\pi^2 + a^2 - \sigma\mathscr{P}_C)\gamma_0 = 0\,. \tag{67.8 c}$$

The linear homogeneous equations (67.8) have a solution for the constants w_0, θ_0 and γ_0 if and only if the values of $\mathscr{R}$ and a^2 are selected so as to make the determinant of the coefficients of these constants vanish. This condition can be written as

$$\begin{aligned}0=&\mathscr{R}^2a^2(\alpha^2-1)+(n^2\pi^2+a^2)^3\\&-\sigma\left\{\frac{\mathscr{R}^2a^2(\mathscr{P}_T\alpha^2-\mathscr{P}_C)}{(n^2\pi^2+a^2)}+(n^2\pi^2+a^2)^2(1+\mathscr{P}_T+\mathscr{P}_C)\right\}\\&+\sigma^2(n^2\pi^2+a^2)(\mathscr{P}_T\mathscr{P}_C+\mathscr{P}_T+\mathscr{P}_C)-\sigma^3\mathscr{P}_T\mathscr{P}_C\,.\end{aligned}\tag{67.9}$$

To find the stability limit, we consider the marginal case $\mathrm{re}(\sigma)=0$ and find the real and imaginary part of (67.9).

$$0=\mathscr{R}^2a^2(\alpha^2-1)+(n^2\pi^2+a^2)^3-(\mathrm{im}(\sigma))^2(\mathscr{P}_C\mathscr{P}_T+\mathscr{P}_C+\mathscr{P}_T)(n^2\pi^2+a^2)\tag{67.10}$$

and

$$0=\mathrm{im}(\sigma)\left\{\frac{\mathscr{R}^2a^2(\mathscr{P}_T\alpha^2-\mathscr{P}_C)}{(n^2\pi^2+a^2)}+(n^2\pi^2+a^2)^2(1+\mathscr{P}_T+\mathscr{P}_C)-(\mathrm{im}(\sigma))^2\mathscr{P}_T\mathscr{P}_C\right\}.\tag{67.11}$$

Putting $\mathrm{im}(\sigma)=0$ in (67.10) we find that

$$\mathscr{R}^2(1-\alpha^2)=(n^2\pi^2+a^2)^3/a^2=27\pi^4/4\,,\tag{67.12}$$

where the minimizing values are $n=1$ and $a^2=\pi^2/2$.

If $\mathrm{im}(\sigma)\neq 0$, then we find after elimination of $\mathrm{im}(\sigma)$ between (67.10) and (67.11) that

$$\mathscr{R}^2=\left(\frac{1+\tau\mathscr{P}_T}{1+\mathscr{P}_T}\right)\frac{\mathscr{C}^2}{\tau^2}+\left(\frac{1}{\tau}+1\right)\left(\frac{1}{\mathscr{P}_C}+1\right)\frac{(n^2\pi^2+a^2)^3}{a^2}\,,\tag{67.13}$$

where $\tau=\mathscr{P}_C/\mathscr{P}_T$, so that the values $n=1$ and $a^2=\pi^2/2$ are again minimizing. The equation $\mathscr{P}_C$ (67.10)—(67.11) $(n^2\pi^2+a^2)/\mathrm{im}(\sigma)$ leads to

$$\mathscr{P}_C(n^2\pi^2+a^2)(\mathscr{P}_T+1)(\mathrm{im}(\sigma))^2=a^2\mathscr{R}^2\alpha^2\left(1-\frac{1}{\tau}\right)-(n^2\pi^2+a^2)^3\left(\frac{1}{\mathscr{P}_C}+\frac{1}{\tau}\right).$$

Since $(\mathrm{im}(\sigma))^2>0$ by hypothesis, it is necessary that

$$\tau>1\,,\qquad \mathscr{R}^2\alpha^2>\mathscr{C}_*^2=\frac{27\pi^4(1+1/\mathscr{P}_T)}{4(\tau-1)}\,.\tag{67.14}$$

If (67.14) holds, instability for small disturbances appears first as an oscillation with $\mathscr{R}^2=\mathscr{R}_L^2$ given by (67.13), $\mathrm{re}\,\sigma(\mathscr{R}_L)=0$ and $\mathrm{im}\,\sigma(\mathscr{R}_L)\neq 0$. When one or both of the inequalities (67.14) fails, $\mathrm{im}\,\sigma(\mathscr{R}_L)=0$ and linear instability occurs as exponential growth without oscillations and with

$\mathscr{R}_L^2-\mathscr{C}^2=27\pi^4/4$ given by (67.12).

We note here that we can always have (67.14) when $\mathscr{P}_T>0$ is fixed and $\mathscr{P}_C\to\infty$ ($\tau\to\infty$). Then, one finds from (67.13) that

$$\mathscr{R}_L^2=27\pi^4/4=\mathscr{R}_B^2\,.$$

We have shown that the rest state of a fluid which is heated and salted from below will give way to oscillations when the parameters have the values specified in the third inequality of (67.5).

The same formulas show that one can have convection when the fluid is gravitationally stable. If it is gravitationally stable, then the density gradient for the conduction-diffusion solution is negative

$$0>\mathbf{e}_z\cdot\nabla\rho=\rho_0\left\{-\alpha_T\frac{dT}{dx_3}+\alpha_C\frac{dC}{dx_3}\right\}=\rho_0\alpha_C\left|\frac{dC}{dx_3}\right|\left\{\frac{\mathscr{R}^2\tau}{\mathscr{C}^2}-1\right\}.$$

Since for large $\mathscr{C}$, (67.13) shows that oscillations can occur when

$$\frac{\mathscr{R}^2\tau}{\mathscr{C}^2}=\frac{1+\tau\mathscr{P}_T}{\tau(1+\mathscr{P}_T)}\leqslant 1\,,$$

the system can be unstable when gravitationally stable.

§ 68. Heated and Salted below: Energy Stability Analysis

The following theorem can be proved for the general problem A; but the instability aspect of the theorem assumes (67.3).

Theorem IX.2: *The constant-gradient solution for heated and salted below is monotonically and globally stable when $\mathscr{R}<\mathscr{R}_B$, and it is unstable when $\mathscr{P}_C\to\infty$ and $\mathscr{R}>\mathscr{R}_B$.*

Proof: This theorem is a consequence of the evolution inequality (57.1) with $\mathbf{U}=0$, $\boldsymbol{\eta}=\boldsymbol{\eta}_T=\boldsymbol{\eta}_C=-\mathbf{e}_z$ and

$$\frac{1}{\tilde{\mathscr{R}}_{\mathscr{E}}}=\min_{\lambda_T,\lambda_C}\max_{\mathbf{H}}\{(1+\lambda_T)\langle w\theta\rangle+\alpha(\lambda_C-1)\langle w\gamma\rangle\}/\mathscr{D}\left[\mathbf{u},\sqrt{\lambda_T}\theta,\sqrt{\lambda_C}\gamma\right]. \tag{68.1}$$

Given λ_C the maximum value in $\mathbf{H}$ of (68.1) must have $(\lambda_C-1)\langle w\gamma\rangle\geqslant 0$. Hence we must have $\lambda_C=1$ at the minimum of (68.1) over λ_C. With $\lambda_C=1$ the maximum of (68.1) has $\gamma=0$. When $\gamma\equiv 0$, by (58.11), $\lambda_T=1$ and

$$\mathscr{R}_{\mathscr{E}}=\mathscr{R}_L(0)\equiv\mathscr{R}_B\,.$$

This means that $\mathscr{E}(t)=\frac{1}{2}\langle|\mathbf{u}|^2+\mathscr{P}_T\theta^2+\lambda_C\mathscr{P}_C\gamma^2\rangle$, defined for any *solution* of the IBVP for the disturbance, decays at least as fast as a bounding exponential of the form

$$\mathscr{E}(0)\exp\{-\Lambda(1-\mathscr{R}/\mathscr{R}_B)\} \quad \text{when} \quad \mathscr{R}<\mathscr{R}_B\,.$$

When $\mathscr{R}>\mathscr{R}_B$, a disturbance can be found such that $\mathscr{E}(t)$ will increase for a time. Ordinarily, if $\mathscr{R}-\mathscr{R}_B>0$ is not too large, this momentary increase will die away. But when $\mathscr{P}_C\to\infty$ then, by virtue of (67.3), this disturbance will not die away but will persist.

The reader's attention is drawn to the fact that the functional (68.1) does not depend on the Prandtl numbers for heat ($\mathscr{P}_T$) or salt ($\mathscr{P}_C$). It follows that the criterion of Theorem IX.2 which is attained when $\mathscr{P}_C\to\infty$, is the best possible criterion for global stability which is independent of $\mathscr{P}_C$ and $\mathscr{P}_T$. At the same time the criterion gives both necessary and sufficient conditions for global, monotonic stability.

To have a better result for global stability, it will be necessary to inject a Prandtl number dependence into the global analysis.

§ 69. Heated and Salted below: Generalized Energy Analysis

The analysis here starts from the energy identities (56.4—7). The "coupling" identity is the new element which will allow us to inject the Prandtl numbers into an energy analysis. We will treat problem B; for this problem the surface terms in (56.4—7) all vanish.

The first step is to form the evolution equation

$$\frac{d}{dt}\{\hat{\mathscr{E}}+\hat{\mathscr{I}}\}=-\langle|\nabla\mathbf{u}|^2+|\nabla\phi|^2\rangle+\frac{\mathscr{R}}{\lambda_T}\left(1+\lambda_T^2-\frac{2\lambda_T\lambda_C\alpha}{1+\tau}\right)\langle w\phi\rangle\,. \tag{69.1}$$

This equation is the linear combination

$$(56.4)+\lambda_T^2\,(56.5)+\lambda_C^2\,(56.6)-2\lambda_T\lambda_C\,(56.7)\,\mathscr{P}_C/(1+\tau) \tag{69.2}$$

of energy identities (56.4—7) with the real coupling constants λ_T and λ_C, and in it

$$\hat{\mathscr{E}}(t)=\frac{1}{2}\left\langle|\mathbf{u}|^2+\frac{\tau\mathscr{P}_T}{1+\tau}\phi^2\right\rangle,$$

$$\hat{\mathscr{I}}(t)=\frac{1}{2}\,\frac{\mathscr{P}_T}{1+\tau}\langle\psi^2\rangle\,,$$

$$\phi(\mathbf{x},t)=\lambda_T\theta-\lambda_C\gamma\,,$$

$$\psi(\mathbf{x},t)=\lambda_T\theta-\tau\lambda_C\gamma\,,$$

where we have chosen only those values of λ_T and λ_C for which

$$\frac{1}{\lambda_T}+\lambda_T-\frac{2\alpha\lambda_C}{1+\tau}=\alpha\left(\frac{1}{\lambda_C}-\lambda_C\right)+\frac{2\lambda_T\tau}{1+\tau}. \tag{69.3}$$

In forming a linear combination of the identities (56.4—7), it is possible to choose three independent coupling constants. Two of these three parameters have been chosen so that the disturbance fields γ and θ appear only in the combination ϕ on the right of (69.1).

For problem B we can prove

Theorem IX.3: *Let*

$$A^2=\mathscr{R}^2-\mathscr{C}^2<\mathscr{R}_B^2 \tag{69.4}$$

when $\mathscr{C}/\mathscr{R}<1\leqslant 1/\tau$, *and let*

$$A=(\tau^2-1)^{-1/2}(\tau\mathscr{R}-\mathscr{C})<\mathscr{R}_B \tag{69.5}$$

when $\mathscr{C}/\mathscr{R}\geqslant 1/\tau$ *where* $\tau\geqslant 1$, *then*

$$\frac{d}{dt}\{\hat{\mathscr{E}}+\hat{\mathscr{J}}\}\leqslant -\xi^2(1-A/\mathscr{R}_B)\hat{\mathscr{E}}(t) \tag{69.6}$$

and $\hat{\mathscr{E}}(t)$ *tends to zero in the following sense:*

$$\lim_{t\to\infty}\int_0^t\hat{\mathscr{E}}(t)\,dt<\infty\,. \tag{69.7}$$

The criterion (69.4) *is necessary and sufficient for stability in the sense of* (69.7).

Proof: Let

$$2A\equiv\mathscr{R}\left(\frac{1}{\lambda_T}+\lambda_T-\frac{2\lambda_C\alpha}{1+\tau}\right).$$

Then we can write Eq. (69.1) as

$$\frac{d}{dt}\{\hat{\mathscr{E}}+\hat{\mathscr{J}}\}=-\langle|\nabla\mathbf{u}|^2+|\nabla\phi|^2\rangle\left\{1-\frac{2A\langle w\phi\rangle}{\langle|\nabla\mathbf{u}|^2\rangle+\langle|\nabla\phi|^2\rangle}\right\},$$

which by virtue of (67.2) is

$$\leqslant -\langle|\nabla\mathbf{u}|^2+|\nabla\phi|^2\rangle(1-A/\mathscr{R}_B)\,,$$

and since there exists $\xi^2>0$ such that $\langle|\nabla\mathbf{u}|^2+|\nabla\phi|^2\rangle\geqslant\xi^2$, we can continue the evolution inequality as

$$\leqslant -\xi^2(1-A/\mathscr{R}_B)\hat{\mathscr{E}}\,.$$

Let $b^2=\xi^2(1-A/\mathscr{R}_B)$ and integrate the last inequality

$$\hat{\mathscr{E}}(t)-\hat{\mathscr{E}}(0)+\hat{\mathscr{J}}(t)-\hat{\mathscr{J}}(0)\leqslant -b^2\int_0^t\hat{\mathscr{E}}(t')\,dt'\,. \tag{69.8}$$

Since $\hat{\mathscr{E}}(0)$ and $\hat{\mathscr{J}}(0)$ are bounded and $\hat{\mathscr{E}}(t)$ and $\hat{\mathscr{J}}(t)$ cannot be negative, it follows from (69.8) that both $\hat{\mathscr{E}}(t)$ and $\hat{\mathscr{J}}(t)$ are bounded uniformly in t. So too, then, must the integral $\int_0^t\hat{\mathscr{E}}(t')\,dt'$ be bounded uniformly in t. Since $\hat{\mathscr{E}}(t)$ is integrable, the statement (69.7) holds when $A<\mathscr{R}_B$.

We next show that λ_T and λ_C can be selected so that A has the values given on the left of (69.4) and (69.5). We seek values λ_T and λ_C which satisfy (69.3) and give the largest possible value (stability limit) on the right of the criterion

$$\mathscr{R}<2\mathscr{R}_B\bigg/\left(\frac{1}{\lambda_T}+\lambda_T-\frac{2\lambda_C\alpha}{1+\tau}\right). \tag{69.9}$$

This problem of the "optimum" stability boundary is an ordinary maximum problem. To solve it, we consider (69.3) to define a function $\lambda_C(\lambda_T)$ and set the total derivative with respect to λ_T on the right of (69.9) to zero. Then besides (69.3), we must have

$$(\lambda_T^2-1)(1+\lambda_C^2)(1+\tau)^2-4\tau\lambda_T^2\lambda_C^2=0\,. \tag{69.10}$$

We next find a continuous solution of (69.3, 10) on two branches. On the first branch $\alpha<1\leqslant 1/\tau$, and

$$\lambda_T=(\sqrt{1-\alpha^2\tau^2}-\tau\sqrt{1-\alpha^2})/(1-\tau)\,,$$
$$\lambda_C=(\sqrt{1-\alpha^2\tau^2}-\sqrt{1-\alpha^2})/\alpha(1-\tau)\,.$$

From (69.9)

$$\mathscr{R}<\mathscr{R}_B/\sqrt{1-\alpha^2}\,,$$

proving (69.4). On the second branch $\alpha\geqslant 1/\tau$, $\tau>1$,

$$\lambda_T=\lambda_C=\left(\frac{\tau+1}{\tau-1}\right)^{1/2},$$

and

$$\mathscr{R}<\frac{\mathscr{R}_B^2(1-\tau^{-2})}{(1-\alpha/\tau)}\,,$$

proving (69.5).

The last statement of the theorem follows from (69.4) and (67.4) taken together. The theorem is now proved.

We want to draw attention to the special value

$$\mathscr{C}^2=\mathscr{C}_1^2=\mathscr{R}_B^2/(\tau^2-1)$$

at which the stability limits (69.4) and (69.6) are joined. On the left of $\mathscr{C}_1$, there can be no subcritical instability. Here the global question about the onset of convection is completely and finally settled.

How good is the criterion (67.5) for $\mathscr{C} > \mathscr{C}_1$? It is, of course, not possible to answer this without knowing much more about the solutions to the nonlinear IBVP, and this knowledge is beyond our reach. But for the special problem C we have, besides (67.5), results of Sani (1965), who treats the nonlinear problem by a perturbation method, and of Veronis (1964, 1968) who treats the nonlinear problem C by modal truncation and a computer (see Fig. IX.1). Both authors find subcritical instabilities in the region $\mathscr{C} > \mathscr{C}_1$ and $\tau > 1$. In this respect, Sani's result is of particular interest since the perturbation method allows a complete sweep of the parameters for motions with small amplitudes. It is to be expected that these results, like (67.5), will depend on both $\mathscr{P}_T$ and $\mathscr{P}_C$. But Sani, by the perturbation method, also finds the same critical value $\mathscr{C}_1$, and his approximate method indicates the existence of two-dimensional convective solutions for all $\mathscr{C} > \mathscr{C}_1$. The result is all the more interesting, because this particular perturbation result, unlike the others, does not depend on separate values of $\mathscr{P}_T$ and $\mathscr{P}_C$ but only on their ratio.

We should call the energy result given by Theorem IX.3 optimal if it could be shown that the negation of the criterion (69.5) implied instability, not for all values $(\mathscr{P}_T, \mathscr{P}_C)$ such that $\tau > 1$, but for all values $\tau > 1$ and any value, say, of $\mathscr{P}_T$. The reason is that (69.1) already shows that the energy criterion could depend on $\mathscr{P}_T$ and $\mathscr{P}_C$ only in the ratio τ. Certainly this limit, given the correctness of Sani's calculation for the special problem C, is attained in the problem C when $\mathscr{C} = \mathscr{C}_1$ and possibly for all problems of class B.

But even without the nonlinear perturbation result, there is a sense in which (69.5) can be shown to be an optimal result. Consider (67.5) at large $\mathscr{C}$ when $\tau \to 1$ and $\mathscr{P}_T$ is fixed and small. Problem C is then unstable when

$$\mathscr{R}^2 > \mathscr{C}^2/\tau^2 .$$

But (69.5) guarantees stability in the same limit when

$$\mathscr{R}^2 < \mathscr{C}^2/\tau^2 .$$

It cannot be said that (69.5) is the global limit, except in the above-mentioned case and for $\mathscr{C} \leqslant \mathscr{C}_1$. But the possibility that it is, or is close to, the true global limit $\mathscr{R}_G$ has not been excluded.

Finally, we take note of the fact that the criteria (69.4, 5) do imply global stability but not monotonic stability for $\hat{\mathscr{E}}(t)$. On the other had, these criteria do guarantee monotonic stability for $\hat{\mathscr{E}}(t) + \hat{\mathscr{J}}(t)$, but it has not been shown that this monotonic stability will imply exponential decay.

Fig. 69.1 is a schematic sketch of the energy (stability) limit (69.5) and the linear (instability) limit for the layer between free surfaces when $\tau > 1$. The top line is the instability limit, and the heavy black line is the stability limit. Both limits have kinks; for the instability limit, the position of kinks and the slope of the line when $\mathscr{C} > \mathscr{C}^*$ depend on both $\mathscr{P}_T$ and $\mathscr{P}_C$, but the stability limit depends only on $\mathscr{P}_C/\mathscr{P}_T$. The two limits coincide when $\mathscr{C}^2 < \mathscr{C}_1^2 = \mathscr{R}_B^2/(\tau^2-1)$. For larger $\mathscr{C}$, the most one can expect from the one (τ) parameter family of energy limits (69.5) is coincidence, say, when τ is fixed for some $\mathscr{P}_T$. For τ fixed and $\mathscr{P}_T \to 0$, the instability and stability limits do coincide both at $\mathscr{C} = \mathscr{C}_1$ and for large $\mathscr{C}$.

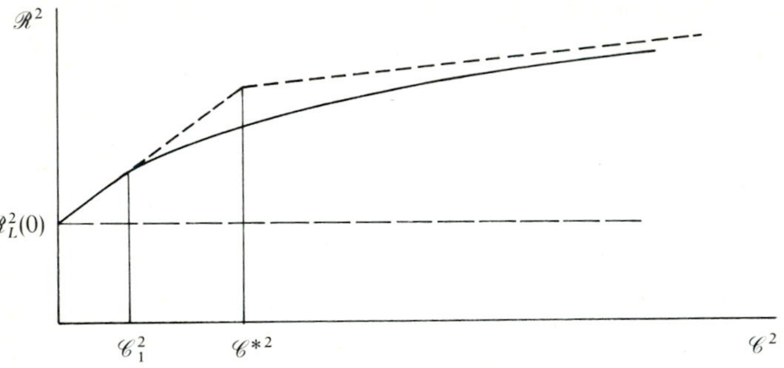

Fig. 69.1: Stability-instability boundaries for heated and salted below (Joseph, 1970)

Only the region between the top two lines of Fig. 69.1 is open to subcritical convection. The perturbation calculation of Sani (1965) indicates that two-dimensional solutions with small amplitudes fill the shaded region near the linear limit for $\mathscr{C} > \mathscr{C}_1$, where $\mathscr{C}_1$ is the above-mentioned energy value.

In the region of energy stability $\mathscr{R} < \mathscr{R}_B$, the decay of a stable disturbance is exponentially rapid.

Exercise 69.1: Salt water is confined in a layer of height l bounded above and below by rigid planes on which the temperature and salt concentration is prescribed. The velocity, temperature and salt concentration are governed by (54.2, 3, 4) with $Q_T = Q_C = 0$. Derive the following energy identities governing fluctuations $(\mathbf{u}, \theta, \gamma)$ in statistically stationary turbulent convection in the layer

$$\alpha_T g\langle w\theta\rangle - \alpha_C g\langle w\gamma\rangle = \mu\langle|\nabla\mathbf{u}|^2\rangle,$$

$$\langle[\overline{w\gamma} - \langle w\gamma\rangle]^2\rangle + \frac{\kappa_C\langle w\gamma\rangle}{l}[C(l) - C(0)] + \kappa_C^2\langle|\nabla\gamma|^2\rangle = 0,$$

$$\langle[\overline{w\theta} - \langle w\theta\rangle]^2\rangle + \frac{\kappa_T\langle w\theta\rangle}{l}[T(l) - T(0)] + \kappa_T^2\langle|\nabla\theta|^2\rangle = 0,$$

and

$$\kappa_C^{-1}\langle\overline{\theta w}[\overline{\gamma w} - \langle\gamma w\rangle]\rangle + \kappa_T^{-1}\langle\overline{\gamma w}[\overline{\theta w} - \langle\theta w\rangle]\rangle + [C(l) - C(0)]\langle\theta w\rangle/l$$
$$+ [T(l) - T(0)]\langle\gamma w\rangle/l + (\kappa_C + \kappa_T)\langle\nabla\theta\cdot\nabla w\rangle = 0.$$

Formulate a variational problem relating the heat and salt transport across the layer to the prescribed boundary values.

Addendum for Chapter IX: Generalized Energy Theory of Stability for Hydromagnetic Flows

Another interesting problem in which a correlation integral plays a role arises in the study of stability of hydromagnetic flow.

The governing equations of motion for a viscous fluid with constant density ρ and finite conductivity σ flowing in a magnetic field are (see Chandrasekhar, 1961):

$$\frac{\partial \mathbf{U}}{\partial t}+\mathbf{U}\cdot\nabla\mathbf{U}=\frac{\mu}{\rho}\mathbf{B}\cdot\nabla\mathbf{B}-\frac{1}{\rho}\nabla\left(p+\frac{\mu}{2}|\mathbf{B}|^2\right)+\nu\nabla^2\mathbf{U}\,, \tag{IX.1a}$$

$$\frac{\partial \mathbf{B}}{\partial t}+\mathbf{U}\cdot\nabla\mathbf{B}=\mathbf{B}\cdot\nabla\mathbf{U}+\frac{1}{\sigma\mu}\nabla^2\mathbf{B} \tag{IX.1b}$$

and

$$\nabla\cdot\mathbf{U}=\nabla\cdot\mathbf{B}=0 \tag{IX.1c}$$

where $\mathbf{B}$ is the magnetic flux density, μ is the magnetic permeability and, as before, ν, $\mathbf{U}$ and p are viscosity, velocity and pressure. From (IX.1b) one finds that $\partial\nabla\cdot\mathbf{B}/\partial t=\frac{1}{\sigma\mu}\nabla^2\nabla\cdot\mathbf{B}$. The condition $\operatorname{div}\mathbf{B}=0$ is automatically guaranteed for solutions of (IX.1b) which have $\operatorname{div}\mathbf{B}=0$ at time zero and on the boundary S of $\mathscr{V}$ at all times. Disturbances $\mathbf{u}$ and $\mathbf{b}$ of $\mathbf{U}$ and $\mathbf{B}$ satisfy

$$\frac{\partial \mathbf{u}}{\partial t}+\mathbf{U}\cdot\nabla\mathbf{u}+\mathbf{u}\cdot\nabla\mathbf{U}+\mathbf{u}\cdot\nabla\mathbf{u}=\frac{\mu}{\rho}(\mathbf{b}\cdot\nabla\mathbf{B}+\mathbf{b}\cdot\nabla\mathbf{b}+\mathbf{B}\cdot\nabla\mathbf{b})$$

$$-\frac{1}{\rho}\nabla\left[\delta p+\frac{\mu}{2}(|\mathbf{B}^*|^2-|\mathbf{B}^2|)\right]+\nu\nabla^2\mathbf{u} \tag{IX.2a}$$

where $\mathbf{B}^*=\mathbf{B}+\mathbf{b}$,

$$\frac{\partial \mathbf{b}}{\partial t}+\mathbf{U}\cdot\nabla\mathbf{b}+\mathbf{u}\cdot\nabla\mathbf{B}+\mathbf{u}\cdot\nabla\mathbf{b}=\mathbf{B}\cdot\nabla\mathbf{u}+\mathbf{b}\cdot\nabla\mathbf{u}+\mathbf{b}\cdot\nabla\mathbf{U}+\frac{1}{\sigma\mu}\nabla^2\mathbf{b}\,, \tag{IX.2b}$$

$$\nabla\cdot\mathbf{u}=\nabla\cdot\mathbf{b}=0 \tag{IX.2c}$$

and

$$\mathbf{u}=\mathbf{b}=0 \quad \text{on } S\,. \tag{IX.2d}$$

Following Rionero (1967, 1968A) and Carmi and Lalas (1970) we derive the energy equations:

$$\tfrac{1}{2}\frac{d}{dt}\langle|\mathbf{u}|^2\rangle=-\langle\mathbf{u}\cdot\mathbf{D}\cdot\mathbf{u}+\nu\nabla\mathbf{u}:\nabla\mathbf{u}\rangle+\frac{\mu}{\rho}\{\langle\mathbf{B}\cdot\nabla\mathbf{b}\cdot\mathbf{u}+\mathbf{b}\cdot\nabla\mathbf{b}\cdot\mathbf{u}+\mathbf{b}\cdot\nabla\mathbf{B}\cdot\mathbf{u}\rangle\} \tag{IX.3a}$$

and

$$\tfrac{1}{2}\frac{d}{dt}\langle|\mathbf{b}|^2\rangle=\langle\mathbf{b}\cdot\mathbf{D}\cdot\mathbf{b}\rangle-\left\langle\frac{1}{\sigma\mu}\nabla\mathbf{b}:\nabla\mathbf{b}\right\rangle+\langle\mathbf{B}\cdot\nabla\mathbf{u}\cdot\mathbf{b}+\mathbf{b}\cdot\nabla\mathbf{u}\cdot\mathbf{b}-\mathbf{u}\cdot\nabla\mathbf{B}\cdot\mathbf{b}\rangle \quad \text{(IX.3b)}$$

where the angle brackets designate volume-averaged integrals and, as before, $\mathbf{D}=\mathbf{D}[\mathbf{U}]$ is the stretching tensor for $\mathbf{U}$. In carrying out the integration, the integral $\langle\nabla\cdot\mathbf{A}_1\rangle=0$ which is added on the right of (IX.3a) and the integral $\langle\nabla\cdot\mathbf{A}_2\rangle=0$ which added on the right of (IX.3b) has been carried to the boundary by the divergence theorem. The normal components of the vector fields

$$\mathbf{A}_1=\nu\nabla\tfrac{1}{2}|\mathbf{u}^2|-\mathbf{u}\left[\tfrac{1}{2}|\mathbf{u}|^2+\frac{\delta p}{\rho}+\frac{\mu}{2\rho}(|\mathbf{B}^*|^2-|\mathbf{B}|^2)\right]$$

and

$$\mathbf{A}_2=\frac{1}{\sigma\mu}\nabla\tfrac{1}{2}|\mathbf{b}|^2-\mathbf{u}\tfrac{1}{2}|\mathbf{b}|^2$$

vanish on S.

The reader's attention is drawn to the fact that some of the cubic nonlinearities in the disturbance, which arise from the quadratic terms $\mathbf{u}$ and $\mathbf{b}$ in (IX.2), do not integrate to zero; that is, though

$$\langle\mathbf{u}\cdot(\mathbf{u}\cdot\nabla)\mathbf{u}\rangle=\langle\mathbf{b}\cdot(\mathbf{u}\cdot\nabla)\mathbf{b}\rangle=0$$

the terms

$$\langle\mathbf{b}\cdot\nabla\mathbf{b}\cdot\mathbf{u}\rangle \quad \text{and} \quad \langle\mathbf{b}\cdot\nabla\mathbf{u}\cdot\mathbf{b}\rangle$$

are not necessarily zero and they appear in the energy balances (IX.3a) and (IX.3b).

There is one linear combination of (IX.3a) and (IX.3b) in which the cubic nonlinearities subtract out: thus,

$$\begin{aligned}\tfrac{1}{2}\frac{d}{dt}\left\langle\frac{\rho}{\mu}|\mathbf{u}|^2+|\mathbf{b}|^2\right\rangle&=-\frac{\rho}{\mu}\langle\mathbf{u}\cdot\mathbf{D}\cdot\mathbf{u}\rangle+\langle\mathbf{b}\cdot\mathbf{D}\cdot\mathbf{b}\rangle+2\langle\mathbf{b}\cdot\boldsymbol{\Omega}_B\cdot\mathbf{u}\rangle\\&\quad-\nu\frac{\rho}{\mu}\langle\nabla\mathbf{u}:\nabla\mathbf{u}\rangle-\frac{1}{\sigma\mu}\langle\nabla\mathbf{b}:\nabla\mathbf{b}\rangle\end{aligned} \quad \text{(IX.4)}$$

where

$$\boldsymbol{\Omega}_B=\text{antisymmetric part of }\nabla\mathbf{B}.$$

This is the energy identity considered by Rionero (1967, 1968A) and by Carmi and Lalas (1970).

There are four fundamental measures of the basic flow; the stretching tensor $\mathbf{D}$, the vorticity tensor $\boldsymbol{\Omega}_U$, the symmetric part of the dyadic gradient of magnetic flow $\mathbf{D}_B$ and the antisymmetric part $\boldsymbol{\Omega}_B$ of the same tensor; that is

$$\nabla\mathbf{U}=\mathbf{D}+\boldsymbol{\Omega}_U \quad \text{and} \quad \nabla\mathbf{B}=\mathbf{D}_B+\boldsymbol{\Omega}_B.$$

Of the four measures only **D** and $\boldsymbol{\Omega}_B$ appear in the identity (IX.4).

A dimensionless form for the energy equations follows by dividing

$$[\mathbf{x}, t, \mathbf{u}, b, \mathbf{B}] \quad \text{by} \quad [l, l/U_0, U_0, U_0\sqrt{\rho\mu}, U_0\sqrt{\rho\mu}]$$

where U_0 and B_0 are scales for the velocity and the magnetic field. The change of variables gives rise to the following dimensionless parameters: $R = U_0 l/\nu$, the Reynolds number; $\mathscr{P}_m = \mu\sigma\nu$, the magnetic Prandtl number, and $A = B_0/U_0(\rho\mu)^{1/2}$, the Alfven number. We will now work only with dimensionless variables which, are also designated as $\mathbf{u}, \mathbf{b}, \mathbf{D}, \mathbf{B}$. In dimensionless variables we have

$$\frac{d\mathscr{E}}{dt} = -\{\langle \mathbf{u}\cdot\mathbf{D}\cdot\mathbf{u}\rangle - \langle \mathbf{b}\cdot\mathbf{D}\cdot\mathbf{b}\rangle\} + 2\langle \mathbf{b}\cdot\boldsymbol{\Omega}_B\cdot\mathbf{u}\rangle$$

$$-\frac{1}{R}\langle \nabla\mathbf{u}:\nabla\mathbf{u}\rangle - \frac{1}{R_m}\langle \nabla\mathbf{b}:\nabla\mathbf{b}\rangle \qquad \text{(IX.5)}$$

where

$$\mathscr{E} = \tfrac{1}{2}\langle |\mathbf{u}|^2 + |\mathbf{b}|^2\rangle, \quad \text{and} \quad R_m = R\mathscr{P}_m$$

and Ω_B is the antisymmetric part of ∇B. Here the magnetic Reynolds number R_m is the ratio of the convection rate to the diffusion rate of the magnetic field. Large R_m implies a thin boundary layer in which dissipation occurs. Outside this region the magnetic field and the flow are "frozen" together. Small R_m, on the other hand, implies that the total magnetic field of the flow is essentially equal to the imposed one, so that the induced field is small. The magnetic Prandtl number $\mathscr{P}_m$ is a measure of the ratio of the rate of diffusion of vorticity to the rate of diffusion of the magnetic field.

Eq. (IX.5) is homogeneous of degree two in **u** and **b**. It is therefore certain that results following from (IX.5) will be independent of the amplitude of the disturbance. Rionero (1967, 1968A) and Carmi and Lalas (1970) used (IX.5) to prove a theorem of unconditional stability.

There are energy identities besides (IX.3) which are of value in treating the stability of hydromagnetic flows. For example, the three Eqs. (IX.2b) for the components of **b** each give an energy equation. Energy equations for the components of **b** are like the separate Eqs. (IX.3a) and (IX.3b) in that they involve cubic nonlinearities.

One identity, the coupling identity, merits special attention. This identity is formed from the sum $\langle \mathbf{b}\cdot(\text{IX.2a})\rangle + \langle \mathbf{u}\cdot(\text{IX.2b})\rangle$ and leads after integration by parts to

$$\frac{d}{dt}\langle \mathbf{u}\cdot\mathbf{b}\rangle = -2\langle \mathbf{u}\cdot\boldsymbol{\Omega}_U\cdot\mathbf{b}\rangle - \langle \mathbf{u}\cdot\mathbf{D}_B\cdot\mathbf{u}\rangle$$

$$+\frac{\mu}{\rho}\langle \mathbf{b}\cdot\mathbf{D}_B\cdot\mathbf{b}\rangle - \left(\nu + \frac{1}{\sigma\mu}\right)\langle \nabla\mathbf{b}:\nabla\mathbf{u}\rangle. \qquad \text{(IX.6)}$$

The coupling identity is striking because it depends on the basic flow only through the measures $\mathbf{D}_B$ and $\boldsymbol{\Omega}_U$ of the "strain rate" of the magnetic flux and the vorticity of the basic motion; these measures are completely absent from the energy identity (IX.4).

A linear combination of the energy identity (IX.4) and the coupling identity (IX.6) forms the basis for generalized energy analysis of the type considered in § 69.

After making (IX.1) dimensionless, we may form a linear combination

$$\frac{d\mathscr{E}_\lambda}{dt} = RI_\lambda - D_\lambda , \tag{IX.7}$$

where

$$\begin{aligned}
&\mathscr{E}_\lambda = (\phi+\lambda)\mathscr{E} + 2\langle \mathbf{u}\cdot\mathbf{b}\rangle ,\\
&\phi = \frac{1}{\mathscr{P}_m} + 1, \qquad \lambda > 0 ,\\
&I_\lambda = (\phi+\lambda) I_1 + 2I_2 ,\\
&I_1 = -\langle \mathbf{u}\cdot\mathbf{D}\cdot\mathbf{u}\rangle + \mathscr{P}_m\langle \mathbf{b}\cdot\mathbf{D}\cdot\mathbf{b}\rangle + 2\mathscr{P}_m\langle \mathbf{b}\cdot\boldsymbol{\Omega}_B\cdot\mathbf{u}\rangle ,\\
&I_2 = -2\langle \mathbf{u}\cdot\boldsymbol{\Omega}_U\cdot\mathbf{b}\rangle - \langle \mathbf{u}\cdot\mathbf{D}_B\cdot\mathbf{u}\rangle + \mathscr{P}_m\langle \mathbf{b}\cdot\mathbf{D}_B\cdot\mathbf{b}\rangle ,\\
&D_\lambda = \lambda\langle |\nabla\mathbf{u}|^2 + |\nabla\mathbf{b}|^2 + \phi\langle |\nabla(\mathbf{u}+\mathbf{b})|^2\rangle .
\end{aligned}$$

We remark that for all fixed values of $\phi > 1$ and $\lambda > 0$ there exist values

$$\frac{1}{R_\lambda} = \max_H \frac{I_\lambda}{D_\lambda} \quad \text{and} \quad \frac{1}{\Lambda_\lambda} = \max_H \frac{\mathscr{E}_\lambda}{D_\lambda} . \tag{IX.8 a, b}$$

These numbers define a stability limit and a decay constant in the energy stability theorem which is to be proved below. It is first necessary to establish a preliminary result.

Lemma: $\mathscr{E}_\lambda \geq 0$ *for all* $\lambda > 0$.

To prove this, we note that

$$\begin{aligned}
\tfrac{1}{2}(\phi+\lambda)(|\mathbf{u}|^2 + \mathscr{P}_m|\mathbf{b}|^2) + 2\mathbf{u}\cdot\mathbf{b} &\geq \tfrac{1}{2}(\phi+\lambda)2\mathscr{P}_m^{1/2}|\mathbf{u}\cdot\mathbf{b}| - 2|\mathbf{u}\cdot\mathbf{b}|\\
&= \mathscr{P}_m^{1/2}\left\{\left(1-\frac{1}{\mathscr{P}_m^{1/2}}\right)^2 + \lambda\right\}|\mathbf{u}\cdot\mathbf{b}| \geq 0 .
\end{aligned}$$

With this preliminary aside we may now establish the following theorem.

Energy theorem for hydromagnetic flow: Let

$$R < R_\lambda \tag{IX.9}$$

for any $\lambda>0$. *Then*

$$\mathscr{E}_\lambda(t)<\mathscr{E}_\lambda(0)\exp\left\{-\Lambda_\lambda t\left[1-\frac{R}{R_\lambda}\right]\right\}. \tag{IX.10}$$

Proof: We may write (IX.7) as

$$\frac{d\mathscr{E}_\lambda}{dt}=-D_\lambda\left\{-R\frac{I_\lambda}{D_\lambda}+1\right\}\leqslant -D_\lambda\left\{1-\frac{R}{R_\lambda}\right\}$$

where we have used (IX.8 a) in forming the last inequality. If $R<R_\lambda$, by (IX.8 b) we have

$$\frac{d\mathscr{E}_\lambda}{dt}<-\Lambda_\lambda\mathscr{E}_\lambda\left\{1-\frac{R}{R_\lambda}\right\}$$

and (IX.10) follows by integration.

The largest R domain of stability is associated with the "energy" $\tilde{\lambda}>0$ *where* $\tilde{\lambda}$ *is value of* λ *for which*

$$\frac{1}{R_{\tilde{\lambda}}}=\sup_{\lambda>0}\frac{1}{R_\lambda}.$$

It is easy to prove that the initial condition which solves the maximum problem (IX.8 a) is also the one which makes $\mathscr{E}_{\tilde{\lambda}}$ increase initially at the smallest R.

Computations for the criterion $R<R_{\tilde{\lambda}}$ for particular flows have not yet been carried out.

The energy stability theorems of Rionero (1967, 1968A) and of Carmi and Lalas (1970) were rediscovered by Bhattacharya and Jain (1972). Rionero (1971B) has given energy stability theorems for magnetohydrodynamics in the presence of a Hall effect. Magnetohydrodynamic stability with respect to weighted metrics and boundary perturbations as well as initial perturbations has been discussed by Galdi (1973). Rionero (1968B) and Lalas and Carmi (1972) consider the global stability of hydromagnetic flows with thermal convection and Patil and Rudraiah (1973) consider this same problem when the fluid fills a porous material. The coupling identity and (IX.10) are due to Joseph (1972).

Chapter X

Two-Sided Bifurcation into Convection

Suppose $\mathscr{R}(\varepsilon)$ is the bifurcation curve for steady convection, $\mathscr{R}$ is the square root of the Rayleigh number, ε is the amplitude and $\mathscr{R}_0=\mathscr{R}(0)=\mathscr{R}_L$ is the critical value for the stability of conduction. There are two possibilities: $\mathscr{R}_1=0$ where $\mathscr{R}_1\equiv\mathscr{R}'(0)$, as in the Bénard problem, or $\mathscr{R}_1\neq 0$, as in the problems of generalized convection to be considered here. Since ε may have positive and negative values and $\mathscr{R}-\mathscr{R}_L=\varepsilon\mathscr{R}_1+O(\varepsilon^2)$, the bifurcation is two-sided when $\mathscr{R}_1\neq 0$. Two-sided bifurcation implies that one side is subcritical and the other supercritical. Hence, we get subcritical bifurcation when $\mathscr{R}_1\neq 0$. This chapter is about $\mathscr{R}_1$.

Two-sided bifurcation into convection can occur in the many problems which generalize the Bénard problem. For example, generalized convection starts as a subcritical bifurcation of the rest state when the material properties (viscosity, specific heat, thermal diffusivity) vary (Palm, 1960; Segel and Stuart, 1962; Busse, 1962, 1967); when the Boussinesq equation of state $\rho(T)$ is nonlinear (Busse, 1962, 1967; Veronis, 1963); when there is a free surface and it is allowed to deflect (Davis and Segel, 1968); when surface tension drives the convection (Scanlon and Segel, 1967), and when heat sources are present (Krishnamurti, 1968A, B; Schwiderski and Jarnagin, 1972).

In this chapter we shall consider fluids whose viscosity varies with temperature. Variations in temperature are induced by internal heating and external heating from below. The chapter starts in § 70 with a discussion of the Darcy-Oberbeck-Boussinesq (DOB) equations. These equations are possibly the mathematically simplest of the nonlinear equations of fluid mechanics which appear to predict observed flows correctly. The energy and linear theory of stability is applied to the conduction solution of the DOB equations in § 71 and some preliminary results about two-sided bifurcation are given in § 72. The problem of bifurcation and stability at a multiple eigenvalue is considered in § 72 and again, in more detail, in § 73, § 74 and in the Addendum to Chapter X. The use of the DOB equations simplifies the calculations; it allows us to study convection in bounded containers with insulated sidewalls by an elementary separation of variables. This separation leads in § 73 to criteria for two-sided bifurcation at simple and multiple eigenvalues. These criteria are applied to various containers of simple shape in § 73. In § 74 we consider the problem of two-sided bifurcation into convection in spherical shells; and, in § 75, the problem of stability and bifurcation in a container of fluid heated internally and from below. The popular topic of cellular convection in fluid layers is treated in § 77, but only briefly, and from a special point of view. We are concerned there with the relation of convection in infinite layers to convection in shallow containers, with an examination of the way various authors define cellular convection, and with an explanation of the important difference in the unique determination of the sign of the motion arising from the analysis of stability of supercritical branches of two-sided convection.

Notation: Several different Rayleigh numbers are used in this chapter:

$$\mathscr{R}^2 = \frac{\alpha g \Delta T l^3}{\nu\kappa}, \qquad \mathscr{R} = \sqrt{\mathscr{R}^2} \quad \text{for the OB equations}$$

and

$$R = \frac{\alpha g \Delta T l k'}{\nu\kappa}, \qquad \mathscr{R} = \sqrt{R} \quad \text{for the DOB equations.}$$

Bifurcation curves are given functions: $R = R(\varepsilon)$ and $\mathscr{R} = \mathscr{R}(\varepsilon)$. The values of these functions at the point of bifurcation are critical values: $R(0) \equiv R_0 = R_L$, $\mathscr{R}(0) \equiv \mathscr{R}_0 = \mathscr{R}_L$. The eigenvalues of the spectral problem for conduction are called σ; the eigenvalues of the spectral problem for convection are called γ. At a critical value, $\sigma(\mathscr{R}_0) = 0$.

§ 70. The DOB Equations for Convection of a Fluid in a Container of Porous Material

The laminar flow of fluid in porous materials and turbulence generally are alike in that the complete, detailed description of either motion is both more than one wants and is beyond what one expects of analysis. The source of the difficulty for the case of flow in porous materials is the complexity of the geometry. A mathematical description of suitably-averaged variables is the goal of analysis of both turbulence and flow in porous materials.

The "theory" of porous flow is largely based on a generalization of empirical observations of Darcy (1856). The Darcy "law" is an empirical equation which relates the mass flux m through a porous material to the pressure drop ΔP across it. In a straight cylinder of radius A and length l, this relation has the form

$$\frac{\Delta P}{l} = \frac{\mu m}{k' \rho A} \tag{70.1}$$

where k' is the permeability expressed in units of length squared. This relation is generalized to steady slow flow in homogeneous isotropic materials by the following vector differential equation:

$$-\rho \mathbf{g} + \nabla P = -\frac{\mu}{k'} \mathbf{U}_m, \qquad \nabla \cdot \mathbf{U}_m = 0. \tag{70.2a, b}$$

Here, $-\rho\mathbf{g}$ is the body force (gravity) per unit volume and $\mathbf{U}_m$ the seepage velocity vector[1].

Eqs. (70.2) should be compared with the Navier-Stokes equations linearized for slow flow,

$$-\rho\mathbf{g}+\nabla p=\mu\nabla^2\mathbf{U}\,,\qquad \nabla\cdot\mathbf{U}=0\,. \tag{70.3a, b}$$

It is seen that the Darcy law replaces the term

$$\mu\nabla^2\mathbf{U}=\text{div (viscous part of the stress tensor)} \tag{70.4}$$

with the term

$$-\mu\mathbf{U}_m/k'\,.$$

This term is sometimes regarded as giving the force which the porous solid exerts on the fluid. Eq. (70.4) is unaffected by the addition to $\mathbf{U}$ of a rigid-body velocity field. It depends only on relative velocities. Analogously, we interpret $\mathbf{U}_m$ to be a velocity measured relative to axes fixed in the porous solid.

In the same way that (70.2) may be said to represent (70.3) for slow steady flow, it is useful, but not correct, to assume that,

$$\rho\frac{d\mathbf{U}}{dt}=-\nabla P+\rho\mathbf{g}-\frac{\mu}{k'}\mathbf{U}_m\,. \tag{70.5}$$

For problems of thermal convection the assumptions about the form of left side of (70.5) are not important since the effects of acceleration are generally very small. The pore-average velocity $\mathbf{U}$ and the seepage velocity $\mathbf{U}_m$ are assumed to be defined at each point of $\mathscr{V}$ and are related by the equation

$$\mathbf{U}_m=\hat{\phi}\mathbf{U}\,, \tag{70.6}$$

where the porosity $\hat{\phi}$ is given by

$$\hat{\phi}=\frac{\text{volume of voids}}{\text{total volume}}\,,\qquad 0<\hat{\phi}<1\,.$$

[1] The seepage velocity is defined at each point in the porous material; equally at points in the solid and in the fluid. Since the average of the product $\mathbf{U}\cdot\nabla\mathbf{U}$ is not the product of averages it is hardly a surprise that this form of the convective acceleration is not appropriate for the DOB equations. Indeed, the appearance of the highest order spatial derivative on the left side of (70.5) alters the fundamental analytic properties of this equation in a quite unacceptable way (see Beck, 1972). In point of fact experiments on rectilinear flow (Forcheimer (1901), Ward (1964), Beavers and Sparrow (1969)) all give a drag proportional to the square of the seepage velocity. It has been argued (cf. Irmay, 1958) that the vector correction for weakly nonlinear porous flow should also take form in a quadratic drag proportional to $\mathbf{U}|\mathbf{U}|$.

In homogeneous isotropic materials $\hat{\phi}$ is a pure constant, but in nonhomogeneous material $\hat{\phi}$ may depend on position. For very fluffy foam metal materials, $\hat{\phi}$ is nearly one and in beds of packed spheres, $\hat{\phi}$ is in the range 0.25—0.50.

For flows driven by buoyancy we shall use the Boussinesq approximation (§ 59) and replace the density with a constant value ρ_0 on the left side of (70.5) and by $\rho=\rho_0[1-\alpha(T-T_0)]$ on the right. The equation

$$\rho_0\frac{d\mathbf{U}}{dt}=-\nabla p-\rho_0\alpha\mathbf{g}(T-T_0)-\frac{\mu}{k'}\mathbf{U}_m \tag{70.7}$$

follows after removing the hydrostatic pressure (we use the symbol p again for the reduced pressure).

The energy equation in permeable materials is obtained from an enthalpy balance over the fluid (see Exercise 54.1)

$$\int_{\mathscr{V}_f}(\rho C_0)_f\left(\frac{\partial T}{\partial t}+\mathbf{U}\cdot\nabla T\right)d\mathscr{V}_f=\int_{\mathscr{V}_f}\nabla\cdot(k_f\nabla T)d\mathscr{V}_f$$

and over the solid

$$\int_{\mathscr{V}_s}(\rho C_0)_s\frac{\partial T}{\partial t}d\mathscr{V}_s=\int_{\mathscr{V}_s}\nabla\cdot(k_s\nabla T)d\mathscr{V}_s\,.$$

The integration is transferred to the common volume $\mathscr{V}=\mathscr{V}_f\cup\mathscr{V}_s$ by the Jacobian relation

$$\frac{d\mathscr{V}_f}{d\mathscr{V}}=\hat{\phi}\quad\text{and}\quad\frac{d\mathscr{V}_s}{d\mathscr{V}}=1-\hat{\phi}\,.$$

Then, upon adding, one finds that

$$\int_{\mathscr{V}}\left\{(\rho_0C_0)_m\frac{\partial T}{\partial t}+(\rho_0C_0)_f\hat{\phi}\mathbf{U}\cdot\nabla T-\nabla\cdot(k_m\nabla T)\right\}d\mathscr{V}=0$$

where the subscript m stands for the fluid-solid mixture, e.g.,

$$k_m=\hat{\phi}k_f+(1-\hat{\phi})k_s\,. \tag{70.8}$$

In this way we arrive at the energy equation

$$(\rho C_0)_m\frac{\partial T}{\partial t}+(\rho_0C_0)_f\hat{\phi}\mathbf{U}\cdot\nabla T=\nabla\cdot(k_m\nabla T)\,. \tag{70.9}$$

This completes our heuristic derivation of the DOB equations.

The dynamic equation (70.7) can be further simplified. Let l be the scale of length, l^2/κ be the scale of time where $\kappa=k_m/(\rho_0C_0)_f$, ΔT be the scale of temperature and κ/l be the velocity scale. In the dimensionless variables we can write (70.7) as

$$B\left[\frac{\partial \mathbf{U}}{\partial t}+(\mathbf{U}\cdot\nabla)\mathbf{U}\right]=-\nabla P-\hat{\phi}\mathbf{U}+R\mathbf{e}_z\left(T-\frac{T_0}{\Delta T}\right) \tag{70.10a}$$

where $B^{-1}=l^2\nu/k'\kappa$ is the Darcy-Prandtl number and $R=\alpha g\Delta T l k'/\nu\kappa$ is the Rayleigh-Darcy number. We note that in porous sands the Darcy permeability coefficient k' is $0(10^{-8}\text{ cm}^2)$ or smaller. Even very porous fiber metals have values of k' no larger than 10^{-4} cm^2. It follows that the value B is commonly so very small that it is reasonable to set it to zero from the outset. Then, collecting the results we have

$$0=-\nabla P-\mathbf{U}_m+R\mathbf{e}_z\left(T-\frac{T_0}{\Delta T}\right) \tag{70.10b}$$

$$\frac{\partial T}{\partial \tau}+\mathbf{U}_m\cdot\nabla T=\nabla^2 T \tag{70.10c}$$

where $\tau=(C_0\rho_0)_f t/(C_0\rho)_m$ and, for homogeneous materials,

$$\nabla\cdot\mathbf{U}_m=0\,. \tag{70.10d}$$

At the boundaries of the fluid saturated porous material we shall regard the normal velocity and the temperature as prescribed. It is not consistent with the lowering of the order of the equations leading to (70.10) from (70.4) to impose a stronger condition on the velocity (cf. Beck, 1972); in particular, the order of the equations will not generally accommodate solutions with prescribed conditions on three components of the velocity at the boundary. In a real porous material the fluid will stick to a solid wall, but the effect of the wall is confined to a boundary layer whose size is measured in pore diameters. Since the wall friction does not sensibly influence the interior motions it is reasonable to replace the true wall with a frictionless wall.

If there are internal sources of heat $\hat{Q}$ in the fluid and the ratio, μ/k', of the viscosity to the permeability is temperature dependent, Eqs. (70.10) become

$$-R\mathbf{e}_z\left(T-\frac{T_0}{\Delta T}\right)+\hat{f}(T)\mathbf{U}_m=-\nabla P \tag{70.11a}$$

$$\frac{\partial T}{\partial \tau}+\mathbf{U}_m\cdot\nabla T=\nabla^2 T+\hat{Q} \tag{70.11b}$$

where

$$\hat{f}(T)=\frac{\mu(T)}{k'(T)}\frac{k'(T_0)}{\mu(T_0)}>0$$

and T_0 is the reference temperature.

We are going to study the following problem: An impermeable container Ω is filled with a porous material saturated with fluid satisfying (70.11 a, b) and (70.10 d). The container is heated from below and the fluid is heated internally. The container Ω is sketched in Fig. 70.1. The boundary $\partial\Omega$ of Ω is the intersection of horizontal planes at $z=0$ and $z=1$ and the boundary of a vertical cylinder of arbitrary cross-section $\mathscr{A}$. The sidewall of the cylinder is called S. The outward normal on $\partial\Omega$ is $\mathbf{N}$ and on S, $\mathbf{N}=\mathbf{n}$. The temperature outside the top wall is $T_0/\Delta T$ and outside the bottom wall it is $1+T_0/\Delta T$.

Fig. 70.1: An impermeable container Ω of constant cross-section A is filled with a fluid-saturated porous material heated from below and internally. The ratio of the viscosity to the permeability depends on the temperature

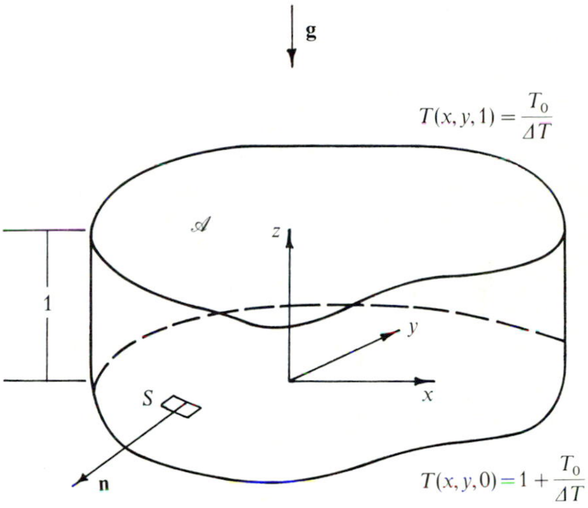

The configuration described in the previous paragraph supports a unique motionless solution, $T=-G(z;\xi)\equiv -G(z)$, the conduction solution, given by (59.7) with $T'=\Delta T$, where

$$G'=(1+\xi g(z)), \tag{70.12}$$

and

$$G''=\xi g'=\hat{Q}.$$

It is assumed that the conditions at the side wall are compatible with the conduction solution. The conditions are compatible in the most interesting case of an insulating sidewall.

Disturbances of the conduction solution are $\mathbf{U}$, P (again) and $\mathscr{R}^{-1}\Theta$ where $\mathscr{R}=\sqrt{R}$:

$\mathbf{U}_m=\mathbf{U}$ with components (U,V,W) relative to cartesian coordinates (x,y,z) and

$$T=-G(z;\xi)+\mathscr{R}^{-1}\Theta.$$

The equations governing disturbances are

$$-\mathbf{e}_z\mathscr{R}\Theta+\hat{f}(-G+\mathscr{R}^{-1}\Theta)\mathbf{U}=-\nabla P\,, \tag{70.13 a}$$

$$\frac{\partial\Theta}{\partial\tau}+\mathbf{U}\cdot\nabla\Theta-\mathscr{R}G'W-\nabla^2\Theta=0\,, \tag{70.13 b}$$

$$(\mathbf{U},\,\Theta)\in\mathscr{B} \tag{70.13 c}$$

where $\mathscr{B}$ is a set of solenoidal vectors $\mathbf{U}$ and temperature disturbances Θ satisfying the boundary conditions. Of course $\mathbf{U}\cdot\mathbf{N}=0$ on $\partial\Omega$, but a variety of boundary conditions depending the resistivity of $\partial\Omega$ can be generated for Θ. For the moment, we shall leave a wide scope for boundary conditions and prescribe that $\mathbf{N}\cdot\nabla\Theta+h(\mathbf{x})\Theta=0$ on $\partial\Omega$ where $h(\mathbf{x})$ is a prescribed function which describes the heat resistivity of the boundary at $\mathbf{x}\in\partial\Omega$. Hence,

$$\mathscr{B}=[\mathbf{U},\,\Theta\colon \operatorname{div}\mathbf{U}=0,\ \mathbf{U}\cdot\mathbf{N}=\mathbf{N}\cdot\nabla\Theta+h(\mathbf{x})\Theta=0 \text{ on } \partial\Omega]\,.$$

In a later application, we shall consider a special case

$$\mathscr{B}_I=[\mathbf{U},\,\Theta\colon \operatorname{div}\mathbf{U}=0,\ \mathbf{U}\cdot\mathbf{N}=0 \text{ on } \partial\Omega,\ \mathbf{n}\cdot\nabla\Theta=0 \text{ on } S,\ \Theta=0 \text{ on } z=0,\,1]\,.$$

The subscript I in $\mathscr{B}_I$ is a reminder that the side wall is insulated.

The boundary value problem (70.13) may be written as

$$\frac{\partial}{\partial\tau}\mathbf{M}\cdot\mathbf{Q}+\mathbf{L}\cdot\mathbf{Q}+\mathbf{N}(\mathbf{Q})\cdot\mathbf{Q}=-\mathbf{d}P\,,\qquad \mathbf{Q}\in\mathscr{B} \tag{70.14}$$

where

$\mathbf{Q}$ is a four-component vector with components $(U,\,V,\,W,\,\Theta)$,
$\mathbf{d}$ is a four-component gradient with component $(\partial_x,\,\partial_y,\,\partial_z,\,0)$,
$\mathbf{M}$ is a 4×4 diagonal matrix with components $(0,\,0,\,0,\,1)$,
$\mathbf{N}(\mathbf{Q})$ is a 4×4 diagonal nonlinear matrix operator with components $(F,\,F,\,F,\,\mathbf{U}\cdot\nabla)$ and

$$F(\Theta)=\hat{f}(-G+\mathscr{R}^{-1}\Theta)-f\,,\qquad F(0)=0,\qquad \hat{f}(-G)\equiv f(z)\equiv f \tag{70.15}$$

and

$$\mathbf{L}=\begin{bmatrix} f & 0 & 0 & 0\\ 0 & f & 0 & 0\\ 0 & 0 & f & -\mathscr{R}\\ 0 & 0 & -\mathscr{R}G' & -\nabla^2 \end{bmatrix}. \tag{70.16}$$

A general property of convection in a container is that the vertical component W of velocity has a zero area-average:

$$\iint_{\mathscr{A}} W\,dx\,dy = 0\,. \tag{70.17}$$

To prove (70.17) note that $\nabla\cdot\mathbf{U} = \nabla_2\cdot\mathbf{U} + \partial_z W$ where $\nabla_2 = \mathbf{e}_x\partial_x + \mathbf{e}_y\partial_y$. Then by the divergence theorem

$$\begin{aligned} 0 &= \iint_{\mathscr{A}} \nabla\cdot\mathbf{U}\,dx\,dy = \iint_{\mathscr{A}} \nabla_2\cdot\mathbf{U}\,dx\,dy + \frac{d}{dz}\iint_{\mathscr{A}} W\,dx\,dy \\ &= \int_{\partial\mathscr{A}} \mathbf{n}\cdot\mathbf{U} + \frac{d}{dz}\iint_{\mathscr{A}} W\,dx\,dy = \frac{d}{dz}\iint_{\mathscr{A}} W\,dx\,dy\,. \end{aligned} \tag{70.18}$$

Since $W=0$ on $z=0,1$, (70.18) implies (70.17).

There are also a number of interesting properties of convection in containers with insulated side-walls. *Let S be the impermeable, insulated vertical side-wall surrounding Ω. Then*

(I) $\left.\dfrac{\partial W}{\partial n}\right|_S = 0\,.$ (70.19)

(II) *The vertical component of the shear stress vanishes on S.* (70.20)

(III) *Suppose S is composed entirely of vertical planar faces. Let l be the arc length on the line of intersection of any plane $z=\text{const}$ and a planar face. Then the horizontal component of the shear stress vanishes on S; that is,*

$$S_{nl} = 2\mu\mathbf{n}\cdot\mathbf{D}\cdot\mathbf{t} = 0 \tag{70.21}$$

where $\mathbf{t}$ *is the unit tangent vector along the line l.*

Proof: The three components of the curl of (70.13a) may be written as

$$\mathbf{e}_z\cdot\operatorname{curl}\hat{f}\,\mathbf{U} = \mathbf{e}_z\cdot(\nabla\Theta\wedge\mathbf{U})\hat{f}'\mathscr{R}^{-1} + \hat{f}\,\mathbf{e}_z\cdot\operatorname{curl}\mathbf{U} = 0\,, \tag{70.22}$$

$$\partial_x(\hat{f}\,W - \mathscr{R}^{-1}\Theta) - \partial_z(\hat{f}\,U) = 0\,, \tag{70.23}$$

and

$$\partial_y(\hat{f}\,W - \mathscr{R}^{-1}\Theta) - \partial_z(\hat{f}\,V) = 0\,. \tag{70.24}$$

We may combine (70.23) and (70.24):

$$\nabla_2(\hat{f}\,W - \mathscr{R}^{-1}\Theta) - \partial_z(\hat{f}\,\mathbf{e}_x U + \hat{f}\,\mathbf{e}_y V) = 0. \tag{70.25}$$

Also

$$\nabla_2\hat{f}\,W = \hat{f}\,\nabla_2 W + \mathscr{R}^{-1}\hat{f}'W\nabla_2\Theta\,.$$

Since $\mathbf{n}$ is independent of z, $\mathbf{n}\cdot\mathbf{U}=\mathbf{n}\cdot(\mathbf{e}_x U+\mathbf{e}_y V)=0$ on S and $\mathbf{n}\cdot\nabla_2\Theta|_S=0$, we find from the scalar product of $\mathbf{n}$ and (70.25) that

$$\mathbf{n}\cdot\nabla_2 W\equiv\frac{\partial W}{\partial n}=0$$

on S, proving (70.19).

The vertical component of the shear stress is proportional to

$$\mathbf{n}\cdot\mathbf{D}\cdot\mathbf{e}_z=n_i\left(\frac{\partial \mathbf{U}_i}{\partial x_j}+\frac{\partial \mathbf{U}_j}{\partial x_i}\right)\delta_{j3} \tag{70.26}$$

where $(x_1, x_2, x_3)=(x, y, z)$. Evaluating (70.26) we find that, on S_1,

$$\frac{\partial \mathbf{n}\cdot\mathbf{U}}{\partial z}+\frac{\partial W}{\partial n}=\mathbf{n}\cdot\mathbf{D}\cdot\mathbf{e}_z=0\,.$$

If S is composed of planar faces then, without loss of generality, we may consider a typical face with normal x. On this face $U=\partial\Theta/\partial x=0$ for all y and z and

$$S_{nl}=S_{xy}=\mu\left(\frac{\partial U}{\partial y}+\frac{\partial V}{\partial x}\right)=\mu\frac{\partial V}{\partial x}\,.$$

Evaluating (70.22) on the planar face we find

$$\mathscr{R}^{-1}\hat{f}'\left[V\frac{\partial\Theta}{\partial x}-U\frac{\partial\Theta}{\partial y}\right]+\hat{f}\left(\frac{\partial V}{\partial x}-\frac{\partial U}{\partial y}\right)=\hat{f}\frac{\partial V}{\partial x}=0\,,$$

proving (70.21).

If S is not composed of planar faces and there are arcs of nonzero length, with curvature κ, on the lines of intersection of S and the planes $z=\text{const}$, then

$$S_{nl}=2\mu\kappa U_l \quad \text{on } S \tag{70.27}$$

where U_l is the component of velocity tangent to the arc with curvature κ. The horizontal component of the shear stress does not vanish on a curved surface S if $U_l\neq 0$. To prove this we decompose vectors on S into orthogonal components $(\mathbf{n}, \mathbf{t}, \mathbf{e}_z)$. Then

$$2D_{nl}=\mathbf{t}\cdot\frac{\partial\mathbf{U}}{\partial n}+\mathbf{n}\cdot\frac{\partial\mathbf{U}}{\partial l}=\frac{\partial U_l}{\partial n}-\mathbf{U}\cdot\frac{\partial\mathbf{t}}{\partial n}+\frac{\partial U_n}{\partial l}-\mathbf{U}\cdot\frac{\partial\mathbf{n}}{\partial l} \tag{70.28}$$

where l is the arc length on the curve of intersection of S and the plane $z=\text{const}$.

Eq. (70.22) may be written as

$$\hat{f}'\mathscr{R}^{-1}\mathbf{e}_z\cdot\left\{\left(\mathbf{n}\frac{\partial\Theta}{\partial n}+\mathbf{t}\frac{\partial\Theta}{\partial l}\right)\wedge(\mathbf{t}U_l+\mathbf{n}U_n)\right\}+\hat{f}\mathbf{e}_z\cdot\left\{\left(\mathbf{n}\frac{\partial}{\partial n}+\mathbf{t}\frac{\partial}{\partial l}\right)\wedge(\mathbf{t}U_l+\mathbf{n}U_n)\right\}=0\,. \tag{70.29}$$

On the plane curve of intersection $U_n=0$, $\partial\Theta/\partial n=0$, $\partial\mathbf{t}/\partial n$ is proportional to $\mathbf{n}$,

$$\frac{\partial\mathbf{t}}{\partial l}=\kappa\mathbf{n}\quad\text{and}\quad\frac{\partial\mathbf{n}}{\partial l}=-\kappa\mathbf{t}\,.$$

Hence (70.28) reduces to

$$2D_{nl}=\frac{\partial U_l}{\partial n}+\kappa U_l \tag{70.30}$$

and (70.29) reduces to

$$\frac{\partial U_l}{\partial n}-\kappa U_l=0\,. \tag{70.31}$$

Combining (70.30) and (70.31) we prove (70.27).

The DOB equations, like Euler's equations of hydrodynamics, do not allow one to impose conditions on the shear stress or on the normal derivatives of velocity when $\mathbf{U}\cdot\mathbf{n}|_S=0$ is prescribed. Property (I) would never be expected to hold for solutions of Euler's equations. Moreover property (I) does not hold generally for the DOB equations; it is forced onto the dynamics as an effect of buoyancy at an insulated side-wall. The vanishing of the shear stress S_{nz} follows easily as a consequence of (I). In general, when S is curved, the other shear stress S_{nl} does not vanish, but when the walls are flat—in polygonal container, S_{nl} does vanish and the side-wall S is free of shear stresses. In this case the container is completely isolated at the side-wall: the side-walls bar the transport of mass, heat and momentum. It is natural to wonder if cells in cellular convection are truly isolated as in containers with planar side-walls S.

Exercise 70.1: Show that

$$S_{r\theta}=-2\mu U_\theta/r$$

on the side-wall b of a container of circular cross-section.

Exercise 70.2: Prove that

$$\mathbf{n}\cdot\nabla P=0$$

on an impermeable vertical side-wall.

§ 71. The Spectral Problem, the Adjoint Spectral Problem and the Energy Theory of Stability

To obtain the spectral problem for the stability of the conduction solution $(\mathbf{Q}, P)=(0, 0)$ we linearize (70.14) and set $(\mathbf{Q}, P)=e^{-\sigma\tau}(\mathbf{q}, p)$. Then,

$$-\sigma\mathbf{M}\cdot\mathbf{q}+\mathbf{L}\cdot\mathbf{q}=-\mathbf{d}p\,, \qquad \mathbf{q}\in\mathscr{B}\,. \tag{71.1}$$

When $G'=1$ $(\xi=0)$, $\mathbf{L}$ is a symmetric operator. Then every eigenvalue σ of $\mathbf{L}$ is real-valued. We assume that σ is real-valued when $\xi\neq 0$. In this case each and every eigenvalue σ of (71.1) is also an eigenvalue of the adjoint problem

$$-\sigma\mathbf{M}\cdot\mathbf{q}^*+\mathbf{L}^T\cdot\mathbf{q}^*=-\mathbf{d}p^*\,, \qquad \mathbf{q}^*\in\mathscr{B} \tag{71.2}$$

where $\mathbf{L}^T$ is the transpose of $\mathbf{L}$.

The eigenvalues $\sigma=\sigma(\mathscr{R})$ are all positive when $\mathscr{R}$ is sufficiently small. As $\mathscr{R}$ is increased one of the eigenvalues becomes negative and the conduction solution loses stability. The value $\mathscr{R}=\mathscr{R}_L$ for which $\sigma(\mathscr{R})$ first vanishes is the critical value and $\sigma'(\mathscr{R}_L)<0$. When $\sigma<0$ conduction is unstable. When $\sigma>0$ conduction is stable to small disturbances but may be unstable to large disturbances. The energy theory of stability is used to find conditions for stability to large disturbances.

The energy identities following from (70.13) are

$$-\mathscr{R}\langle W\Theta\rangle+\langle\hat{f}(-G+\mathscr{R}^{-1}\Theta)|\mathbf{U}|^2\rangle=0 \tag{71.3 a}$$

and

$$\tfrac{1}{2}\frac{d}{d\tau}\langle\Theta^2\rangle-\mathscr{R}\langle G'W\Theta\rangle+\langle|\nabla\Theta|^2\rangle+\langle h\Theta^2\rangle_{\partial\Omega}=0 \tag{71.3 b}$$

where $\langle\cdot\rangle_{\partial\Omega}\equiv\dfrac{1}{\Omega}\int_{\partial\Omega}$,

and

$$\langle\cdot\rangle=\int_\Omega\cdot\,d\Omega\Big/\int_\Omega d\Omega$$

is a volume-averaged integral.

We may write (71.3) as a single equation with a coupling parameter $\lambda>0$ to be chosen to optimize the criterion for stability

$$\tfrac{1}{2}\frac{d\langle\Theta^2\rangle}{d\tau}=\mathscr{R}\langle(G'+\lambda)W\Theta\rangle-\langle|\nabla\Theta|^2+\lambda\hat{f}(-G+\mathscr{R}^{-1}\Theta)|\mathbf{U}|^2\rangle-\langle h\Theta^2\rangle_{\partial\Omega}. \tag{71.4}$$

Clearly, $\langle \Theta^2 \rangle$ decreases when $\mathscr{R} < \mathscr{R}_{\mathscr{E}}$ where

$$\frac{1}{\mathscr{R}_{\mathscr{E}}} = \inf_{\lambda>0} \sup_{[\mathbf{U},\Theta]\in\mathscr{B}} \frac{\langle (G'+\lambda) W\Theta \rangle}{\langle |\nabla\Theta|^2 + \lambda \hat{f}(-G+\mathscr{R}^{-1}\Theta)|\mathbf{U}|^2 \rangle + \langle h\Theta^2 \rangle_{\partial\Omega}} \tag{71.5}$$

It is instructive to consider (71.5) in three special cases.

(i) Bénard problem for convection in a porous media (Westbrook, 1969): In this case $G' = \hat{f} = 1$ and $\mathscr{R}_{\mathscr{E}} = \mathscr{R}_L$. Subcritical instabilities are impossible.

(ii) The material properties are constant, $\hat{f} = 1$, but there may be heat sources, $\xi \neq 0$. Then one may show that the supremum and infimum in (71.5) are attained,

$$\frac{1}{\mathscr{R}_{\mathscr{E}}} = \min_{\lambda>0} \max_{[\mathbf{U},\Theta]\in\mathscr{B}} \frac{\langle (G'+\lambda) W\Theta \rangle}{\langle |\nabla\Theta|^2 + \lambda |\mathbf{U}|^2 \rangle + \langle h\Theta^2 \rangle_{\partial\Omega}} \tag{71.6}$$

and that

$$\mathscr{R}_{\mathscr{E}} \leqslant \mathscr{R}_L \tag{71.7}$$

with equality only when $\xi = 0$. The good properties arise from the fact the energy equation with $\hat{f} = 1$ is homogeneous of degree two and the production and dissipation integrals are both quadratic in the components of $\mathbf{Q}$. It is possible to obtain many results comparing $\mathscr{R}_{\mathscr{E}}$ and $\mathscr{R}_L$ without solving explicit problems (see Exercises of § 76).

(iii) There are no heat sources but material properties depend on the temperature. In this case $G' = 1$ and we may rewrite (71.4) as

$$\tfrac{1}{2}\frac{d}{d\tau}\langle \theta^2 \rangle = \mathscr{R}(1+\lambda)\langle w\theta \rangle - \langle |\nabla\theta|^2 + \lambda \hat{f}(-G+\varepsilon\mathscr{R}^{-1}\theta)|\underline{u}|^2 \rangle - \langle h\theta^2 \rangle_{\partial\Omega} \tag{71.8}$$

where we have put $\mathbf{U} = \varepsilon\mathbf{u}$ and $\Theta = \varepsilon\theta$. The presence of ε in $\hat{f}$ shows Eq. (71.8) is not homogeneous. It is therefore impossible, without extra conditions, to deduce global criteria of stability, independent of the amplitude ε, from (71.8). This difficulty in energy analysis arises from material nonlinearities which, unlike inertial nonlinearities, do not wash out of the overall energy balance.

An extra condition which is needed to deduce global stability from (71.8) is provided if $\hat{f}$ is bounded from below by some positive constant $\underline{\hat{f}}$. The existence of this constant might be supposed to stem from nature; viscosity is positive, but the expressions we use to represent $\hat{f}$ are usually unbounded when $-\infty < \varepsilon < \infty$ is allowed to take on all values. It is an interesting fact that the temperature, $-G + \mathscr{R}^{-1}\Theta$, on which $\hat{f}$ depends can be bounded from above and below by the maximum principle for parabolic equations when the initial and boundary values of T are suitably prescribed (see Exercise 71.2).

Exercise 71.1: Prove the inequality (71.7).

Exercise 71.2 (Rabinowitz, 1973): Find bounds on the values of the total $T = -G(z;0) + \mathscr{R}^{-1}\Theta$ in terms of the values of T given on the parabolic boundary (*Hint*: see Exercise 40.2).

§ 72. Two-Sided Bifurcation

The general theory of bifurcation at a simple real-valued eigenvalue σ has been given in § 15 of Chapter II. Now we apply this theory to special problems of steady convection. At the end of this section we shall develop elements in the theory of bifurcation at an eigenvalue of higher multiplicity. The bifurcation problems to be considered are defined by

$$\mathbf{L}\cdot\mathbf{q}+\mathbf{N}(\varepsilon\mathbf{q})\cdot\mathbf{q}=-\mathbf{d}p\,,\qquad \mathbf{q}\in\mathscr{B}\,,\qquad \varepsilon=\langle\mathbf{Q}\cdot\mathbf{q}_0\rangle \tag{72.1}$$

where $P=\varepsilon p$, $\mathbf{Q}=\varepsilon\mathbf{q}(x,y,z;\varepsilon)$, $\mathbf{q}_0=\mathbf{q}(x,y,z;0)$ and $\mathbf{q}$ is a four-component vector with components (u, v, w, θ).

The defining condition for the amplitude ε is useful for discussing the sign of the motion. This condition implies that $\langle|\mathbf{q}_0|^2\rangle=1$. To fix the sign of $\mathbf{q}_0$, we choose some point (x_0, y_0, z_0) in the interior of Ω at which $w_0=\mathbf{e}_z\cdot\mathbf{q}_0\neq 0$ and, since the sign of $\mathbf{q}_0(x_0, y_0, z_0)$ is not determined by normalization we may take $w_0(x_0,y_0,z_0)>0$. Then, to lowest order, $\varepsilon w_0(x_0,y_0,z_0)$ gives the sign of the vertical velocity, which is upwards when $\varepsilon>0$ and downwards when $\varepsilon<0$. Since the area average (70.17) of w vanishes, the fluid must rise in some parts of the container and sink in others.

(a) Simple Eigenvalue

We now suppose that $\sigma(\mathscr{R}_L)=0$ is a simple isolated eigenvalue of $\mathbf{L}_0$ where $\mathbf{L}_0$ is the operator defined by (70.16) when $\mathscr{R}=\mathscr{R}_L\equiv\mathscr{R}_0$. Equivalently, we may put $\sigma=0$ and regard $\mathscr{R}_L$ as the eigenvalue parameter. The bifurcation theory developed in § 15 applies. There is a unique solution of (72.1)

$$\mathbf{q}(x,\ y,\ z;\ \varepsilon),\ \ p(x,\ y,\ z;\ \varepsilon),\ \ \mathscr{R}(\varepsilon)$$

analytic in ε, which bifurcates from conduction when $\mathscr{R}=\mathscr{R}(0)=\mathscr{R}_0$. The bifurcating solution is subcritical if

$$\mathscr{R}(\varepsilon)-\mathscr{R}_0=\varepsilon\mathscr{R}_1+\varepsilon^2\mathscr{R}_2+\dots$$

is negative and is supercritical if $\mathscr{R}(\varepsilon)-\mathscr{R}_0$ is positive. Since ε may take on both signs, the bifurcation is two-sided when $\mathscr{R}_1\neq 0$.

To obtain a formula for $\mathscr{R}_1$, we first note that when $\varepsilon=0$ (72.1) reduces to

$$\mathbf{L}_0\cdot\mathbf{q}_0=-\mathbf{d}p,\qquad \mathbf{q}_0\in\mathscr{B}\,,\qquad \langle|\mathbf{q}_0|^2\rangle=1\,. \tag{72.2}$$

Differentiating (72.1) with respect to ε at $\varepsilon=0$ we find that

$$\mathbf{L}_0\cdot\mathbf{q}_1+\mathbf{L}_1\cdot\mathbf{q}_0+\mathbf{N}_1(\mathbf{q}_0)\cdot\mathbf{q}_0=-\mathbf{d}p_1\,,\qquad \mathbf{q}\in\mathscr{B}\,,\qquad \langle\mathbf{q}_0\cdot\mathbf{q}_1\rangle=0 \tag{72.3}$$

where

$$\mathbf{L}_1 = -\mathscr{R}_1\hat{\mathbf{L}}, \qquad \hat{\mathbf{L}} = \begin{bmatrix} 0 & 0 & 0 & 0 \\ 0 & 0 & 0 & 0 \\ 0 & 0 & 0 & 1 \\ 0 & 0 & G' & 0 \end{bmatrix}$$

and $\mathbf{N}_1(\mathbf{q}_0)$ is a 4×4 diagonal matrix with components $(\mathscr{R}_0^{-1}\hat{f}'\theta_0, \mathscr{R}_0^{-1}\hat{f}'\theta_0, \mathscr{R}_0^{-1}\hat{f}'\theta_0, \mathbf{u}_0\cdot\nabla)$ where $\hat{f}'$ is the derivative of $\hat{f}(\eta)$ evaluated at $\eta=-G$.

Eq. (72.3) may be solved if and only if $\mathscr{R}_1(\mathbf{q}_0)$ is given by

$$-\mathscr{R}_1\langle \mathbf{q}^*\cdot\hat{\mathbf{L}}\cdot\mathbf{q}_0\rangle + \langle \mathbf{q}^*\cdot\mathbf{N}_1(\mathbf{q}_0)\cdot\mathbf{q}_0\rangle = 0 \tag{72.4}$$

where $\mathbf{q}^*\equiv\mathbf{q}_0^*$ (to simplify notation) is the adjoint eigenfunction of (71.2) when $\sigma(\mathscr{R}_0)=0$. It follows from (72.4) that

$$\varepsilon\mathscr{R}_1(\varepsilon\mathbf{q}_0) = \varepsilon\frac{\langle \mathbf{q}^*\cdot\mathbf{N}_1(\varepsilon\mathbf{q}_0)\cdot\mathbf{q}_0\rangle}{\langle \mathbf{q}^*\cdot\hat{\mathbf{L}}\cdot\mathbf{q}_0\rangle} = -\varepsilon\frac{\langle \mathbf{q}^*\cdot\mathbf{N}_1(-\varepsilon\mathbf{q}_0)\cdot\mathbf{q}_0\rangle}{\langle \mathbf{q}^*\cdot\hat{\mathbf{L}}\cdot\mathbf{q}_0\rangle} = -\varepsilon\mathscr{R}_1(-\varepsilon\mathbf{q}_0). \tag{72.5}$$

The value of $\mathscr{R}_1$ determines the sign of the motion. Recall that without loss of generality $w_0(x_0, y_0, z_0)>0$. Suppose now that $\mathscr{R}_1$ is calculated from (72.4). $\mathscr{R}_1$ is unique because the sign of $\mathbf{q}_0$ is fixed by our sign convention for w_0. Suppose further that $\mathscr{R}_1>0$. Then the subcritical branch $\varepsilon\mathscr{R}_1<0$ has $\varepsilon<0$ and $W(x_0, y_0, z_0)\sim\varepsilon w(x_0, y_0, z)<0$. The fluid at (x_0, y_0, z_0) on the subcritical branch is sinking and (72.5) shows that the fluid is rising at the same point on the supercritical branch.

Eq. (72.4) may be written as

$$-\mathscr{R}_1\langle w^*\theta_0 + w_0\theta^* G'\rangle + \mathscr{R}_0^{-1}\langle \hat{f}'\theta_0\mathbf{u}^*\cdot\mathbf{u}_0\rangle + \langle \theta^*\mathbf{u}_0\cdot\nabla\theta\rangle = 0. \tag{72.6}$$

When $\xi=0$, then $G'=1$, $\mathbf{q}^*=\mathbf{q}_0$, $\langle\theta_0(\mathbf{u}\cdot\nabla)\theta_0\rangle=0$ and

$$\mathscr{R}_1 = \mathscr{R}_0^{-1}\langle \hat{f}'\theta_0|\mathbf{u}_0|^2\rangle / 2\langle w_0\theta_0\rangle. \tag{72.7}$$

Now $\mathbf{q}_0$ does not depend explicitly on $\hat{f}'$ and the values of θ_0 do not depend strongly on whether $\hat{f}$ is a decreasing or increasing function. Hence (72.7) suggests that the sign of $\mathscr{R}_1$ is associated uniquely with the sign of $\hat{f}'$. This means that $\mathscr{R}_1$ will have one value in fluid whose viscosity increases with temperature, $\hat{f}'>0$, and the other sign in the other case. This fact has interesting physical consequences which we shall discuss in § 77. When $\xi=0$ and $\hat{f}=1$, $\mathscr{R}_1=0$ and the bifurcation is supercritical (see Exercise 76.1). When $\mathscr{R}_1$ is known we may solve (72.3) for $\mathbf{q}_1$ and proceed sequentially to higher orders.

(b) Multiple Eigenvalues

For simplicity we shall consider the case when $\mathbf{L}_0$ has two eigenfunctions for the one eigenvalue $\gamma(\mathscr{R}_0)=0$. Call the two eigenfunctions $\boldsymbol{\phi}$ and $\boldsymbol{\psi}$ so that

$\operatorname{curl}\mathbf{L}_0\cdot\boldsymbol{\phi}=\operatorname{curl}\mathbf{L}_0\cdot\boldsymbol{\psi}=0$. Then, there are two adjoint eigenfunctions $\boldsymbol{\phi}^*$ and $\boldsymbol{\psi}^*$ such that $\operatorname{curl}\mathbf{L}_0^T\cdot\boldsymbol{\phi}^*=\operatorname{curl}\mathbf{L}_0^T\cdot\boldsymbol{\psi}^*=0$. The vector

$$\mathbf{q}_0=A_1\boldsymbol{\phi}+A_2\boldsymbol{\psi} \tag{72.8}$$

is said to span the null space of $\mathbf{L}_0$, but the constants A_1 and A_2 are undetermined at zeroth order. The vectors $\boldsymbol{\phi}$ and $\boldsymbol{\psi}$ are linearly independent and may be selected so as to satisfy conditions of orthonormality

$$\langle\boldsymbol{\phi}\cdot\boldsymbol{\psi}\rangle=0\,,\qquad \langle|\boldsymbol{\phi}|^2\rangle=\langle|\boldsymbol{\psi}|^2\rangle=1\,.$$

Then $\langle|\mathbf{q}_0|^2\rangle=1=A_1^2+A_2^2=A_1^2\,(1+\mu^2)$ where $\mu=A_2/A_1$. Only certain values of μ are compatible with solutions of (72.3). These values may be determined by solvability conditions (72.3) and (72.4). We find that

$$\begin{aligned}&\mathbf{L}_0\cdot\mathbf{q}_1-\mathscr{R}_1(A_1\hat{\mathbf{L}}\cdot\boldsymbol{\phi}+A_2\hat{\mathbf{L}}\cdot\boldsymbol{\psi})+\mathbf{N}_1(A_1\boldsymbol{\phi}+A_2\boldsymbol{\psi})\cdot(A_1\boldsymbol{\phi}+A_2\boldsymbol{\psi})=-\mathbf{d}p_1\,,\\&\mathbf{q}_1\in\mathscr{B}\,,\qquad\langle\mathbf{q}_0\cdot\mathbf{q}_1\rangle=0\,.\end{aligned} \tag{72.9}$$

(72.9) is solvable only if the inhomogeneous terms are orthogonal to $\boldsymbol{\phi}^*$ and $\boldsymbol{\psi}^*$. There are then two equations:

$$\begin{aligned}&\mathscr{R}_1(1+\mu^2)^{1/2}\{\langle\boldsymbol{\phi}^*\cdot\hat{\mathbf{L}}\cdot\boldsymbol{\phi}\rangle+\mu\langle\boldsymbol{\phi}^*\cdot\hat{\mathbf{L}}\cdot\boldsymbol{\psi}\rangle\}\\&\quad=\langle\boldsymbol{\phi}^*\cdot\mathbf{N}_1(\boldsymbol{\phi})\cdot\boldsymbol{\phi}\rangle+\mu\langle\boldsymbol{\phi}^*\cdot\mathbf{N}_1(\boldsymbol{\psi})\cdot\boldsymbol{\phi}\rangle+\mu\langle\boldsymbol{\phi}^*\cdot\mathbf{N}_1(\boldsymbol{\phi})\cdot\boldsymbol{\psi}\rangle+\mu^2\langle\boldsymbol{\phi}^*\cdot\mathbf{N}_1(\boldsymbol{\psi})\cdot\boldsymbol{\psi}\rangle\end{aligned} \tag{72.10a}$$

and

the same equation with $\boldsymbol{\psi}^*$ replacing $\boldsymbol{\phi}^*$. (72.10b)

These equations may be solved for

$$\mathscr{R}_1(1+\mu)^{1/2}=\frac{a+\mu b+\mu^2 c}{d+\mu e}=\frac{f+\mu g+\mu^2 h}{k+\mu l}\,.$$

The last equation implies that

$$(k+\mu l)(a+\mu b+\mu^2 c)=(d+\mu e)(f+\mu g+\mu^2 h)\,. \tag{72.11}$$

(72.11) is a cubic equation in $\mu=A_2/A_1$ and it has either one real and two conjugate complex roots or three real roots. When the roots μ are known, we find $A_1=(1+\mu^2)^{-1/2}$ and $A_2=\mu(1+\mu^2)^{-1/2}$. In the case of three distinct real roots there are three initiating vectors

$$\mathbf{q}_0=\begin{bmatrix}\mathbf{q}_{01}\\\mathbf{q}_{02}\\\mathbf{q}_{03}\end{bmatrix}=\begin{bmatrix}(1+\mu_1^2)^{-1/2}\;(\boldsymbol{\phi}+\mu_1\boldsymbol{\psi})\\(1+\mu_2^2)^{-1/2}\;(\boldsymbol{\phi}+\mu_2\boldsymbol{\psi})\\(1+\mu_3^2)^{-1/2}\;(\boldsymbol{\phi}+\mu_3\boldsymbol{\psi})\end{bmatrix},$$

three values $\mathscr{R}_{11}$, $\mathscr{R}_{12}$ and $\mathscr{R}_{13}$ and three branches of solutions bifurcate from the eigenvalue $\sigma(\mathscr{R}_0)=0$ of multiplicity two. In the other case only one solution bifurcates. When no root of 72.11 is repeated, either one solution or three solutions bifurcate from a double eigenvalue.

When $\mathscr{R}_1=0$ the determination of the number of bifurcating branches and the allowed values of A_1 and A_2 must be determined at order two. A more complete discussion of these matters is given in the addendum to this chapter.

(c) Stability of Bifurcating Solutions at Eigenvalues of Higher Multiplicity

To obtain the spectral problem for the bifurcating solution we substitute

$$Q=\varepsilon\mathbf{q}+\tilde{\mathbf{q}}e^{-\gamma\tau}$$

into (70.14) and find that

$$-\gamma\mathbf{M}\cdot\tilde{\mathbf{q}}+\mathbf{L}\cdot\tilde{\mathbf{q}}+\mathbf{N}(\varepsilon\mathbf{q})\cdot\tilde{\mathbf{q}}+\mathbf{N}(\tilde{\mathbf{q}})\cdot\varepsilon\mathbf{q}=-\mathbf{d}\pi \tag{72.12}$$

where $\tilde{\mathbf{q}}\in\mathscr{B}$. When $\varepsilon=0$, (72.12) reduces to

$$-\gamma_0\mathbf{M}\cdot\tilde{\mathbf{q}}_0+\mathbf{L}_0\cdot\tilde{\mathbf{q}}_0=-\mathbf{d}\pi_0\,. \tag{72.13}$$

We suppose that $\mathscr{R}_0$ is as small as possible. Then $\gamma_0\geqslant 0$. The most dangerous case is when $\gamma_0=0$. Then $\tilde{\mathbf{q}}_0$ is a linear combination of eigenfunctions in the null space of $\mathbf{L}_0$. In general the bifurcating flow is also a linear combination of eigenfunctions in the null space of $\mathbf{L}_0$. Thus

$$\begin{aligned}\mathbf{q}_0&=\textstyle\sum_{n=1}^{N}A_n\hat{\mathbf{q}}_n\,,\\ \tilde{\mathbf{q}}_0&=\textstyle\sum_{n=1}^{N}B_n\hat{\mathbf{q}}_n\,.\end{aligned} \tag{72.14}$$

There are also N independent eigenfunctions $\mathbf{q}_n^*$ of $\mathbf{L}_0^T$.

Notation: The subscripts on $\hat{\mathbf{q}}_n$, $\mathbf{q}_n^*$ and $\mathbf{q}_n$ have different meanings:

$\hat{\mathbf{q}}_1, \hat{\mathbf{q}}_2, \ldots, \hat{\mathbf{q}}_N$ are the independent eigenfunctions of $\mathbf{L}_0$

$\mathbf{q}_1^*, \mathbf{q}_2^*, \ldots, \mathbf{q}_N^*$ are the independent eigenfunctions of $\mathbf{L}_0^T$

$\mathbf{q}_l^*=(u_l^*, \theta_l^*) \qquad (l=1, 2, \ldots, N)$

whereas

$$\mathbf{q}_n=\frac{1}{n!}\partial^n\mathbf{q}/\partial\varepsilon^n\big|_{\varepsilon=0} \qquad (n\geqslant 0)\,.$$

Let us suppose that some solution of the bifurcation problem is given; that is, a set of coefficients $\{A_n\}$, and a value $\mathscr{R}_1$ and the function $\mathbf{q}_0$ have been determined.

The coefficients A_n and associated slopes $\mathscr{R}_1$ are obtained from the equations ($n=1, 2, \ldots, N$)

$$\mathscr{R}_1\langle \mathbf{q}_n^* \cdot \hat{\mathbf{L}} \cdot \mathbf{q}_0\rangle = \langle \mathbf{q}_n^* \cdot \mathbf{N}_1(\mathbf{q}_0) \cdot \mathbf{q}_0\rangle$$

and some normalizing condition for the A_n.

Given $N+1$ values $\{A_n, \mathscr{R}_1\}$ we may state the problem for γ_1 and $\tilde{\mathbf{q}}_1$ as

$$-\gamma_1 \mathbf{M} \cdot \tilde{\mathbf{q}}_0 - \mathscr{R}_1 \hat{\mathbf{L}} \cdot \tilde{\mathbf{q}}_0 + \mathbf{L} \cdot \tilde{\mathbf{q}}_1 + \mathbf{N}_1(\mathbf{q}_0) \cdot \tilde{\mathbf{q}}_0 + \mathbf{N}_1(\tilde{\mathbf{q}}_0) \cdot \mathbf{q}_0 = -\mathbf{d}\pi_1, \tag{72.15}$$

(72.15) is solvable for $\mathbf{q}_1 \in \mathscr{B}$ if

$$-\gamma_1\langle \mathbf{q}_n^* \cdot \mathbf{M} \cdot \tilde{\mathbf{q}}_0\rangle - \mathscr{R}_1\langle \mathbf{q}_n^* \cdot \hat{\mathbf{L}} \cdot \tilde{\mathbf{q}}_0\rangle + \langle \mathbf{q}_n^* \cdot \mathbf{N}_1(\mathbf{q}_0) \cdot \tilde{\mathbf{q}}_0\rangle + \langle \mathbf{q}_n^* \cdot \mathbf{N}_1(\tilde{\mathbf{q}}_0) \cdot \mathbf{q}_0\rangle = 0 \tag{72.16}$$

for $n=1, 2, \ldots, N$. If we supplement (72.16) with a normalizing condition we have $N+1$ equations for the $N+1$ values $\{B_n, \gamma_1\}$. Since (72.15) is a linear problem, N solutions typically branch from an eigenvalue of multiplicity N.

Eq. (72.16) may be written as

$$\begin{aligned}\gamma_1\langle \theta_n^* \tilde{\theta}\rangle + \mathscr{R}_1\langle w_n^* \tilde{\theta} + G' \tilde{w} \theta_n^*\rangle = & \frac{1}{\mathscr{R}_0}\langle f'(\theta_0 \tilde{\mathbf{u}} \cdot \mathbf{u}_n^* + \tilde{\theta} \mathbf{u}_0 \cdot \mathbf{u}_n^*)\rangle \\ & + \langle \theta_n^*(\mathbf{u}_0 \cdot \nabla \tilde{\theta} + \tilde{\mathbf{u}} \cdot \nabla \theta_0)\rangle\end{aligned} \tag{72.17}$$

where $\mathbf{u}_n^*$ and θ_n^* are the components of $\mathbf{q}_n^*$. The stability problem for hexagonal convection may be formulated in a set of functions for which $N=2$. We shall apply (72.17) to study hexagonal convection in hexagonal containers (§ 73c) and in horizontally infinite fluid layers (§ 77).

Exercise 72.1: Find a formula giving $\mathscr{R}_2$ in terms of lower order quantities.

Exercise 72.2: Define $\mathscr{R}_\mathscr{E}(\varepsilon)$ by (71.5) and the constraint $\langle w\theta\rangle = \varepsilon^2$. Show that

$$\mathscr{R}_\mathscr{E}(0) = \mathscr{R}_0 \quad \text{and} \quad \frac{d\mathscr{R}_\mathscr{E}(0)}{d\varepsilon} = \mathscr{R}_1 .$$

Exercise 72.3: Assume that $\sigma(\mathscr{R}_0) = 0$ is a simple eigenvalue of $\mathbf{L}_0$. Show that

$$\frac{d\sigma(\mathscr{R}_0)}{d\mathscr{R}} = -\langle \mathbf{q}^* \cdot \hat{\mathbf{L}} \cdot \mathbf{q}_0\rangle / \langle \mathbf{q}^* \cdot \mathbf{M} \cdot \mathbf{q}_0\rangle$$

§ 73. Conditions for the Existence of Two-Sided Bifurcation

We turn now to the problem of bifurcation in a container with insulated side-walls. This is a physically interesting but mathematically special boundary condition at the container wall. The conditions which we shall derive for this special problem have a wider applicability deriving from the fact that the conditions

$$\mathbf{n}\cdot\nabla\Theta=\mathbf{n}\cdot\mathbf{U}=0 \quad \text{on } S \tag{73.1}$$

which hold at the side-wall of an insulated container are identical to the conditions which hold at the boundary of certain cells in cellular convection.

For simplicity we shall now assume that the top and bottom of the container are made of perfectly conducting materials. Then Θ and W vanish on $z=0, 1$.

Recalling that $f(z)\equiv\hat{f}(-G)$ we may rewrite Eqs. (72.2) as

$$f\mathbf{u}_0-\mathscr{R}_0\mathbf{e}_z\theta_0=-\nabla p_0\,, \tag{73.2a}$$

$$\mathscr{R}_0G'w_0+\nabla^2\theta_0=0\,, \tag{73.2b}$$

$$w_0=\theta_0=0 \quad \text{on} \quad z=0,1\,, \tag{73.2c}$$

$$\mathbf{u}_0\cdot\mathbf{n}=\mathbf{n}\cdot\nabla\theta_0=0 \quad \text{on } S. \tag{73.2d}$$

The adjoint eigenvalue problem is

$$f\mathbf{u}^*-\mathbf{e}_z\mathscr{R}_0G'\theta_0=-\nabla p^* \tag{73.3a}$$

$$\mathscr{R}_0w^*+\nabla^2\theta^*=0 \tag{73.3b}$$

$$w^*=\theta^*=0 \quad \text{on} \quad z=0,1 \tag{73.3c}$$

$$\mathbf{u}^*\cdot\mathbf{n}=\mathbf{n}\cdot\nabla\theta^*=0 \quad \text{on } S\,. \tag{73.3d}$$

It is necessary to add that w_0 and w_* both have a zero area-average.

Noting now that $G(z;\xi)$ is independent of x and y we may verify that $\mathbf{e}_z\cdot\operatorname{curl}\mathbf{u}_0=\mathbf{e}_z\cdot\operatorname{curl}\mathbf{u}^*=0$. Since $\mathbf{u}_0$ and $\mathbf{u}^*$ are solenoidal we may introduce poloidal potentials (see Appendix B)

$$\mathbf{u}_0=\boldsymbol{\delta}\chi_0\,, \qquad \mathbf{u}^*=\boldsymbol{\delta}\chi^*$$

where

$$\boldsymbol{\delta}=\nabla\frac{\partial}{\partial z}-\mathbf{e}_z\nabla^2=\nabla_2\frac{\partial}{\partial z}-\mathbf{e}_z\nabla_2^2=\operatorname{curl}^2\mathbf{e}_z\,.$$

Applying the operator $\mathbf{e}_z\cdot\operatorname{curl}^2$ to (73.2a) and (73.3a) we find that

$$-f\,\nabla^2w_0-f'\partial_zw_0+\mathscr{R}_0\nabla_2^2\theta_0=0$$

and

$$-f\,\nabla^2w^*-f'\partial_zw^*+\mathscr{R}_0G'\nabla_2^2\theta^*=0$$

where $f'=df/dz$. Since $w_0=-\nabla_2^2\chi_0$ and $w^*=-\nabla_2^2\chi^*$

$$f\nabla^2\chi_0+f'\partial_z\chi_0+\mathscr{R}_0\theta_0=H_1(x,y,z)$$

and

$$f\nabla^2\chi^*+f'\partial_z\chi^*+\mathscr{R}_0G'\theta^*=H_2(x,y,z)$$

where $\nabla_2^2H_1=\nabla_2^2H_2=0$. To find the boundary values of H_1 and H_2 we apply $\mathbf{n}\cdot\nabla_2$ and utilize (73.2c, 2d, 3c, 3d) to deduce that

$$\left.\frac{\partial H_1}{\partial n}\right|_S=\left.\frac{\partial H_2}{\partial n}\right|_S=0.$$

It follows that H_1 and H_2 are functions of z alone. Such functions of z alone may be absorbed into χ without changing the velocities. It follows that we may put $H_1=H_2=0$ without loss of generality. Then $\mathbf{u}_0$, $\mathbf{u}^*$, θ_0 and θ^* may be found as solutions of the eigenvalue problems

$$f\nabla^2\chi_0+f'\partial_z\chi_0+\mathscr{R}_0\theta_0=0, \tag{73.4a}$$

$$-\mathscr{R}_0G'\nabla_2^2\chi_0+\nabla^2\theta_0=0, \tag{73.4b}$$

$$\chi_0=\theta_0=0 \quad \text{on} \quad z=0,1, \tag{73.4c}$$

$$\mathbf{n}\cdot\nabla\partial_z\chi_0=\mathbf{n}\cdot\nabla\theta_0=0 \quad \text{on } S \tag{73.4d}$$

and

$$f\nabla^2\chi^*+f'\partial_z\chi^*+\mathscr{R}_0G'\theta_0=0, \tag{73.5a}$$

$$-\mathscr{R}_0\nabla_2^2\chi^*+\nabla^2\theta^*=0, \tag{73.5b}$$

$$\chi^*=\theta^*=0 \quad \text{on} \quad z=0,1, \tag{73.5c}$$

$$\mathbf{n}\cdot\nabla\partial_z\chi^*=\mathbf{n}\cdot\nabla\theta^*=0 \quad \text{on } S. \tag{73.5d}$$

The eigenvalue problem (73.4a) and (73.5a) can be solved by separation of variables:

$$\begin{bmatrix}\chi_0\\ \theta_0\\ \chi^*\\ \theta^*\end{bmatrix}=\phi(x,y)\begin{bmatrix}\bar{\bar{\chi}}(z)\\ \bar{\bar{\theta}}(z)\\ \bar{\bar{\chi}}^*(z)\\ \bar{\bar{\theta}}^*(z)\end{bmatrix} \tag{73.6}$$

where ϕ is called a planform function. The function ϕ is an eigenfunction with eigenvalue a^2 of the plane Laplacian

$$\nabla_2^2\phi+a^2\phi=0, \tag{73.7a}$$

$$\mathbf{n}\cdot\nabla_2\phi=0 \tag{73.7b}$$

$$\iint_{\mathscr{A}}\phi\,dx\,dy=0. \tag{73.7c}$$

The zero average condition (73.7c) guarantees that w_0 and w^* will have zero area-averages. In fact (73.7c) is implied by (73.7a) and (73.7b) for all nonconstant solutions ϕ. The separation (73.6) arises when one tries to solve (73.4) and (73.5) as a Fourier series in the eigenfunctions of (73.7) with z dependent Fourier coefficients. The same separation will not work in containers filled with OB fluid; for these $\mathbf{u}=0$ on S and the representations (73.6) are insufficiently general. The separation (73.6) will not work when the side-walls are conducting and $\theta=0$ on S. This boundary condition leads to a more complicated problem (see Exercises 76.4 and 76.5). However, the separation does hold when $\partial_z\theta+h\theta=0$ with a constant but different value of h on the horizontal walls at $z=0, 1$.

The overbar functions of z satisfy ordinary differential equations

$$f(D^2-a^2)\bar{\bar{\chi}}+f'D\bar{\bar{\chi}}+\mathscr{R}_0\bar{\bar{\theta}}=0\,, \tag{73.8a}$$

$$\mathscr{R}_0G'a^2\bar{\bar{\chi}}+(D^2-a^2)\bar{\bar{\theta}}=0\,, \tag{73.8b}$$

$$\bar{\bar{\chi}}=\bar{\bar{\theta}}=0 \quad \text{on} \quad z=0, 1 \tag{73.8c}$$

and

$$f(D^2-a^2)\bar{\bar{\chi}}^*+f'D\bar{\bar{\chi}}^*+\mathscr{R}_0G'\bar{\bar{\theta}}=0\,, \tag{73.9a}$$

$$\mathscr{R}_0a^2\bar{\bar{\chi}}^*+(D^2-a^2)\bar{\bar{\theta}}^*=0\,, \tag{73.9b}$$

$$\bar{\bar{\chi}}^*=\bar{\bar{\theta}}^*=0 \quad \text{on} \quad z=0, 1 \tag{73.9c}$$

where $D^2=d^2/dz^2$. (73.9) is adjoint to (73.8). These two systems determine the same eigenvalues $\mathscr{R}_0(a^2)$. $\mathscr{R}_0(a^2)>0$ is a continuous (analytic) function of $a^2>0$ which tends to infinity as $a^2\to\infty$ or $a^2\to 0$.

There are, in general, a countably infinite set of eigenvalues $a^2=a_l^2$ of (73.7) and a countably infinite set of eigenvalues $\mathscr{R}_0(a_l^2)$ for each one of the eigenvalues a_l^2. In the typical problem $f(z)$ and $G'(z;\xi)$ are positive when $0\leqslant z\leqslant 1$. Then *every eigenvalue* $\mathscr{R}_0(a_l^2)$ *is a simple eigenvalue of* (73.8). The proof of this result is developed as an application (d) of the theory of oscillation kernels in App. D of Vol. 1. For a given value a_l^2 we understand, in a loose notation, that $\mathscr{R}_0(a_l^2)$ is smallest of the positive eigenvalues of (73.8). The main interest is in the minimum of these smallest values

$$\mathscr{R}_0(\tilde{a}_l^2)=\min_{a_l^2}\mathscr{R}_0(a_l^2)\,. \tag{73.10}$$

Eq. (73.10) selects the critical planform eigenvalue and eigenfunction ϕ_l and determines the nodal lines of convection with small amplitudes.

We turn now to the problem of bifurcation and assume that a^2 is a simple eigenvalue of (73.7). To evaluate how the separation affects $\mathscr{R}_1$ we first compute

$$\begin{aligned}\theta_0\mathbf{u}_0\cdot\mathbf{u}^* &= \theta_0\left[\frac{\partial^2\chi_0}{\partial x\partial z}\frac{\partial^2\chi^*}{\partial x\partial z}+\frac{\partial^2\chi_0}{\partial y\partial z}\frac{\partial^2\chi^*}{\partial y\partial z}+\nabla_2^2\chi_0\nabla_2^2\chi^*\right]\\ &=\bar{\bar{\theta}}D\bar{\bar{\chi}}D\bar{\bar{\chi}}^*\phi|\nabla_2\phi|^2+a^4\bar{\bar{\theta}}\bar{\bar{\chi}}\bar{\bar{\chi}}^*\phi^3\\ &=\tfrac{1}{2}\bar{\bar{\theta}}D\bar{\bar{\chi}}D\bar{\bar{\chi}}^*\nabla_2\phi\cdot\nabla_2\phi^2+a^4\bar{\bar{\chi}}\bar{\bar{\chi}}^*\bar{\bar{\theta}}\phi^3\end{aligned}$$

and

$$\theta^*\mathbf{u}_0\cdot\nabla\theta_0=\theta^*\left[\frac{\partial^2\chi_0}{\partial x\partial z}\frac{\partial\theta_0}{\partial x}+\frac{\partial^2\chi_0}{\partial y\partial z}\frac{\partial\theta_0}{\partial y}-\nabla_2^2\chi_0\frac{\partial\theta_0}{\partial z}\right]$$
$$=\tfrac{1}{2}\bar{\bar{\theta}}^*\bar{\bar{\theta}}D\bar{\bar{\chi}}\nabla_2\phi\cdot\nabla_2\phi^2+a^2\bar{\bar{\theta}}^*\bar{\bar{\chi}}D\bar{\bar{\theta}}\phi^3\,.$$

We note next that

$$\iint_{\mathscr{A}}\nabla_2\phi\cdot\nabla_2\phi^2=\iint_{\mathscr{A}}\nabla_2\cdot[\phi^2\nabla_2\phi]-\iint_{\mathscr{A}}\phi^2\nabla_2^2\phi$$
$$=\int_S\phi^2\mathbf{n}\cdot\nabla\phi+a^2\iint_{\mathscr{A}}\phi^3$$
$$=a^2\iint_{\mathscr{A}}\phi^3\,.$$

Finally, we evaluate $\mathscr{R}_1$ as given by (72.6).

$$\mathscr{R}_1=M(a^2)\iint_{\mathscr{A}}\phi^3dxdy/\iint_{\mathscr{A}}\phi^2dxdy \tag{73.11}$$

where

$$M(a^2)=\frac{\int_0^1[\mathscr{R}_0^{-1}\bar{\bar{\theta}}f'(\frac{1}{2}D\bar{\bar{\chi}}D\bar{\bar{\chi}}^*+a^2\bar{\bar{\chi}}\bar{\bar{\chi}}^*)+\bar{\bar{\theta}}^*(\frac{1}{2}\bar{\bar{\theta}}D\bar{\bar{\chi}}+\bar{\bar{\chi}}D\bar{\bar{\theta}})]dz}{\int_0^1[\bar{\bar{\chi}}^*\bar{\bar{\theta}}+G'\bar{\bar{\chi}}\bar{\bar{\theta}}^*]dz}\,. \tag{73.12}$$

The following theorem specifies necessary and sufficient conditions for two-sided bifurcation at a simple eigenvalue.

Suppose a^2 *is a simple eigenvalue of* (73.7), $M(a^2)\neq 0$ *and*

$$\iint_{\mathscr{A}}\phi^3dxdy/\iint\phi^2dxdy\neq 0\,. \tag{73.13}$$

Then $\mathscr{R}_1\neq 0$ *and the bifurcation is two-sided. If* $\mathscr{R}_1\neq 0$ *and the bifurcation is two-sided, then* (73.13) *holds.* We have already noted that the same theorem holds for problems of cellular convection with or without porous materials. The condition (73.13) contains all previously known results; for instance, it implies previously known results about two-sided bifurcation of cellular convection with hexagonally symmetric plan forms. The quantity $M(a^2)$ depends in a more complicated way on the details of the problem. In the cases computed so far $M(a^2)\neq 0$ and $\mathscr{R}_2>0$ (see Exercises 73.1 and 76.1). The criterion (73.13) may be modified to treat the case when a^2 is an eigenvalue of higher multiplicity (see Exercise 73.2).

Since ϕ has a zero area-average it has positive and negative parts on $\mathscr{A}$ and it is possible for the area-average of the cube of ϕ to vanish. The vanishing of the average of the cube of ϕ requires that ϕ have high degree of symmetry. Such symmetry could be generated only if $\mathscr{A}$ itself has considerable symmetry. It is of interest to compute the value of $\iint_{\mathscr{A}}\phi^3dxdy$ in special cases.

(a) Axisymmetric Convection in Round Containers[2]

The unique solution of

$$\frac{1}{r}\frac{d}{dr}\left(r\frac{d\phi}{dr}\right)+a^2\phi=0\,, \qquad \frac{d\phi(b)}{dr}=0 \tag{73.14}$$

which is bounded at $r=0$ is

$$\phi=J_0(ar) \tag{73.15}$$

where $a_j b=\lambda_j$ are the positive zeros of $J_1(\lambda)$; $J_1(\lambda_j)=0$, $j=1,2,\dots$. Convection starts first when, for the given b, $\mathscr{R}_0(a_j^2)=\mathscr{R}_0(\lambda_j^2/b^2)$ is minimized over the zeros λ_j^2 of J_1. For problems which are not too different from the Bénard problem for a DOB fluid the minimum value of $\mathscr{R}^2$ is near to $4\pi^2$ and is attained when a_j^2 is closest to π^2. a_j^2 is a simple eigenvalue of (73.14). $J_0(a_j r)$ has j nodal lines (circles) inside the circle of radius b. $J_0(a_1 r)$, for example, is positive when $r<\tilde{r}<b$, where $\tilde{r}\simeq 2.4\,b/3.8$, and is negative when $\tilde{r}<r\leqslant b$.

Axisymmetric convection satisfies the condition (73.13) for two-sided bifurcation (see Table 73.1).

Table 73.1: Table of values of the first five roots of $J_1(\lambda_j)=0$ and $I_j=\int_0^1 rJ_0^3(\lambda_j r)\,dr$

j	λ_j	I_j
1	3.832	2.857×10^{-2}
2	7.016	7.001×10^{-3}
3	10.173	3.682×10^{-3}
4	13.324	2.019×10^{-3}
5	16.470	1.379×10^{-3}

(b) Nonaxisymmetric Convection in Round Containers[2]

All of the bounded solutions of

$$\frac{1}{r}\frac{\partial}{\partial r}\left(r\frac{\partial\phi}{\partial r}\right)+\left(a^2-\frac{m^2}{r^2}\right)\phi=0\,, \qquad \frac{\partial\phi(b)}{\partial r}=0 \tag{73.16}$$

are given by

$$J_m(ar)\begin{Bmatrix}\cos m\theta\\ \sin m\theta\end{Bmatrix} \tag{73.17}$$

[2] A theoretical analysis of bifurcation of an OB fluid into axisymmetric convection has been given by Liang, Vidal and Acrivos (1969) and by Charlson and Sani (1975). Convection of OB fluids in round cylinders has been studied in the experiments of Liang, Vidal and Acrivos (1969), Koschmieder (1974), and Stork and Müller (1975).

where $a=a_l$ are the positive roots of

$$J'_m(a_l b)=0\,. \tag{73.18}$$

$a_l^2(m)$ is a double eigenvalue when $m\neq 0$. To study bifurcation when $m\neq 0$ it is necessary to treat the double eigenvalue (see Exercises 73.2 and 5). This follows along the lines laid out in § 72(b) and in the example to be considered under (c) below.

(c) Convection in a Hexagonal Container

The dimensions of the hexagonal container of side d are given in Fig. 73.1.

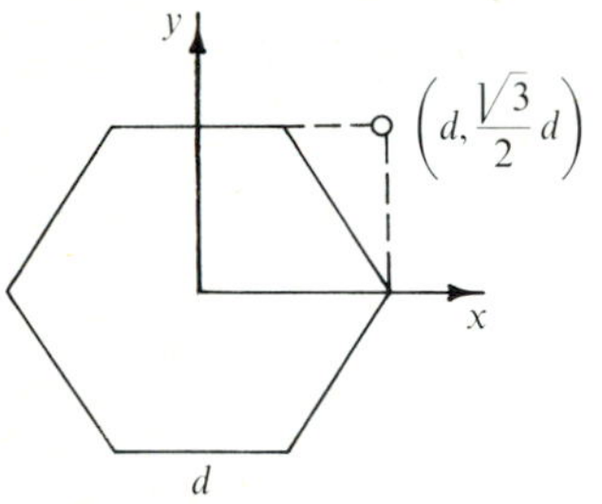

Fig. 73.1: Top view of the hexagonal container

We are going to discuss the special solutions of (73.7) in the hexagonal container which can be embedded in a class of functions of partial hexagonal symmetry. Hexagonally symmetric functions are important in the problem of two-sided bifurcation in cellular convection.

First we shall discuss the functions; then, the solutions. The functions $\Psi(x, y)$ which we shall say have a property of partial hexagonal symmetry are doubly-periodic functions with a period $3d$ in x and a period of $\sqrt{3}d$ in y and which are invariant to rotations of $2\pi/3$. We say that $\Psi(x, y)$ has full hexagonal symmetry if it is a function of partial hexagonal symmetry for which $\Psi(x, -y)=\Psi(-x, y)=\Psi(x, y)$[3]. Our analysis of hexagonally symmetric functions follows Yudovich (1966).

[3] This definition of full hexagonal symmetry leads to a correct functional form for (x,y). However, T.H. Schwab (1972) has pointed out to the author that a mathematically precise definition of a function with hexagonal point symmetry requires $\Psi(x,y)$ to be invariant under the group C_{6v}. (The C_{6v} group has symmetry with regard to six rotations and reflections in six mirror planes.) This definition is preferred over those requiring invariance to certain distinct rotations together with evenness in x and in y. Definitions of the latter type lead to an ambiguity in the selection of the group elements necessary to generate C_{6v} because evenness and invariance under reflections in two perpendicular mirror planes (both elements of C_{6v}) are identical constraints. Also, it is not *a priori* clear that hexagonally symmetric functions must be even in the coordinates. Only by investigation of the elements of C_{6v} can this be foreseen.

Every function of partial hexagonal symmetry has a Fourier series of the form

$$\begin{aligned}\theta(x, y) &= \textstyle\sum_{l=-\infty}^{\infty} \sum_{v=-\infty}^{\infty} \theta_{vl}(x, y) \\ &= \tfrac{1}{3} \textstyle\sum_{l=-\infty}^{\infty} \sum_{v=-\infty}^{\infty} C_{vl} \{ e^{ia/2\,(vx+\sqrt{3}ly)} \\ &\quad + e^{ia/4(-(v+3l)x+\sqrt{3}(v-l)y)} \\ &\quad + e^{ia/4((3l-v)x-\sqrt{3}(v+l)y)} \}\end{aligned} \tag{73.19}$$

where $d=4\pi/3a$ and the summation is over all integers v and l such that $v+l$ is even. The Fourier coefficients in (73.19) are uniquely determined by $\theta(x, y)$ with the orthogonality condition defined by integration over the hexagon. The formula (73.19) can be derived from the relation

$$\theta(x, y) = \tfrac{1}{3}\{\theta(x, y) + \theta(x', y') + \theta(x'', y'')\} \tag{73.20}$$

where primed variables are related to unprimed variables by a rotation of $2\pi/3$, and double-primed to unprimed variables by a rotation of $4\pi/3$. A Fourier series of exponential terms with the required periodicity of $3d$ in x and $\sqrt{3}d$ in y, and Fourier coefficient C_{vl}, are now used to represent the functions θ in (72.12).

The coefficients in (73.19) are not arbitrary but must be chosen to satisfy the invariance requirement $\theta(x, y)=\theta(x', y')=\theta(x'', y'')$. Through orthogonality conditions, one finds that

$$\begin{aligned}C_{vl} &= \textstyle\sum_{n=-\infty}^{\infty} \sum_{m=-\infty}^{\infty} C_{nm}\delta(2v+n+3m)\delta(2l-n+m) \\ &= \textstyle\sum_{n=-\infty}^{\infty} \sum_{m=-\infty}^{\infty} C_{nm}\delta(2v-3m+n)\delta(2l+m+n)\end{aligned} \tag{73.21}$$

where $\delta(0)=1$ and $\delta(m)=0$ when $m \neq 0$. This insures invariance through rotation angles of $2\pi/3$ and completes the proof of the representation (73.19).

Now we want to find eigensolutions of (73.7) which have partial hexagonal symmetry. We find that

$$\nabla_2^2\theta(x, y) = -\frac{a^2}{4} \textstyle\sum_{-\infty}^{\infty} \sum_{-\infty}^{\infty} (v^2+3l^2)\theta_{vl}(x, y) = -a^2\theta(x, y)$$

if and only if

$$v^2+3l^2=4\,.$$

Hence,

$$(v, l) = (\pm 2, 0), (\pm 1, \pm 1)$$

and, by (73.21)

$$C_{11}=C_{1,-1}=C_{-2,0}\,; \quad C_{-1,-1}=C_{-1,1}=C_{20}\,; \text{ all other } C_{vl}=0\,. \tag{73.22}$$

Using (73.22) we find that eigenfunctions of (73.7a) of partial hexagonal symmetry must be of the form

$$\phi(x, y)=\theta(x, y)=\textstyle\sum_{v=1}^{3}\left[C_1 \exp\{i\mathbf{k}_v\cdot\mathbf{r}\}+C_2 \exp\{-i\mathbf{k}_v\cdot\mathbf{r}\}\right] \tag{73.23}$$

where $\mathbf{r}=\mathbf{e}_x x+\mathbf{e}_y y$ and

$$\mathbf{k}_1=(a, 0), \quad \mathbf{k}_2=(-a/2, \sqrt{3}a/2), \quad \mathbf{k}_3=(-a/2, -\sqrt{3}a/2).$$

It is now necessary to verify that (73.23) has a vanishing normal derivative on the walls of the hexagonal container. (73.23) may be written as

$$\begin{aligned}\phi(x, y)&=A_1\phi_1+A_2\phi_2,\\ \phi_1&=\cos ax+2\cos\frac{a}{2}x\cos\frac{\sqrt{3}a}{2}y,\\ \phi_2&=-\sin ax+2\sin\frac{a}{2}x\cos\frac{\sqrt{3}a}{2}y.\end{aligned} \tag{73.24}$$

It is enough to show that $\mathbf{n}\cdot\nabla_2\phi$ vanishes on the walls at $y=\pm\sqrt{3}d/2=\pm 2\pi/\sqrt{3}a$. Then, by invariance to rotations through angles of 120°, $\mathbf{n}\cdot\nabla_2\phi=0$ on all walls of the hexagon. We leave the rest of the demonstration as an exercise (73.3).

It follows from (73.24) that a^2 has a multiplicity of two among functions of partial hexagonal symmetry. Functions ϕ_1 of full hexagonal symmetry are even functions of x and y. It follows that eigensolutions of (73.7) which have full hexagonal symmetry are given by (73.24) with $A_2=0$. The eigenfunction of full hexagonal symmetry is unique for each and every value of a. It may also be represented as

$$\phi(x, y)=A_1\textstyle\sum_{v=-3}^{3}\exp\{i\mathbf{k}_v\cdot r\}. \tag{73.25}$$

It follows from (73.11) that

$$\mathscr{R}_1=A_1 M(a^2) \tag{73.26}$$

and the bifurcation into fully hexagonal convection is two-sided. The factorization theorem for steady bifurcation shows that the supercritical branch is stable and the subcritical branch is unstable.

It is, of course, artificial to restrict attention to the hexagons of full symmetry. Among the functions of partial hexagonal symmetry a^2 has multiplicity two and the result given in Exercise (73.2) holds. Hence,

$$A_1\mathscr{R}_1=M(a^2)\textstyle\int\int_{\mathscr{A}}\phi_1(A_1\phi_1+A_2\phi_2)^2 dxdy\Big/\int\int_{\mathscr{A}}\phi_1^2 dxdy$$

and

$$A_2\mathscr{R}_1=M(a^2)\textstyle\int\int_{\mathscr{A}}\phi_2(A_1\phi_1+A_2\phi_2)^2 dxdy\Big/\int\int_{\mathscr{A}}\phi_2^2 dxdy$$

after carrying out the integrations (73.31) (see Exercise 73.6) we get

$$A_1\mathscr{R}_1 = M(a^2)(A_1^2 - A_2^2)$$

and (73.27)

$$A_2\mathscr{R}_1 = -M(a^2)2A_1A_2\,.$$

There are three solutions of (73.27); three solutions bifurcate from an eigenvalue of multiplicity two:

$$\begin{aligned}
&\text{I:}\ A_2=0\,, \quad \phi = A_1\phi_1\,, \quad \mathscr{R}_{11}=A_1M(a^2)\,;\\
&\text{II:}\ A_2=\sqrt{3}A_1\,, \quad \phi=A_1[\phi_1+\sqrt{3}\phi_2]\,, \quad \mathscr{R}_{12}=-2A_1M(a^2)\,;\\
&\text{III:}\ A_2=-\sqrt{3}A_1\,, \quad \phi=A_1[\phi_1-\sqrt{3}\phi_2]\,, \quad \mathscr{R}_{13}=\mathscr{R}_{12}\,.
\end{aligned}$$

Solution I corresponds to bifurcation from a simple eigenvalue. The constant A_1 is determined by a normalization which is determined by the definition of the amplitude ε.

It is of interest to consider the stability of the three solutions of partial hexagonal symmetry which bifurcate from an eigenvalue of multiplicity two. We follow the method introduced in § 72(c). When f and G' are positive $\mathscr{R}_0(a^2)$ is a simple eigenvalue of the ordinary differential equations (73.8) and (73.9) which arise when (72.13) is solved by the separation of variables (73.6). The functions $\bar{\bar{\chi}}(z)$ and $\bar{\bar{\theta}}(z)$ are then unique.

(d) Stability of Solutions Bifurcating at an Eigenvalue of Multiplicity N

In the next step we may keep the analysis general. We first suppose that ϕ and $\mathscr{R}_1$ are given by analysis of bifurcation; for example, we could study the stability of each of solutions I, II, III of (73.27). We let ϕ and $\mathscr{R}_1$ stand for one of these, but in the general case $\phi=\sum_1^N A_n\phi_n$ where N is the multiplicity of a^2. Then the disturbances $\tilde{\chi}_0$ and $\tilde{\theta}_0$ are also a linear combination of eigenfunctions ϕ_n

$$\tilde{\chi}_0 = \bar{\bar{\chi}}(z)\textstyle\sum_1^N B_n\phi_n = \bar{\bar{\chi}}(z)\tilde{\phi}(x, y)$$

and

$$\tilde{\theta}_0 = \bar{\bar{\theta}}(z)\textstyle\sum_1^N B_n\phi_n = \bar{\bar{\theta}}(z)\tilde{\phi}(x, y)$$

where, in our special case,

$$\tilde{\phi} = B_1\phi_1 + B_2\phi_2\,.$$

The adjoint eigenfunctions are also separable and proportional to ϕ_1 and ϕ_2.

Using these representations we may reduce (72.16) to

$$\begin{aligned}\gamma_1\langle\bar{\bar\theta}^*\bar{\bar\theta}\phi_l\tilde\phi\rangle+\mathcal{R}_1a^2\langle[\bar{\bar\chi}^*\bar{\bar\theta}+G'\bar{\bar\theta}^*\bar{\bar\chi}]\phi_l\tilde\phi\rangle\\=\frac{1}{\mathcal{R}_0}\langle f'[\bar{\bar\theta}D\bar{\bar\chi}D\bar{\bar\chi}^*\phi\nabla_2\phi_l\cdot\nabla_2\tilde\phi+a^4\bar{\bar\theta}\bar{\bar\chi}\bar{\bar\chi}^*\phi\phi_l\tilde\phi]\rangle\\+\frac{1}{\mathcal{R}_0}\langle f'[\bar{\bar\theta}D\bar{\bar\chi}D\bar{\bar\chi}^*\tilde\phi\nabla_2\phi_l\cdot\nabla_2\phi+a^4\bar{\bar\theta}\bar{\bar\chi}\bar{\bar\chi}^*\phi\phi_l\tilde\phi]\rangle\\+\langle\bar{\bar\theta}^*[\bar{\bar\theta}D\bar{\bar\chi}\phi_l\nabla_2\phi\cdot\nabla_2\tilde\phi+a^2D\bar{\bar\theta}\bar{\bar\chi}\phi_l\phi\tilde\phi]\rangle\\+\langle\bar{\bar\theta}^*[D\bar{\bar\chi}\bar{\bar\theta}\phi_l\nabla_2\tilde\phi\cdot\nabla_2\phi+a^2D\bar{\bar\theta}\bar{\bar\chi}\phi\phi_l\tilde\phi]\rangle\end{aligned}$$

where $l=1, 2$. After integrating the planform products by parts, using the boundary condition $\partial\phi_l/\partial n|_s=0$ and the differential equation $\nabla_2^2\phi_l+a^2\phi_l=0$, repeatedly, we may reduce the expression for γ_1 to

$$[\gamma_1I_6+a^2\mathcal{R}_1I_5]\int\int_{\mathcal{A}}\tilde\phi\phi_l=2a^2M(a^2)I_5\int\int_{\mathcal{A}}\phi\tilde\phi\phi_l \tag{73.28}$$

where $l=1, 2, \ldots, N$,

$$I_5=\int_0^1[\bar{\bar\chi}^*\bar{\bar\theta}+G'\bar{\bar\theta}^*\bar{\bar\chi}]dz$$

and

$$I_6=\int_0^1\bar{\bar\theta}^*\bar{\bar\theta}dz\,.$$

In carrying out the reduction to (73.28) we integrated the terms involving ∇_2 by parts using identities like the one above (73.11) and

$$\begin{aligned}2\int\int_{\mathcal{A}}[\phi_n\cdot\nabla_2\phi_j\cdot\nabla_2\phi_l-\tfrac{1}{2}a^2\phi_n\phi_l\phi_j]dxdy&=\int\int_{\mathcal{A}}[\phi_n\nabla_2\phi_j-\phi_j\nabla_2\phi_n]\cdot\nabla_2\phi_ldxdy\\&=-\int\int_{\mathcal{A}}\phi_l\{\phi_n\nabla_2^2\phi_j-\phi_j\nabla_2^2\phi_n\}dxdy\\&=0\end{aligned}$$

which hold for all planform eigenfunctions with eigenvalue a^2.

Eqs. (73.28) are N linear, homogeneous equations in the N unknown coefficients B_l where

$$\tilde\phi=\sum_1^N B_l\phi_l\,.$$

Supposing now that the set $\{\phi_l\}$ has been chosen to form an orthogonal basis,

$$\int\int_{\mathcal{A}}\phi_l\phi_n=C\delta_{ln}$$

where C is a constant, we may reduce (73.28) to

$$[\gamma_1I_6+a^2\mathcal{R}_1I_5]CB_l=2a^2M(a^2)I_5\sum_{n=1}^N B_nS_{nl} \tag{73.29}$$

where

$$S_{nl} = S_{ln} = \int\int_{\mathscr{A}} \phi\phi_n\phi_l$$

is a symmetric $N \times N$ matrix. Then (73.29) may be solved for B_l if and only if

$$\det[\lambda\delta_{nl} - S_{nl}] = 0 \tag{73.30}$$

where

$$\lambda = \frac{(\gamma_1 I_6 + a^2\mathscr{R}_1 I_5)C}{2a^2 M(a^2) I_5}$$

is an eigenvalue of S_{nl}; that is, λ is a root of (73.30). There are N real roots $\lambda_1, \lambda_2, \ldots, \lambda_N$ of the polynomial (73.30). Each distinct root λ_l is associated with a distinct value for γ_{1l}. If one of the values γ_{1l} is negative the bifurcating flow is unstable.

(e) Stability of Bifurcating Hexagonal Convection

We shall now apply (73.30) to the problem of stability of the three branches of hexagonal convection arising from (73.27). In this case, $N=2$ and, by computation,

$$\tfrac{3}{2} = \frac{C}{\mathscr{A}} = \frac{1}{\mathscr{A}}\int\int_{\mathscr{A}} \phi_1^2 = \frac{1}{\mathscr{A}}\int\int_{\mathscr{A}} \phi_2^2 = \frac{1}{\mathscr{A}}\int\int_{\mathscr{A}} \phi_1^3 = -\frac{1}{\mathscr{A}}\int\int_{\mathscr{A}} \phi_1\phi_2^2 \tag{73.31a}$$

where $\mathscr{A}$ is the area of the hexagon, and

$$\int\int_{\mathscr{A}} \phi_1\phi_2 = \int\int_{\mathscr{A}} \phi_2^3 = \int\int_{\mathscr{A}} \phi_1^2\phi_2 = 0\,. \tag{73.31b}$$

Using $\phi = A_1\phi_1 + A_2\phi_2$ we find that

$$S_{nl} = \begin{bmatrix} \int\int_{\mathscr{A}} \phi\phi_1^2 & \int\int_{\mathscr{A}} \phi\phi_1\phi_2 \\ \int\int_{\mathscr{A}} \phi\phi_2\phi_1 & \int\int_{\mathscr{A}} \phi\phi_2^2 \end{bmatrix} = \tfrac{3}{2}\mathscr{A} \begin{bmatrix} A_1 & -A_2 \\ -A_2 & -A_1 \end{bmatrix}$$

and

$$\det[\lambda\delta_{nl} - S_{nl}] = \det\begin{bmatrix} \lambda - CA_1 & CA_2 \\ CA_2 & \lambda + CA_1 \end{bmatrix} = \lambda^2 - C^2(A_1^2 + A_2^2) = 0\,. \tag{73.32}$$

Eq. (73.32) leads to values γ_{1i} $(i = 1, 2, 3)$

$$\gamma_{1i} I_6 + a^2\mathscr{R}_{1i} I_5 = \pm 2a^2 M(a^2) I_5 (A_1^2 + A_2^2)^{1/2}$$

corresponding to the three bifurcating solutions of hexagonal symmetry: For example, in case I

$$A_2=0, \quad \mathscr{R}_1=\mathscr{R}_{11}=A_1 M(a^2) \tag{73.33a}$$

and

$$\gamma_{11}=a^2\mathscr{R}_{11}\frac{I_5}{I_6}\begin{cases}1\\-3\end{cases}. \tag{73.33b}$$

The disturbances $\tilde{\phi}=B_1\phi_1+B_2\phi_2$ are given by $B_2=0$, $B_1\neq 0$ for the first alternative in (73.3b) and by $B_2\neq 0$, $B_1=0$ for the second. There are two values of γ_{11} of different sign. It follows that when ε is small, we always have negative values $\gamma=\varepsilon\gamma_{11}+O(\varepsilon^2)$ and the hexagon of full symmetry is unstable on both sides of criticality. If we had artificially restricted disturbances to full hexagonal symmetry, we would be obliged to take $\tilde{\phi}=B_1\phi_1$. Then the stability problem would be a special case of the problem of stability of solutions bifurcating at a simple eigenvalue. We would then find that the supercritical branch of convection is stable and subcritical branch is unstable. This example emphasizes that in the study of stability it is necessary to account for all allowed disturbances.

A similar computation starting from (73.32) shows that the bifurcating solutions II and III of partial hexagonal symmetry are also unstable on both sides of criticality (Exercise 73.4). It would be of interest to compute the stability of bifurcating convection to $O(\varepsilon^2)$. This computation requires the evaluation of the integrals like I_i and could be carried out using the two parameter expansions discussed in § 76.

(f) One-Sided Convection in Containers of Rectangular Cross-Section

Now we shall study (73.7) when the cross-section A of the container Ω is rectangular. Let (x, y) be coordinates measured from the bottom corner on the left. The eigenfunctions of (73.7) are

$$\phi=\cos\frac{m\pi x}{l}\cos\frac{n\pi y}{d}, \quad m^2+n^2\neq 0 \tag{73.34}$$

the eigenvalues are

$$a^2=a^2(m, n)=\pi^2\left(\frac{m^2}{l^2}+\frac{n^2}{d^2}\right). \tag{73.35}$$

The eigenvalue $a^2(m, n)$ is simple unless there is another pair of integers (m', n') such that

$$\frac{m^2}{l^2}+\frac{n^2}{d^2}=\frac{m'^2}{l^2}+\frac{n'^2}{d^2}.$$

Every eigenvalue $a^2(m, n)$ is simple when the ratio l/d of sides of the rectangle is irrational and the minimum value of $a^2(m, n)$ for all m, n with $m^2+n^2\neq 0$ is always simple.

The condition (73.13) for two-sided bifurcation is not satisfied by (73.34); instead we compute

$$\iint_{\mathscr{A}} \phi^3 dxdy=0\,.$$

This shows that $\mathscr{R}_1=0$ and suggests that generalized convection in boxes of porous material with insulated sidewalls is always supercritical.

Moral: There is a natural temptation to generalize from experience with problems for which computed results are known but it is a temptation which one should resist. Explicit computations, like those given above might suggest that in most containers and for most ϕ

$$\iint_{\mathscr{A}} \phi^3 dxdy=0\,.$$

But explicit computations are possible only when $\mathscr{A}$ has a high degree of symmetry. When $\mathscr{A}$ has no such symmetry it is not possible to give exact analytical expressions for the eigenfunctions ϕ. But in such asymmetric containers

$$\iint_{\mathscr{A}} \phi^3 dxdy\neq 0$$

and the bifurcation is two-sided.

(g) The Bénard Problem for a DOB Fluid in a Container

When $\xi=0$ and $f=1$, the problem (73.8) is self-adjoint, $\bar{\bar{\chi}}=\bar{\bar{\chi}}^*$, has constant coefficients and involves only even order derivatives which respect to z which have the property that $\bar{\bar{\chi}}^{(2n)}(1)=0$. It follows that all the solutions $\bar{\bar{\chi}}, \bar{\bar{\theta}}$ are proportional to $\sin n\pi z$ and the minimum value of $\mathscr{R}_0$ is attained for each and every eigenvalue a_l^2 of (73.7) when $n=1$; that is,

$$\mathscr{R}_0(a_l^2)=(\pi^2+a_l^2)^2/a_l^2\,.$$

The critical planform eigenvalue $a_l^2=\tilde{a}_l^2$ is the root of

$$\mathscr{R}_0(\tilde{a}_l^2)=\min_{a_l^2}(\pi^2+a_l^2)^2/a_l^2\geq 4\pi^2\,.$$

Generally $\tilde{a}_l^2$ is the eigenvalue which is closest to π^2.

The nodal properties of small-amplitude convection are determined by the planform eigenfunction. In axisymmetric convection in round containers the planform eigenfunction $\phi_l(r)=J_0(a_l r)$ has l interior zeros. The nodal lines for axisymmetric convection are circles of radius $r_l=\lambda_l/a_l<b$ where the λ_l are the zeros of $J_0(\lambda_l)$ and b is the ratio of the outer radius of the container to its depth.

For nonaxisymmetric convection the critical planform eigenfunction is determined by

$$\mathscr{R}_0(\tilde{a}_l^2(m)) = \min_{a_l^2(m)} (\pi^2 + a_l^2(m))^2 / a_l^2(m).$$

If the eigenvalue $a_l^2(m)$ nearest to π^2 has $m \neq 0$ then the nodal lines of the critical eigenfunction will be composed of circles and radial lines.

Exercise 73.1: Show that $M(a^2) \neq 0$ and $\mathscr{R}_2 > 0$ when $\xi = 0$ and $f = 1$.

Exercise 73.2: Suppose $\mathscr{R}_0(a^2)$ is a simple eigenvalue of (73.8) and a^2 is a double eigenvalue of (73.7). Show that

$$A_1 \mathscr{R}_1 = M(a^2) \iint_{\mathscr{A}} \phi_1 \phi^2 \, dx dy / \iint_{\mathscr{A}} \phi_1^2 \, dx dy$$

and

$$A_2 \mathscr{R}_1 = M(a^2) \iint_{\mathscr{A}} \phi_2 \phi^2 \, dx dy / \iint_{\mathscr{A}} \phi_2^2 \, dx dy \tag{73.36}$$

where

$$\phi = A_1 \phi_1 + A_2 \phi_2$$

and ϕ_1 and ϕ_2 are orthogonal eigenfunctions on the null space of the operator $\nabla_2^2 + a^2$.

Exercise 73.3: Show that $\mathbf{n} \cdot \nabla_2 \phi = 0$ on the sidewall S of the hexagonal container shown in Fig. 73.1 when ϕ is given by (73.24).

Exercise 73.4: Suppose that the bifurcating solutions are normalized by the definition $\varepsilon = \langle \mathbf{Q} \cdot \mathbf{q}_0 \rangle$ of the amplitude ε. Find the magnitude of the constant A_1 for the three bifurcating solutions I, II and III of hexagonal symmetry. Show that the bifurcating solutions II and III are unstable on both sides of criticality.

Exercise 73.5: Consider the problem of bifurcation and stability in a ring shaped container. First assume that the smallest eigenvalue $\mathscr{R}_0(a^2)$ is associated with an axisymmetric eigenfunction. Show that the bifurcation is two-sided, that subcritical bifurcation is unstable and supercritical bifurcation is stable. Carry out the analysis of bifurcation when the critical eigenfunctions are not axisymmetric. In this last case you will need to consider the problem of bifurcation and stability at a double eigenvalue. Show that the bifurcation into nonaxisymmetric convection is supercritical.

Exercise 73.6: Show that functions of hexagonal symmetry tesselate the plane with identical hexagons. *Hint*: show that the functions (73.24) are periodic with period $\sqrt{3}d$ along three lines separated by 120°. Simplify the integration leading to (73.31) by noting, for example, that

$$\frac{1}{\mathscr{A}} \iint_{\mathscr{A}} \phi_1 \phi_2^2 \, dx dy = \lim_{L \to \infty} \frac{1}{4L^2} \iint_{-L}^{L} \phi_1 \phi_2^2 \, dx dy.$$

Exercise 73.7: Suppose that $\sigma(\mathscr{R}_0) = 0$ is a simple eigenvalue of $\mathbf{L}_0$ and that the loss of stability in the container Ω with insulated side-walls S, is strict, $\sigma'(\mathscr{R}_0) < 0$. Show that

$$\gamma = \sigma'(\mathscr{R}_0) \mathscr{R}_1 \varepsilon + 0(\varepsilon^2)$$

where

$$\sigma'(\mathscr{R}_0) = -a^2 I_5 / I_6. \tag{73.37}$$

Under what conditions does (73.37) hold when $\sigma(\mathscr{R}_0) = 0$ is an eigenvalue of higher multiplicity?

§ 74. Two-Sided Bifurcation between Spherical Shells

In this section and in § 75 we shall consider convection in the presence of heat-sources in fluids having constant material properties. In this section, we shall discuss Busse's analysis of nonlinear convection of an OB fluid between spheres. Mathematically, the problem is interesting because the region between spheres is closed and though the eigenvalues of the two dimensional Laplacian (∇_s^2) on spheres are degenerate, their multiplicity is finite. In the future, this geometry should provide a good subject for application of group-theoretical ideas to the problems of nonlinear convection. Some authors say that the theory of convection between spheres is also an interesting physical theory, but I don't know whether or not such claims are justified.

The theory of convection of DOB fluids in a spherical container with cone-shaped insulated side walls (see Fig. 74.3) is just like the one I developed in § 72 and § 73. This theory is described by the results summarized in the Exercises to this section.

Notation: In this section θ and ϕ stand for the angles used in a spherical coordinate system. We use Φ for the plan-form eigenfunction and Θ for the temperature.

In the analysis to follow we will be interested in the stability of the motionless state of heat conduction in a sphere or spherical annulus (inner radius $=\hat{a}$, outer radius $=\hat{b}$). The assumptions about this problem which make it a model for geo-astrophysical applications have been discussed by Chandrasekhar (1961) and by Busse (1975). The motionless solution can exist if and only if the variations of gravity and temperature in the annulus are purely radial (see § 59 and Exercise 59.3). A linear stability analysis of this motionless solution was given by Chandrasekhar (1961), an energy stability analysis was given by Joseph and Carmi (1966) and a bifurcation analysis was given by Busse (1975). Many of the properties of instability and bifurcation depend only weakly on the details of the gravity and temperature distributions in the annulus and on the precise forms of the boundary conditions.

Without loss of generality we may choose scale factors g, T', $l=\hat{b}$ introduced under (56.1)

$$\begin{aligned} \boldsymbol{\eta} &= \mathbf{g}/g' = -\mathbf{r}\,\mathscr{g}(r)\,, \\ \boldsymbol{\eta}_T &= -\mathbf{r}\,\ell(r)\,, \\ \mathbf{r} &= \mathbf{e}_r r \end{aligned} \tag{74.1}$$

where r is the polar radius measured from the sphere center. The functions $\mathscr{g}(r)$ and $\ell(r)$ are arbitrary; they represent radially symmetric solutions of the equations for the gravitational potential and the temperature distribution (Exercise 59.3). Disturbances of (74.1) are governed by the appropriate reduction of problem (56.3):

$$(\partial_t + \varepsilon \mathbf{u}\cdot\nabla)\begin{bmatrix} \mathbf{u} \\ \mathscr{P}\Theta \end{bmatrix} - \mathbf{u}\cdot\begin{bmatrix} 0 \\ \mathscr{R}\mathbf{r}\,\ell(r) \end{bmatrix} = \begin{bmatrix} -\nabla p + \mathscr{R}\mathbf{r}\,\mathscr{g}(r)\,\Theta + \nabla^2\mathbf{u} \\ \nabla^2\Theta \end{bmatrix} \tag{74.2}$$

and (56.3 b, c, d).

The equations of the linear and energy theory of stability are obtained from (61.1) and (61.2), respectively, with $R=\gamma=\boldsymbol{\eta}_c=0$, $\mathbb{R}=\mathscr{R}$ and $\mathscr{A}_{\mathscr{R}}=1$. The best λ of the energy theory is given by (61.3)

$$\lambda=\lambda_T=\frac{\langle\boldsymbol{\eta}\cdot\mathbf{u}\Theta\rangle}{\langle\boldsymbol{\eta}_T\cdot\mathbf{u}\Theta\rangle}=\frac{\langle g(r)f\Theta\rangle}{\langle \ell(r)f\Theta\rangle} \tag{74.3}$$

where

$$f(r,\theta,\phi)=rw(r,\theta,\phi)=\mathbf{r}\cdot\mathbf{u}$$

and θ and ϕ are polar angles. It is an immediate consequence of the theorem proved in § 61 that $\mathscr{R}_L=\mathscr{R}_{\mathscr{E}}$, $\operatorname{im}\sigma(\mathscr{R}_L)=0$, when $g(r)=\ell(r)$. In this exceptional case subcritical bifurcations are impossible.

It is convenient to eliminate the equation $\operatorname{div}\mathbf{u}=0$ by expressing the velocity in terms of poloidal and toroidal potentials (see Appendix B.6)

$$\mathbf{u}=\operatorname{grad}\left(\frac{\partial(r\chi)}{\partial r}\right)-\mathbf{r}\nabla^2\chi+\mathbf{r}\wedge\operatorname{grad}\Psi\,. \tag{74.4}$$

We next note that

$$\mathbf{r}\cdot\nabla^2\hat{\boldsymbol{\phi}}=x_i\frac{\partial^2\hat{\phi}_i}{\partial x_l\,\partial x_l}=\nabla^2(\mathbf{r}\cdot\hat{\boldsymbol{\phi}})-2\operatorname{div}\hat{\boldsymbol{\phi}}$$

$$\begin{aligned}\mathbf{r}\cdot\operatorname{curl}\operatorname{curl}\hat{\boldsymbol{\phi}}&=-\mathbf{r}\cdot\nabla^2\hat{\boldsymbol{\phi}}+\mathbf{r}\cdot\nabla\operatorname{div}\hat{\boldsymbol{\phi}}=-\nabla^2(\mathbf{r}\cdot\hat{\boldsymbol{\phi}})+2\operatorname{div}\hat{\boldsymbol{\phi}}+\mathbf{r}\cdot\nabla\operatorname{div}\hat{\boldsymbol{\phi}}\\&=-\nabla^2(\mathbf{r}\cdot\hat{\boldsymbol{\phi}})+\frac{1}{r}\partial_r(r^2\operatorname{div}\hat{\boldsymbol{\phi}})=-\frac{1}{r^2}\nabla_s^2(\mathbf{r}\cdot\hat{\boldsymbol{\phi}})+\frac{1}{r}\partial_r(r\operatorname{div}_s\hat{\boldsymbol{\phi}})\end{aligned}$$

where

$$\nabla_s=\mathbf{e}_\theta\frac{\partial}{\partial\theta}+\frac{\mathbf{e}_\phi}{\sin\theta}\,\frac{\partial}{\partial\phi},$$

$$\operatorname{div}_s\hat{\boldsymbol{\phi}}=\frac{1}{\sin\theta}\,\frac{\partial(\sin\theta\,\hat{\phi}_\theta)}{\partial\theta}+\frac{1}{\sin\theta}\,\frac{\partial\hat{\phi}_\phi}{\partial\phi}$$

and

$$\nabla_s^2=\frac{1}{\sin\theta}\,\frac{\partial}{\partial\theta}\sin\theta\frac{\partial}{\partial\theta}+\frac{1}{\sin^2\theta}\,\frac{\partial^2}{\partial\phi^2}.$$

Application of the operator $\mathbf{r}\cdot\operatorname{curl}\operatorname{curl}$ to the first of Eqs. (74.2) then leads to

$$\begin{aligned}&\partial_t\nabla^2 f-\varepsilon\mathbf{r}\cdot\operatorname{curl}^2(\mathbf{u}\cdot\nabla\mathbf{u})=\mathscr{R}g(r)\nabla_s^2\Theta+\nabla^4 f,\\&\mathscr{P}(\partial_t+\varepsilon\mathbf{u}\cdot\nabla)\Theta=\mathscr{R}\ell(r)f+\nabla^2\Theta\,.\end{aligned} \tag{74.5}$$

We next write the equations which govern the spectral problem of the linearized theory of stability

$$\begin{bmatrix} \nabla^4 & \mathscr{R}g(r)\nabla_s^2 \\ \mathscr{R}\ell(r)\nabla_s^2 & \nabla^2\nabla_s^2 \end{bmatrix} \begin{bmatrix} f \\ \Theta \end{bmatrix} = 0 \tag{74.6}$$

where we have assumed that at criticality $\sigma(\mathscr{R}_L)=0$.

The equations which govern the energy theory of stability are

$$\begin{bmatrix} \nabla^4 & \frac{1}{2}\mathscr{R}\left\{\frac{g(r)}{\sqrt{\lambda}}+\sqrt{\lambda}\ell(r)\right\}\nabla_s^2 \\ \frac{1}{2}\mathscr{R}\left\{\frac{g(r)}{\sqrt{\lambda}}+\sqrt{\lambda}\ell(r)\right\}\nabla_s^2 & \nabla^2\nabla_s^2 \end{bmatrix} \begin{bmatrix} f \\ \Theta \end{bmatrix} = 0. \tag{74.7}$$

The variables $f=\mathbf{u}\cdot\mathbf{r}$ and Θ are now expanded into a complete set of spherical harmonics

$$\begin{bmatrix} f \\ \Theta \end{bmatrix} = \sum \begin{bmatrix} \bar{\bar{f}}_l(r) \\ \bar{\bar{\Theta}}_l(r) \end{bmatrix} Y_l^m(\theta,\phi) \tag{74.8}$$

where

$$Y_l^m(\theta,\phi) = P_l^m(\cos\theta)\begin{Bmatrix} \sin m\phi \\ \cos m\phi \end{Bmatrix}, \qquad 0\leqslant m\leqslant l, \qquad l=1,2,\ldots \tag{74.9}$$

are spherical harmonics, and P_l^m are associated Legendre polynomials,

$$\left.\begin{aligned} &\nabla_s^2 Y_l^m(\theta,\phi) = -l(l+1)\, Y_l^m(\theta,\phi) \\ &\text{and} \\ &\nabla^2 \bar{\bar{f}}(r)\, Y_l^m(\theta,\phi) = Y_l^m(\theta,\phi)\, \mathscr{D}_l^2 \bar{\bar{f}} \end{aligned}\right\} \tag{74.10}$$

where

$$\mathscr{D}_l^2 = \frac{d^2}{dr^2} + \frac{2}{r}\frac{d}{dr} - \frac{l(l+1)}{r^2}.$$

Inserting (74.8) into (74.6) and (74.7) we find, using (74.9) and (74.10), that

$$\begin{bmatrix} \mathscr{D}_l^4 & -\mathscr{R}g(r)\,l(l+1) \\ -\mathscr{R}\ell(r)\,l(l+1) & -l(l+1)\mathscr{D}_l^2 \end{bmatrix} \begin{bmatrix} \bar{\bar{f}}_l(r) \\ \bar{\bar{\Theta}}_l(r) \end{bmatrix} = 0 \tag{74.11}$$

for the linear stability problem and

$$\begin{bmatrix} \mathscr{D}_l^4 & -l(l+1)\frac{1}{2}\mathscr{R}\left\{\frac{g(r)}{\sqrt{\lambda}}+\sqrt{\lambda}\ell(r)\right\} \\ -l(l+1)\frac{1}{2}\mathscr{R}\left\{\frac{g(r)}{\sqrt{\lambda}}+\sqrt{\lambda}\ell(r)\right\} & -l(l+1)\mathscr{D}_l^2 \end{bmatrix} \begin{bmatrix} \bar{\bar{f}}(r) \\ \bar{\bar{\Theta}}(r) \end{bmatrix} = 0 \tag{74.12}$$

for the energy stability problem. These problems are completely specified when boundary conditions on $\bar{\bar{f}}$ and $\bar{\bar{\Theta}}$ are prescribed; for example

$$\bar{\bar{f}}(1)=\bar{\bar{f}}(\eta)=\bar{\bar{f}}'(1)=\bar{\bar{f}}'(\eta)=\bar{\bar{\Theta}}(1)=\bar{\bar{\Theta}}(\eta)=0 \tag{74.13}$$

on rigid conducting surfaces.

The eigenvalues of (74.11) are designated as

$$\mathscr{R}=\mathscr{R}_L(l,\eta)\,.$$

There are $2l+1$ eigenfunctions belonging to each eigenvalue $\mathscr{R}_L(l,\eta)$. These eigenfunctions are given by (74.9) with $m=0, \pm 1, \pm, \ldots, \pm l$. The smallest critical value of linear theory is given by

$$\mathscr{R}_L(\eta)=\min_{l=1,2,\ldots}\mathscr{R}_L(l,\eta)\,.$$

The eigenvalues of (74.12) are designated as

$$R=R_\lambda(l,\eta)\,.$$

The energy stability limit is given by

$$\mathscr{R}_{\mathscr{E}}(\eta)=\max_{\lambda>0}\min_{l=1,2,\ldots}\mathscr{R}_\lambda(l,\eta)\,.$$

Table 74.1: Critical Rayleigh numbers for energy and linear stability theory (see Joseph and Carmi, 1966 and Chandrasekhar, 1961, p. 250)

	η	minimizing l	Estimated λ (74.14) with $\bar{r}=\bar{\bar{r}}=(1+\eta)/2$	Best λ	$\mathscr{R}_{\mathscr{E}}^2\times 10^{-3}$	$\mathscr{R}_L^2\times 10^{-3}$
$\ell=1$	0.2	2	1.67	1.69	7.52	7.57
	0.3	2	1.54	1.52	11.67	11.71
$g=\frac{1}{r}$	0.4	3	1.43	1.42	19.35	19.31
	0.5	4	1.33	1.34	37.07	37.12
	0.6	6	1.25	1.25	84.37	84.44
	0.8	12	1.11	1.11	1129.0	1129.0
$\ell=1$	0.2	2	4.63	6.57	2.125	2.270
	0.3	2	3.65	4.60	4.174	4.320
$g=\frac{1}{r^3}$	0.4	3	2.92	3.34	8.605	8.786
	0.5	4	2.37	2.55	19.69	19.93
	0.6	6	1.95	2.00	52.38	52.74
	0.8	12	1.36	1.38	908.4	909.8
$\ell=1$	0.2	2	1.52	1.71	7.586	7.707
	0.3	2	1.38	1.48	12.49	12.57
$g=\left(6+\frac{1}{r^3}\right)/7$	0.4	3	1.27	1.31	21.26	21.31
	0.5	4	1.20	1.22	40.93	40.98
	0.6	6	1.14	1.14	92.42	92.46
	0.8	12	1.05	1.05	1191.00	1191.00

The principal eigenfunctions $\bar{\bar{f}}(r)$ and $\bar{\bar{\Theta}}(r)$ of (74.11) and (74.12) subject to (74.13) (or to other boundary conditions, see Appendix D) are one-signed in the spherical annulus $(\eta<r<1)$ provided only that the gravity function $g(r)$ and the heat source function $\ell(r)$ are one-signed. In this case the system (74.12), (74.13) can be reduced to an integral equation with an oscillation kernel and the λ which makes $\mathcal{R}_\lambda(l,\eta)$ maximum may be estimated from (74.3) and the mean value theorem as

$$\lambda=\frac{\langle g(r)\, f\, \Theta\rangle}{\langle \ell(r)\, f\, \Theta\rangle}=\frac{\int_\eta^1 g(r)\,\bar{\bar{f}}\,\bar{\bar{\Theta}}\,dr}{\int_\eta^1 \ell(r)\,\bar{\bar{f}}\,\bar{\bar{\Theta}}\,dr}=\frac{g(\bar{r})}{\ell(\bar{\bar{r}})} \tag{74.14}$$

where $\bar{r}$ and $\bar{\bar{r}}$ are mean values of r in the interval $(\eta,1)$. For the problems considered by Joseph and Carmi (1966) the estimate $\bar{r}=\bar{\bar{r}}=(1+\eta)/2$ gives a fairly accurate estimate of λ. Some representative comparisons of the energy and linear stability limits when the boundary conditions are given by (74.13) are exhibited in Table 74.1.

The problem of bifurcation into spherical convection has been studied by Busse (1975). This problem involves perturbing an eigenvalue of multiplicity $2l+1$ corresponding to the $2l+1$ eigenfunctions (74.8, 9) and it may be studied by the methods developed in this chapter. The number of nonlinear solutions which bifurcate and the forms of motion which are allowed by the nonlinear equations cannot be obtained from linear theory. To study these questions we develop the solution in a series of powers of ε

$$\begin{bmatrix}\chi\\ \psi\\ \Theta\\ \mathcal{R}\end{bmatrix}=\begin{bmatrix}\chi_0\\ 0\\ \Theta_0\\ \mathcal{R}_0\end{bmatrix}+\begin{bmatrix}\chi_1\\ \psi_1\\ \Theta_1\\ \mathcal{R}_1\end{bmatrix}\varepsilon+O(\varepsilon^2) \tag{74.15}$$

where χ, ψ and Θ are functions of position, $f_n=-\nabla_s^2\chi_n$ and $\mathcal{R}_0=\mathcal{R}_L(l,\eta)$. The parameter ε may be defined by any one of a number of equivalent conventions. The representations (74.15) are now inserted into (74.5) and they lead to a sequence of perturbation problems

$f_0(r,\theta,\phi;\eta)$, $\quad\Theta_0(r,\theta,\phi;\eta)$ and $\mathcal{R}_0(l,\eta)$ satisfy (74.6);

$$\left.\begin{aligned}&\nabla^4 f_1+\mathcal{R}_L g(r)\nabla_s^2\Theta_1=-\mathbf{r}\cdot\mathrm{curl}^2(\mathbf{u}_0\cdot\nabla\mathbf{u}_0)-\mathcal{R}_1 g(r)\nabla_s^2\Theta_0\\ &\text{and}\\ &\nabla^2\Theta_1+\mathcal{R}_L\ell(r)f_1=\mathcal{P}\mathbf{u}_0\cdot\nabla\Theta_0-\mathcal{R}_1\ell(r)f_0\end{aligned}\right\} \tag{74.16}$$

where

$$\mathbf{u}_0=\mathrm{grad}\frac{\partial(r\chi_0)}{\partial r}-\mathbf{r}\nabla^2\chi_0=-\frac{\mathbf{e}_r}{r}\nabla_s^2\chi_0+\frac{1}{r}\nabla_s\chi_0\,. \tag{74.17}$$

All solutions of (74.16) and the boundary conditions may be formed from linear combinations of the $2l+1$ eigenfunctions (74.9) belonging to $\mathscr{R}_L(l,\eta)=\mathscr{R}_0$.

$$\begin{bmatrix} f_0 \\ \Theta_0 \end{bmatrix} = \begin{bmatrix} \bar{\bar{f}}_{0l}(r) \\ \bar{\bar{\Theta}}_{0l}(r) \end{bmatrix} \sum_{m=-l}^{l} [\alpha_m \exp(\mathrm{i}m\phi)]\, P_l^m(\cos\theta) \tag{74.18}$$

where $\bar{\bar{f}}_{0l}(r)$ and $\bar{\bar{\Theta}}_{0l}(r)$ are assumed to be the unique solution of (74.11). The coefficients α_m and β_m in (74.18) are indeterminate in the linear theory.

We next show that solvability conditions for (74.16) determine a number $M'(l)$ of sets of $2l+2$ values ($\alpha_{m\nu}, \beta_{m\nu}, \mathscr{R}_{1\nu}$; $\nu=1,2,\ldots,M'$). Thus, from the number $2l+1$ of linearly independent eigensolutions of (74.6) we get $M'(l)$ bifurcating branches. Typically, $M'\neq 2l+1$ and M' may be larger or smaller than $2l+1$, depending on conditions. (Recall that in (74.9) we found that $M'(2)$ was typically 1 or 3 and $M'(2)=2$ in degenerate cases.) Busse finds that for spherical convection, $M'<2l+1$.

To determine solvability conditions for (74.17) we introduce the adjoint spectral problem

$$\begin{bmatrix} \nabla^4 & \mathscr{R}_0 \ell(r) \nabla_s^2 \\ \mathscr{R}_0 g(r) \nabla_s^2 & \nabla^2 \nabla_s^2 \end{bmatrix} \begin{bmatrix} f^* \\ \Theta^* \end{bmatrix} = 0 \tag{74.19}$$

which together with self-adjoint boundary conditions form the adjoint spectral problem. Solutions of this problem are in the form

$$\begin{bmatrix} f^* \\ \Theta^* \end{bmatrix} = \begin{bmatrix} \bar{\bar{f}}_l^*(r) \\ \bar{\bar{\Theta}}_l^*(r) \end{bmatrix} Y_l^m(\theta,\phi) \equiv \begin{bmatrix} f_m^* \\ \Theta_m^* \end{bmatrix} \qquad (m=0,\pm 1,\ldots,\pm l)\,. \tag{74.20}$$

There are therefore $2l+1$ eigensolutions of the adjoint spectral problem. Eqs. (74.16) together with boundary conditions can be solved if and only if the right hand side of (76.16) is orthogonal to each of the $2l+1$ eigenfunctions (74.20):

$$\begin{aligned} &\langle f_m^* \mathbf{r}\cdot \mathrm{curl}^2(\mathbf{u}_0\cdot\nabla\mathbf{u}_0)\rangle + \mathscr{R}_1 \langle g(r) f_m^* \nabla_s^2 \Theta_0 + \ell(r)\, \Theta_m^* \nabla_s^2 f_0\rangle \\ &= \mathscr{P}\langle \Theta_m^* \mathbf{u}_0\cdot\nabla\Theta_0\rangle \qquad (m=0,\pm 1,\ldots,\pm l)\,. \end{aligned} \tag{74.21}$$

Eqs. (74.21) are $2l+1$ nonlinear (bilinear) equations in the $2l+2$ values $(\alpha_m, \beta_m, \mathscr{R}_1)$. The coefficients α_m and β_m enter (74.21) through the representations (74.18). The $2l+1$ equations among $2l+2$ constants determine all but one constant. This constant may be fixed by any convenient normalization ($2l+1$ equations + normalization = $2l+2$ equations). Because the $2l+2$ equations are not linear, they may have $M'(l)$ real solutions $[\alpha_1, \alpha_2, \ldots, \alpha_l, \beta_1, \beta_2, \ldots, \beta_l, \mathscr{R}_1]$; that is, there may be $M'(l)$ sets of constants

$$[\alpha_{1\nu}, \alpha_{2\nu}, \ldots, \alpha_{l\nu}, \beta_{1\nu}, \beta_{2\nu}, \ldots, \mathscr{R}_{1\nu}], \qquad \nu=1,2,\ldots,M'\,. \tag{74.22}$$

In this case $M'(l)$ solutions bifurcate from $2l+1$ solutions and $M'(l)$ need not be equal to $2l+1$.

Busse (1975) has studied (74.21) for values of $l<\infty$. He evaluates the part of the integrals in (74.21) which depend on spherical harmonics; the integrals over r are not evaluated (see Exercise 74.5). The results of Busse's analysis are general because they do not require that one first solve (74.11). He shows (Exercise 74.8) that

$$\mathscr{R}_1=0 \quad \text{when } l \text{ is odd.}$$

He speculates that the convection is supercritical when l is odd. He is able to solve for the coefficients (74.22) when l is even and less than seven. For these cases $\mathscr{R}_1 \neq 0$ and the bifurcation is two-sided. He finds that $M'(l)<2l+1$ in the cases calculated by him. When $l=4$ and $l=6$, $M'(l)>1$. Hence, there can be more than one plan form with eigenvalue $a^2=l(l+1)$ in two-sided bifurcating convection in spheres. According to Busse, the forms of convection which can exist when $l=4$ and $l=6$ exhaust the symmetries of Platonic bodies, except for the tetrahedron.

No stability results are given by Busse. Stability results at $O(\varepsilon)$ may be computed without details of the radial variations of eigenfunctions, as in § 73, but results at $O(\varepsilon^2)$ depend on these details. Busse speculates that among the M' solutions corresponding to a given value of $a^2=l(l+1)$, with bifurcating curves $\mathscr{R}(\varepsilon, v)$, $v=1,2,\ldots,M'$ the stable branch is the one for which the $\min_\varepsilon \mathscr{R}(\varepsilon, v)=\mathscr{R}(\varepsilon^*, v)$, is smallest; that is, the stable branch is identified as the one for which $v=\tilde{v}$ where

$$\mathscr{R}(\varepsilon^*, \tilde{v})=\min_{v=1,2,\ldots,M'} \mathscr{R}(\varepsilon^*, v).$$

He suggests that this distinguished branch is the one for which

$$\varepsilon\mathscr{R}_1(0, \tilde{v})=\min_{v=1,2,\ldots} \varepsilon\mathscr{R}_1(0, v).$$

Of course, such a branch would be unstable when $\varepsilon<\varepsilon_*$ and, if stable, stable only when $\varepsilon>\varepsilon_*$ (see Fig. 77.1).

In Exercises (74.1—6) we develop a bifurcation theory for convection in porous materials in closed containers bounded by spherical caps and insulated side walls. The theory is the analogue to that which was developed in § 73. Here, and in § 73, the theory which is developed appears to generalize to problems of cellular solutions of the DOB and OB equations in which there are no side-walls. The bounded containers considered here are formed from the intersection of a spherical shell (inner radius r_1 and outer radius r_2) and a cone whose generators are radius vectors from the sphere center. The container volume is designated as Ω and $\partial\Omega=S\cup\mathscr{A}(r_2)\cup\mathscr{A}(r_1)$ where S is the side wall and $\mathscr{A}(r)$ is the cross-section (the intersection of a sphere of radius r and the cone). The construction of the container is such that $\partial\mathscr{A}(r)$ is on S and is described by a function $f(\theta, \phi)$, independent of r (see Fig. 74.3).

We shall consider the bifurcation into convection of the conduction solution (74.1). Since the container boundary $\partial\Omega$ is impermeable the normal component of $\mathbf{u}$ vanishes on $\partial\Omega$. Since S is insulating, $\partial\Theta/\partial n|_S=0$. The results in the exer-

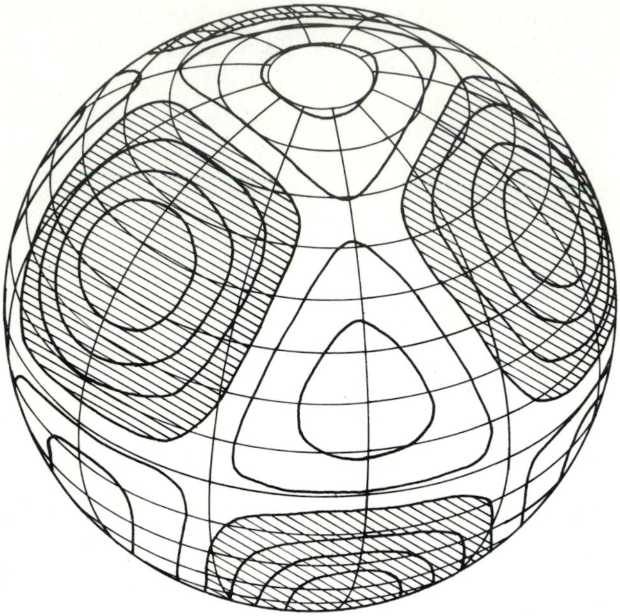

Fig. 74.1: Convection in a spherical shell, $l=4$ (Busse, 1975)

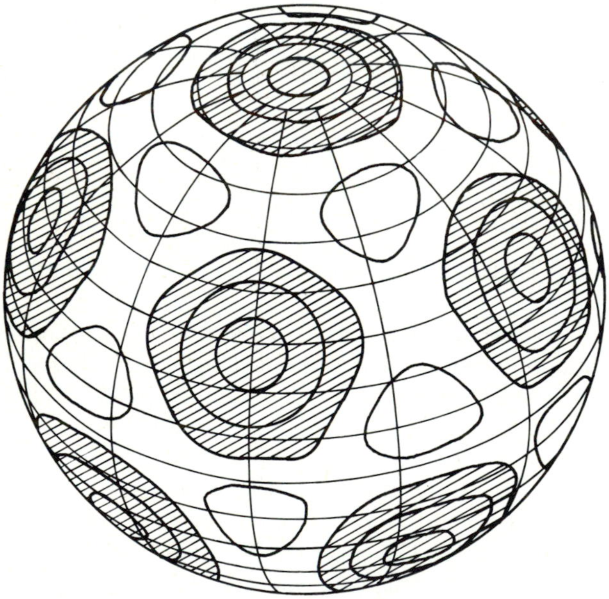

Fig. 74.2: Convection in a spherical shell, $l=6$ (Busse, 1975)

cises hold for many different prescribed conditions for Θ on $\mathscr{A}(r_1)$ and $\mathscr{A}(r_2)$. In general, any linear, homogeneous condition $\mathscr{L}\Theta=0$ will do, provided that $\mathscr{L}$ is independent of θ and ϕ. To be definite we shall assume conducting caps with $\Theta=0$ at $r=r_1$ and $r=r_2$.

Fig. 74.3: Container Ω formed from intersecting a spherical shell and a cone. The container is filled with a fluid-saturated porous material. The boundary $\partial\Omega$ is impermeable. The sidewall S is insulated

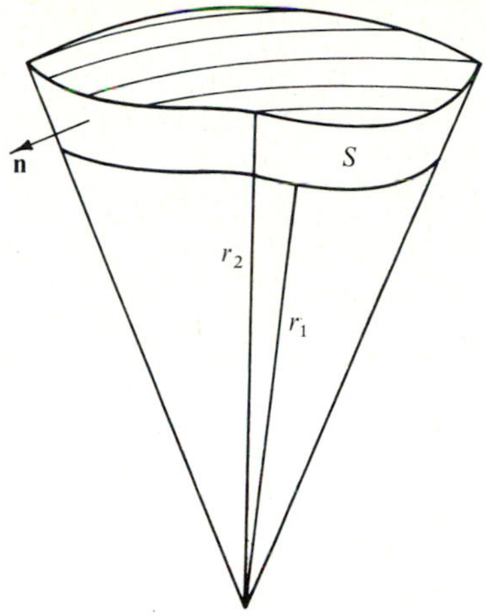

Exercise 74.1: Show that the DOB equations (70.10) for bifurcation of conduction into convection in the container Ω of Fig. 74.3 may be written as

$$(\partial_\tau + u\cdot\nabla)\begin{bmatrix}0\\ \Theta\end{bmatrix} - \mathscr{R}\,\mathbf{u}\cdot\begin{bmatrix}0\\ \mathbf{r}\ell(r)\end{bmatrix} = \begin{bmatrix}-\nabla p + \mathscr{R}\mathbf{r}\,g(r)\Theta - \mathbf{u}\\ \nabla^2\Theta\end{bmatrix}. \tag{74.23}$$

Show that the radial vorticity of solutions of (74.23) vanishes. Conclude that the velocity $\mathbf{u}$ is poloidal

$$\mathbf{u} = \boldsymbol{\delta}\chi = -\mathbf{e}_r\frac{1}{r^2}\nabla_s^2\chi + \frac{1}{r^2}\nabla_s\frac{\partial r\chi}{\partial r} \tag{74.24}$$

and that (74.23) may be written as

$$\nabla^2\chi + \mathscr{R}\,g(r)\Theta = 0\,, \tag{74.25a}$$

$$\frac{\partial\Theta}{\partial t} + \boldsymbol{\delta}\chi\cdot\nabla\Theta = -\mathscr{R}\ell(r)\nabla_s^2\chi + \nabla^2\Theta \tag{74.25b}$$

where on the side walls S of the container

$$\frac{\partial\chi}{\partial n} = \frac{\partial\Theta}{\partial n} = 0\,. \tag{74.25c}$$

and

$$\chi = \Theta = 0 \quad \text{at} \quad r_1 \text{ and } r_2\,. \tag{74.25d}$$

Exercise 74.2: Expand steady bifurcating solutions

$$\begin{bmatrix}\chi\\ \Theta\\ \mathscr{R}-\mathscr{R}_0\end{bmatrix} = \begin{bmatrix}\chi_0\\ \Theta_0\\ \mathscr{R}_1\end{bmatrix}\varepsilon + \begin{bmatrix}\chi_1\\ \Theta_1\\ \mathscr{R}_2\end{bmatrix}\varepsilon^2 + O(\varepsilon^3)\,.$$

Show that

$$\nabla^2\begin{bmatrix}\chi_0\\ \Theta_0\end{bmatrix}+\mathscr{R}_0\begin{bmatrix}0 & g(r)\\ -\ell(r)\nabla_S^2 & 0\end{bmatrix}\begin{bmatrix}\chi_0\\ \Theta_0\end{bmatrix}=0 \qquad (74.26\text{a})$$

$$\frac{\partial\chi_0}{\partial n}=\frac{\partial\Theta_0}{\partial n}=0 \quad \text{on} \quad S\,, \qquad (74.26\text{b})$$

$$\chi_0=\Theta_0=0 \quad \text{at} \quad r_1 \text{ and } r_2 \qquad (74.26\text{c})$$

and

$$\mathscr{R}_1\langle\ell(r)\,\Theta^*\nabla_S^2\chi_0\rangle+\langle\Theta^*(\mathbf{u}_0\cdot\nabla)\,\Theta_0\rangle=0 \qquad (74.26\text{d})$$

where Θ^* and χ^* are solutions of the adjoint problem

$$\nabla^2\begin{bmatrix}\chi^*\\ \Theta^*\end{bmatrix}+\mathscr{R}_0\begin{bmatrix}0 & -\ell(r)\nabla_S^2\\ g(r) & 0\end{bmatrix}\begin{bmatrix}\chi^*\\ \Theta^*\end{bmatrix}=0 \qquad (74.27)$$

and χ^* and Θ^* satisfy the same boundary conditions as χ and Θ.

Exercise 74.3: Show that (74.26) and its adjoint support separable solutions of the form

$$\begin{bmatrix}\chi_0\\ \Theta_0\\ \chi^*\\ \theta^*\end{bmatrix}=\begin{bmatrix}\bar{\bar{\chi}}(r)\\ \bar{\bar{\Theta}}(r)\\ \bar{\bar{\chi}}^*(r)\\ \bar{\bar{\Theta}}^*(r)\end{bmatrix}\Phi(\theta,\phi) \qquad (74.28)$$

where

$$\nabla_S^2\Phi+a^2\Phi=0 \quad \text{and} \quad \left.\frac{\partial\Phi}{\partial n}\right|_S=0\,. \qquad (74.29)$$

Show that

$$\langle\Theta^*(\mathbf{u}_0\cdot\nabla)\,\Theta\rangle=a^2\left\langle\frac{\bar{\bar{\Theta}}}{r}[\bar{\bar{\chi}}\,\bar{\bar{\Theta}}'+\tfrac{1}{2}(r\bar{\bar{\chi}})'\,\bar{\bar{\Theta}}]\,\Phi^3\right\rangle\,. \qquad (74.30)$$

where

$$\mathscr{D}_a^2\begin{bmatrix}\bar{\bar{\chi}}\\ \bar{\bar{\Theta}}\end{bmatrix}+\mathscr{R}_0\begin{bmatrix}0 & g(r)\\ a^2\ell(r) & 0\end{bmatrix}\begin{bmatrix}\bar{\bar{\chi}}\\ \bar{\bar{\Theta}}\end{bmatrix}=0\,, \qquad (74.31)$$

$$\mathscr{D}_a^2\begin{bmatrix}\bar{\bar{\chi}}^*\\ \bar{\bar{\Theta}}^*\end{bmatrix}+\mathscr{R}_0\begin{bmatrix}0 & a^2\ell(r)\\ g(r) & 0\end{bmatrix}\begin{bmatrix}\bar{\bar{\chi}}^*\\ \bar{\bar{\Theta}}^*\end{bmatrix}=0\,,$$

$$\bar{\bar{\chi}}=\bar{\bar{\Theta}}=\bar{\bar{\chi}}^*=\bar{\bar{\Theta}}^*=0 \quad \text{on} \quad r=r_1 \quad \text{and} \quad r=r_2\,,$$

and

$$\mathscr{D}_a^2=\frac{1}{r^2}\frac{d}{dr}\left(r^2\frac{d}{dr}\right)-\frac{a^2}{r^2}\,.$$

Exercise 74.4: Formulate and prove necessary and sufficient conditions for two-sided bifurcation which involves the integral

$$\iint_{\mathscr{A}}\Phi^3\sin\theta\,d\theta\,d\phi \qquad (74.32)$$

where Φ satisfies (74.29). Apart from simple differences in geometry these conditions should coincide with those stated around (73.13)

Exercise 74.5: Suppose that a^2 is a simple eigenvalue of (74.29). Show that

$$\langle \mathbf{u}^*(\mathbf{u}_0 \cdot \nabla)\mathbf{u}_0 \rangle = \langle F(r)\Phi^3 \rangle .$$

Find the form of $F(r)$ in terms of $\bar{\bar{\chi}}(r)$, $\bar{\bar{\chi}}^*(r)$ and a^2.

Exercise 74.6: Show that the conditions for two-sided bifurcation proved in Exercise 74.4 hold for the DOB equations (70.10a) with $B \neq 0$.

Exercise 74.7 (Busse, 1975): Consider the problem of bifurcation into convection when Ω is the full sphere. Show that $\Phi(\theta,\phi) = Y_l^m(\theta,\phi)$ and $a^2 = l(l+1)$ for $l \geqslant 1$; the linear combination

$$\Phi(\theta,\phi) = \textstyle\sum_{m=0}^{l}[\alpha_m \cos m\phi + \beta_m \sin m\phi]\, P_l^m(\cos\theta) \tag{74.33}$$

forms an orthogonal basis for the null space of the operator $\nabla_s^2 + l(l+1)$. The constants α_m and β_m must be selected by solvability conditions. Suppose that $\mathscr{R}_0(a^2) = \mathscr{R}$ is a simple eigenvalue of (74.11). Show that the perturbation problems for χ_1 and Θ_1 are solvable when for each and every $m = 0, 1, \ldots, l$,

$$\mathscr{R}_1 = \frac{\langle \bar{\bar{\Theta}}^*[\bar{\bar{\chi}}\bar{\bar{\Theta}}' + \frac{1}{2}(r\bar{\bar{\chi}})'\,\bar{\bar{\Theta}}]\, Y_l^m(\theta,\phi)\Phi^2 \rangle}{\langle \ell(r)\,\bar{\bar{\chi}}\bar{\bar{\Theta}}^*\,\Phi\, Y_l^m(\theta,\phi)\rangle} = M(l)\frac{\int_0^\pi \int_0^{2\pi} Y_l^m \Phi^2 \sin\theta\, d\theta d\phi}{\int_0^\pi \int_0^{2\pi} Y_l^m \Phi \sin\theta\, d\theta d\phi} \tag{74.34}$$

where Φ is the linear combination (74.33), $Y_l^m(\theta,\phi)$ is $\sin m\phi\, P_l^m(\cos\theta)$ or $\cos m\phi\, P_l^m(\cos\theta)$ and

$$M(l) = \frac{\int_{r_1}^{r_2} \bar{\bar{\Theta}}^*[\bar{\bar{\chi}}\bar{\bar{\Theta}}' \pm \frac{1}{2}(r\bar{\bar{\chi}})'\,\bar{\bar{\Theta}}]\, r^2 dr}{\int_{r_1}^{r_2} \ell(r)\,\bar{\bar{\chi}}\bar{\bar{\Theta}}^* r^2\, dr}$$

depends on l but not on m. The solvability condition (74.34) may be written as

$$\begin{aligned}\mathscr{R}_1 \alpha_m N_{lm} = M(l) \textstyle\int_0^\pi \int_0^{2\pi} \cos m\phi\, P_l^m(\cos\theta \{ \sum_{\nu=0}^{l} \sum_{\mu=0}^{l} [\alpha_\nu \cos\nu\phi + \beta_\nu \sin\nu\phi]\, P_l^\nu(\cos\theta)] \\ \cdot [\alpha_\mu \cos\mu\phi + \beta_\mu \sin\mu\phi]\, P_l^\mu(\cos\theta)\} \sin\theta\, d\theta d\phi\end{aligned} \tag{74.35a}$$

and

$$\mathscr{R}_1 \beta_m N_{lm} = M(l) \quad \text{times the same integral with } \sin m\phi \text{ replacing } \cos m\phi \tag{74.35b}$$

where

$$N_{lm} = \pi \textstyle\int_0^\pi (P_l^m(\cos\theta))^2 \sin\theta\, d\theta .$$

Eqs. (74.35) are $2l+1$ bilinear algebraic equations for the $2l+2$ constants α_m $(m = 0, 1, \ldots l)$, β_m $(m = 1, 2, \ldots l)$ and $\mathscr{R}_1$. Since Φ is the solution of a linear problem the $(2l+2)$nd equation is given by a normalizing condition.

Exercise 74.8 (Busse, 1975): $P_l^m(\cos\theta)$ is $\sin^m\theta$ times a polynomial of degree $l-m$ in $\cos\theta$. Show that

$$\psi_l(\theta,\phi) = \begin{Bmatrix} P_l^m(\cos\theta)\cos m\phi \\ P_l^m(\cos\theta)\sin m\phi \end{Bmatrix} = (-1)^l \psi_l(\pi-\theta, \phi+\pi) . \tag{74.36}$$

When l is odd $\psi_l(\theta,\phi) = -\psi_l(\pi-\theta,\phi+\pi)$ and $\Phi^2(\theta,\phi) = \Phi^2(\pi-\theta,\phi+\pi)$ is even, where Φ is given by (74.33) with l odd. Use (74.36) to show that

$$\mathscr{R}_1 = 0 \quad \text{when } l \text{ is odd.}$$

Exercise 74.9: Eigenvalues which are degenerate in a wide class of solutions can sometimes be made simple in an appropriately restricted class of solutions. Show that $a^2 = l(l+1)$ is a simple eigenvalue of (74.29) among solutions which are unchanged by rotation about the polar axis (axisymmetric solutions). Suppose a^2 is simple is some subset of solutions generated by invariance to group transformations on the sphere. Can anything be concluded about the stability of solutions possessing this same property of invariance?

§ 75. Stability of the Conduction Solution in a Container Heated below and Internally

The conduction solution whose stability is now to be studied is given by (59.6) in terms of the dimensionless temperature gradient $\boldsymbol{\eta}_T = -\mathbf{e}_z(1+\xi g(z))$, $-\frac{1}{2} \leqslant z \leqslant \frac{1}{2}$. The function $g(z)$ gives the derivative of the part of the basic temperature distribution which comes from the heat sources.

A uniform distribution of heat sources Q leads to a parabolic temperature distribution and adds to the linear variation engendered by heating from below. Uniform heat sources have $g(z)=z$ and $\boldsymbol{\eta}_T \cdot \mathbf{e}_z = -(1+\xi z)$; heat sources make the adverse temperature gradient larger where $\xi z > 0$. When $|\xi|$ is large, there is a large effective adverse temperature gradient over the region $\xi z > 0$. Hence, large sources (and sinks) $|\xi|$ are physically destabilizing and convection will become possible when the effective Rayleigh number exceeds a critical value, say R_*,

$$\frac{\alpha g |\xi| l^4}{\nu k} \geqslant R_* .$$

This implies instability when

$$\frac{\alpha g |dT(0)/dx_3| |\xi| l^4}{\nu \kappa |dT(0)/dx_3|} = \frac{\mathscr{R}^2 |\xi|}{|dT(0)/dx_3|} \geqslant R_* .$$

Hence, the critical value of $\mathscr{R}^2$ can be expected to decrease with $1/|\xi|$ at large $|\xi|$ when $|dT(0)/dx_3|$ is fixed (see Exercise 75.4c and Fig. 75.2).

An odd distribution of heat sources $Q(x_3) = -Q(-x_3)$ leads to an even heat source temperature gradient $g(z) = g(-z)$. Suppose $g(z)$ is a positive polynomial of even powers of z. Then $\xi > 0$ is destabilizing for all z and $\xi < 0$ is stabilizing. In this case, the effect of a heat source on stability is opposite to the effect of a heat sink (see Fig. 75.1).

The equations which govern a disturbance of the conduction solution (59.7) are

$$\left(\frac{\partial}{\partial t} + \mathbf{u} \cdot \nabla\right) \mathbf{M} \cdot \mathbf{Q} + \mathbf{L} \cdot \mathbf{Q} + \mathscr{R} \xi \mathbf{F} \cdot \mathbf{Q} = -\mathbf{d}P \tag{75.1}$$

with $\operatorname{div} \mathbf{u} = 0$ and subject to general boundary conditions which for the moment, we take as $\mathbf{u} = \theta = 0$ on S. The four component vectors $\mathbf{Q} = (U, V, W, \Theta)$ and

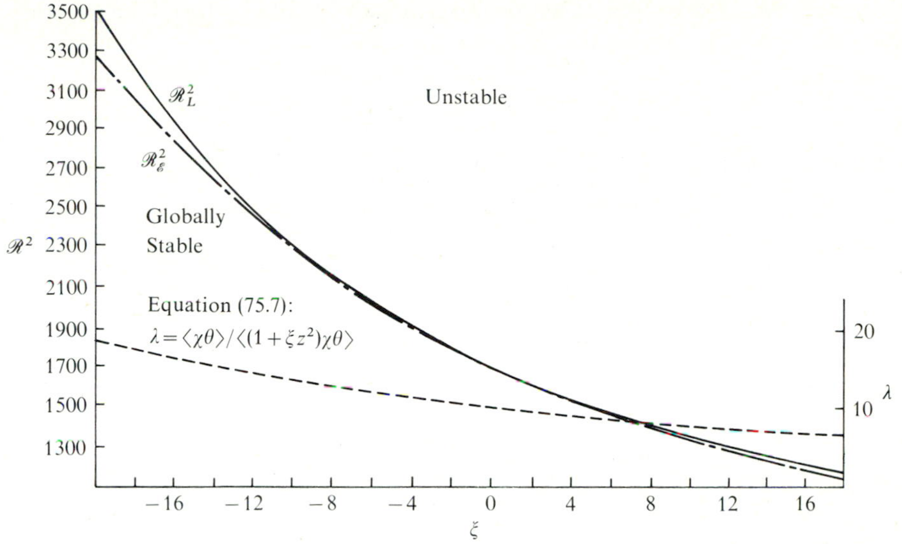

Fig. 75.1: Stability regions for the case $g(z)=z^2$ in which half the channel has heat sources and half has heat sinks. Here the calculation is for conducting rigid surfaces. The picture is the same for the free surface case, but the numbers are different. The size of the unstable region in the channel center increases with ξ

$\mathbf{d}=(\partial_x,\partial_y,\partial_z,0)$. **M** is a 4×4 diagonal matrix with components $[1,1,1,\mathscr{P}]$, **F** is a 4×4 matrix whose only nonzero component $-g(z)$ is in the third row and fourth column and

$$\mathbf{L}=-\begin{bmatrix}\nabla^2 & 0 & 0 & 0\\ 0 & \nabla^2 & 0 & 0\\ 0 & 0 & \nabla^2 & \mathscr{R}\\ 0 & 0 & \mathscr{R} & \nabla^2\end{bmatrix}.$$

The spectral problem for disturbances proportional to $e^{-\sigma t}$ is

$$-\sigma\mathbf{M}\cdot\mathbf{q}+\mathbf{L}\cdot\mathbf{q}+\mathscr{R}\xi\mathbf{F}\cdot\mathbf{q}=-\mathbf{d}p \tag{75.2}$$

We assume $\sigma(R,\xi)$ is real-valued. $\sigma(R,0)$ is real-valued. The adjoint problem is

$$-\sigma\mathbf{M}\cdot\mathbf{q}^*+\mathbf{L}\cdot\mathbf{q}^*+\mathscr{R}\xi\mathbf{F}^T\cdot\mathbf{q}^*=-\mathbf{d}p^* \tag{75.3}$$

where $\mathbf{F}^T$ is the transpose of **F**.

The conduction solution (59.7) is unstable when

$$\mathscr{R}>\mathscr{R}_L \tag{75.4}$$

where $\mathscr{R}_L$ is the first critical value of the Rayleigh number; that is, there is an eigenvalue $\sigma(\mathscr{R}) \leqslant 0$ with equality only when $\mathscr{R} = \mathscr{R}_L$.

The conduction solution (59.7) is monotonically and globally stable with respect to the energy $\mathscr{E}_\lambda(t) = \frac{1}{2}\langle |\mathbf{u}|^2 + \lambda \mathscr{R}\theta^2 \rangle$ for any $\lambda > 0$ when

$$\frac{1}{\mathscr{R}} > \frac{1}{\mathscr{R}_\lambda} = \max_{\mathbf{H}} \frac{\langle [1 + \lambda + \lambda\xi\hat{g}(z)]\, w\theta \rangle}{\langle |\nabla \mathbf{u}|^2 + \lambda |\nabla \theta|^2 \rangle}. \tag{75.5}$$

The largest stability limit for which a decaying energy can be found is called the optimum stability boundary. This boundary is associated with the function $\lambda = \tilde{\lambda}(\xi)$ which is generated as the solution of the problem[4]

$$\mathscr{R}_{\mathscr{E}}(\xi) = \max_{\lambda > 0} \mathscr{R}_\lambda(\xi) = \mathscr{R}_{\tilde{\lambda}(\xi)}(\xi) \tag{75.6}$$

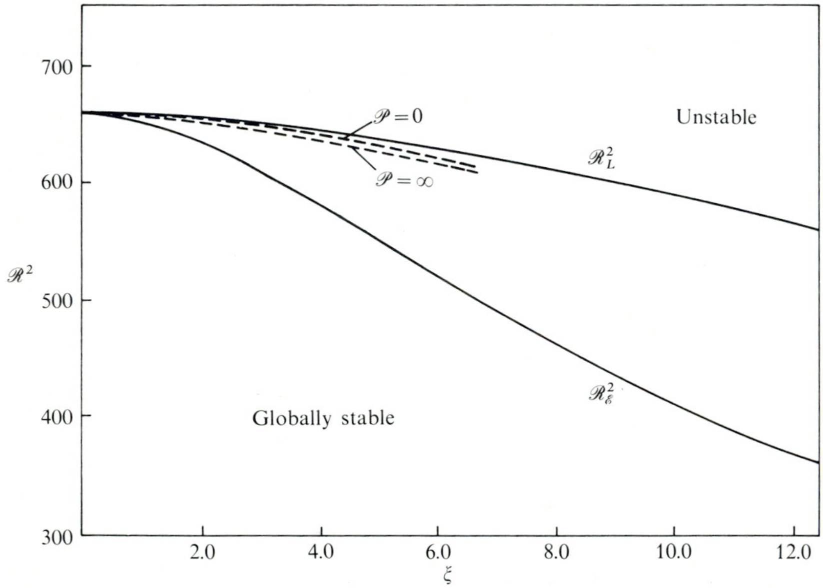

Fig. 75.2: Critical values and bifurcation curves for a fluid layer between zero-shear-stress planes, heated from below and internally (by uniform heat sources with $g(z) = z$). Steady convection will bifurcate from the rest state when the point $(\xi, \mathscr{R}^2)$ is on a bifurcation curve. Two such curves, given by (76.5), (76.7) and (76.8) to $O(\xi^3)$, are shown as dashed lines in the figure. All curves are symmetric with respect to the line $\xi = 0$

Figs. 75.1, 2, 3 show graphs of the curves $\mathscr{R}_L^2(\xi)$ and $\mathscr{R}_{\mathscr{E}}^2(\xi)$ which give the limits for instability and global stability for particular heat source distributions in fluids confined between parallel planes. The working equations for the numerical computations leading to the graphs are given in Exercise (75.9). These

[4] Problem (75.6) was first considered by Joseph and Shir (1966). Homsy (1973, 1974) has developed a related energy theory to discuss the global stability of time-dependent flows.

computations show that there is a region in the $(\xi, \mathscr{R}^2)$ plane, between the curves $\mathscr{R}_{\mathscr{E}}^2(\xi)$ and $\mathscr{R}_L^2(\xi)$, in which steady subcritical motions may exist.

Krishnamurti (1968A, B) has studied the bifurcation of the conduction solution induced by heat sources by a two parameter expansion (ε and ξ) and experimentally. The analysis is discussed in § 76 and an example of the type of calculations which are involved is given in the long Exercise 76.2. In Krishnamurti's experiment the heat sources are created by slowly changing the mean temperature.

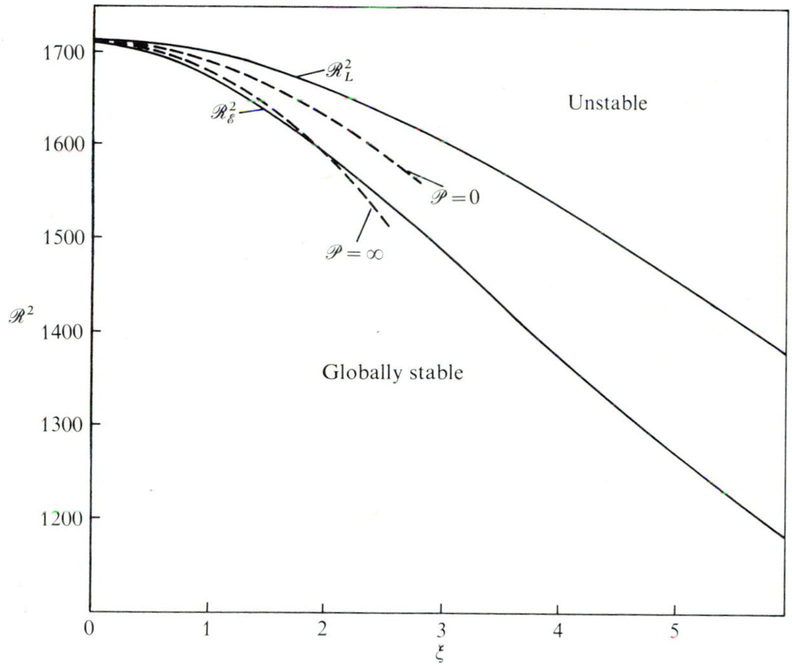

Fig. 75.3: Critical values and bifurcation curves for a fluid layer between rigid walls, heated from below and internally. The figure description is similar to 75.2. The bifurcation curves (dashed) are due to Krishnamurti (1968A). The critical values $\mathscr{R}_L$ of the linear theory of stability are due to Sparrow, Goldstein and Jonsson (1964)

Her η equals our 2ξ. When there are no heat sources, convection in rolls is observed, in agreement with stability predictions of Schlüter, Lortz and Busse (1965), but when $\eta \neq 0$, stable subcritical convection in hexagons is observed: "The critical Rayleigh number predicted for $\eta=0$ is 1465, which is 14% below the critical number for $\eta=0$. This is interpreted as a finite amplitude instability occurring at a Rayleigh number below that predicted by linear theory".

Other experiments on convective instability in internally heated fluid layers have been reported by Tritton and Zarraga (1967), Schwiderski and Schwab (1971) and Kulacki and Goldstein (1972, 1975). These authors do not report observations of subcritical convection.

Remarks about the exercises: Exercises 75.1 through 75.5 are about the energy theory of stability of the conduction solution. Exercises 75.6 and 75.7 are about the spectral problems. The theories are compared in Exercises 75.8 and 75.9.

Exercise 75.1: Prove the energy stability theorem for the conduction solution in a container heated from below and internally. Show that the decay of the energy $\mathscr{E}_{\tilde{\lambda}}$ of any disturbance $\mathbf{Q}$ is global, monotonic and exponential when $\mathscr{R}<\mathscr{R}_{\mathscr{E}}(\xi)$.

Exercise 75.2: Show that the solution $\lambda=\tilde{\lambda}(\xi)$ of (75.6) satisfies the relation

$$\lambda=\tilde{\lambda}(\xi)=\langle\tilde{\chi}\tilde{\theta}\rangle/\langle[1+\xi g(z)]\tilde{\chi}\tilde{\theta}\rangle \tag{75.7}$$

where $\tilde{\mathbf{u}}$ and $\tilde{\theta}$ solve (75.5). (See Appendix B10.)

Exercise 75.3: Use the result of exercise B4.6 or otherwise, to prove that

$$\begin{aligned}\frac{d\mathscr{R}_{\mathscr{E}}}{d\xi}&=-\tilde{\lambda}\mathscr{R}_{\mathscr{E}}^2\langle g\tilde{w}\tilde{\theta}\rangle/\langle|\nabla\tilde{\mathbf{u}}|^2+\tilde{\lambda}|\nabla\tilde{\theta}|^2\rangle\\&=(\tilde{\lambda}-1)\mathscr{R}_{\mathscr{E}}(\xi)/2\xi=-\bar{g}(\xi)\mathscr{R}(\xi)/2(1+\xi\bar{g}(\xi))\end{aligned} \tag{75.8}$$

where

$$\bar{g}(\xi)=\langle g(z)\tilde{w}\tilde{\theta}\rangle/\langle\tilde{w}\tilde{\theta}\rangle\,.$$

Exercise 75.4: (a) Prove that

$$d\xi^2\left\langle w\theta\left\{\left(\frac{d\lambda}{d\xi}\right)^2\frac{\partial^2}{\partial\lambda^2}+2\frac{d\lambda}{d\xi}\frac{\partial^2}{\partial\lambda\,\partial\xi}+\frac{\partial^2}{\partial\xi^2}\right\}(1+\lambda+\lambda\hat{g}(z)\xi)\right\rangle>0\,.$$

Hint: Follow the procedure used to prove B4.3 of Appendix B.

(b) Show that

$$0>\frac{d^2\mathscr{R}_{\mathscr{E}}}{d\xi^2}+\frac{(1-\lambda)}{\xi}\frac{d\mathscr{R}_{\mathscr{E}}}{d\xi}+\frac{\mathscr{R}_{\mathscr{E}}(1-\lambda)}{2\lambda\xi}\frac{d\lambda}{d\xi}\,.$$

(c) Let $\bar{g}(0)=0$. Then $\bar{g}(\xi)$ changes sign at $\xi=0$, is positive for all positive ξ and is negative for all negative ξ. $\mathscr{R}_{\mathscr{E}}(\xi)$ is an increasing function when $\xi<0$, a maximum at $\xi=0$, a decreasing function for $\xi>0$ and such that at large $|\xi|$, $|\xi|\mathscr{R}(\xi)=\text{const.}$

Exercise 75.5: Show that the Euler equations for (75.5) may be written as

$$\nabla^2\mathbf{u}+\frac{\mathscr{R}_\lambda}{2\sqrt{\lambda}}(1+\lambda+\lambda\xi g(z))\mathbf{e}_z\hat{\theta}=\nabla p \tag{75.9 a}$$

and

$$\nabla^2\hat{\theta}+\frac{\mathscr{R}_\lambda}{2\sqrt{\lambda}}(1+\lambda+\lambda\xi g(z))w=0 \tag{75.9 b}$$

where we have substituted $\sqrt{\lambda}\hat{\theta}$ for θ.

Exercise 75.6 (Exchange of stability): The assumption that the eigenvalues $\sigma(\mathscr{R}_L)=i\,\mathrm{im}\,\sigma(\mathscr{R}_L)=0$ are real-valued at criticality is difficult to prove even when the bounded domain is taken as a period cell between parallel plates (see Davis (1969A), DiPrima and Habetler (1969) and Yih (1972A, B) for proofs of exchange of stability in this special case). However, easier proofs (Spiegel in Veronis, 1963) are possible when the boundaries are assumed to be nondeflecting but unable to resist tangential tractions.

(a) Consider problem (75.2) in a plane fluid layer bounded by free nondeflecting surfaces on which $\theta=0$. Suppose all disturbances are almost periodic functions of the horizontal variables. Prove that all eigensolutions of (75.2) which have $\text{re}(\sigma)<0$ also have $\text{im}(\sigma)=0$. *Hint:* note that $\langle\nabla p\cdot\nabla^2\bar{\mathbf{u}}\rangle=0$ where $\bar{\mathbf{u}}=$complex conjugate of $\mathbf{u}$.

(b) Consider the problem of stability of the rest state in an internally heated self-gravitating spherical annulus (Exercise 59.3). Formulate disturbance equations. Show that the conditions

$$\mathbf{u}\cdot\mathbf{r}=\partial^2\mathbf{u}\cdot\mathbf{r}/\partial r^2=0\,, \tag{75.10}$$

hold on free spherical surfaces of constant radius. Here $\mathbf{u}\cdot\mathbf{r}$ is the radial velocity in spherical polar coordinates. Suppose the annulus is bounded by free nondeflecting surfaces on which (75.10) holds and $\theta|_S=0$. Show that the result given under (a) holds for the spherical annulus.

(c) Suppose that the spherical surfaces are allowed to deflect under the action of motions and that these surfaces follow the laws of surface tension described in § 55. Formulate disturbance equations (like those derived in §§ 93 and 94) which govern the stability and bifurcation of the rest state. Under what circumstances is the result described under (b) valid?

Exercise 75.7: Use the spectral problem (75.2) and its adjoint (75.3) to show that

$$-\sigma\mathbf{M}\cdot\mathbf{q}+\partial p^*=\mathscr{R}_\xi\mathscr{F}^T\cdot\mathbf{q}^*+\mathbf{S}\cdot\mathbf{q}^*,\qquad \partial\cdot\mathbf{q}^*=0,\qquad \mathbf{q}^*|_S=0 \tag{75.11}$$

where

$$\frac{\langle w^*\theta+w\theta^*+\xi g w\theta^*\rangle}{\langle\nabla\mathbf{u}:\nabla\mathbf{u}^*+\nabla\theta\cdot\nabla\theta^*\rangle}=\mathscr{I}(\varepsilon),\qquad \left(\mathscr{I}(0)=\frac{1}{\mathscr{R}_L}\right)$$

is a stationary functional in the sense of Exercise B4.6, $\mathbf{u}$ and $\mathbf{u}^*$ are solenoidal vectors vanishing on the boundary S of $\mathscr{V}$, θ and θ^* vanish on S and

$$\mathbf{u}-\mathbf{u}_0=\varepsilon\boldsymbol{\eta}\,,\qquad \mathbf{u}^*-\mathbf{u}_0^*=\varepsilon\boldsymbol{\eta}^*\,,\qquad \theta-\theta_0=\varepsilon\phi\,,\qquad \theta^*-\theta_0^*=\varepsilon\phi^*$$

where the functions with subscripts zero solve the spectral problem when $\mathscr{R}=\mathscr{R}_L$.

Show that

$$\frac{d\mathscr{R}_L}{d\xi}\langle w_0^*\theta_0+w_0\theta_0^*\rangle+\frac{d(\xi\mathscr{R}_L)}{d\xi}\langle g w_0\theta_0^*\rangle=0$$

and

$$\frac{d\mathscr{R}_L}{d\xi}=-\hat{\hat{g}}(\xi)\mathscr{R}_L(\xi)/2(1+\xi g(\xi)) \tag{75.12}$$

where

$$\hat{\hat{g}}(\xi)=\langle\hat{g}w_0\theta_0^*\rangle/\langle w_0\theta_0^*\rangle\,.$$

Exercise 75.8 (Comparison of stability limits and instability limits):

Show that when $\xi=0$, $\mathscr{R}_L(\xi)=\mathscr{R}_{\mathscr{E}}(\xi)$ and

$$\frac{d\mathscr{R}_L(0)}{d\xi}=\frac{d\mathscr{R}_{\mathscr{E}}(0)}{d\xi}\,.$$

Compare the slope formulas (75.8) with (75.12). Under what conditions would you expect $\mathscr{R}_{\mathscr{E}}(\xi)$ and $\mathscr{R}_L(\xi)$ to be curves of similar shape (see Figs. 75.1, 2, 3)?

Exercise 75.9: Show that the vertical vorticity of the velocity field $\mathbf{u}$ satisfying (75.9) or (75.2) must vanish when the fluid is contained between infinite horizontal planes on which the velocity or the tangential components of the stress vanish. Justify the introduction of the poloidal potential χ satisfying (62.2c) or (62.2d). Show that the eigenvalue problems for energy and linear stability theory may be reduced, using the planform hypothesis (62.4a, b) to

$$\begin{bmatrix} L^2\hat{\chi} \\ L\hat{\theta} \end{bmatrix} - \frac{\mathcal{R}_\lambda}{2}\left(\frac{1}{\sqrt{\lambda}} + \sqrt{\lambda} + \sqrt{\lambda}\xi\hat{g}(z)\right)\begin{bmatrix} \hat{\theta} \\ -a^2\hat{\chi} \end{bmatrix} = 0\,, \tag{75.13}$$

$$\begin{bmatrix} L^2\hat{\chi} \\ L\hat{\theta} \end{bmatrix} - \mathcal{R}_L\begin{bmatrix} \hat{\theta} \\ -a^2(1+\xi\hat{g}(z))\hat{\chi} \end{bmatrix} = 0 \tag{75.14}$$

and

$$\hat{\theta} = \hat{\chi} = D\hat{\chi} = 0 \quad \text{or} \quad \hat{\theta} = \hat{\chi} = D^2\hat{\chi} = 0 \tag{75.15}$$

at a rigid or free boundary. The eigenvalues $\mathcal{R}_\lambda(a^2, \xi)$ and $\mathcal{R}_l(a^2, \xi)$ can be generated by numerical integration and

$$\mathcal{R}_{\mathscr{E}} = \max_{\lambda>0} \min_{a^2>0} \mathcal{R}_\lambda(a^2, \xi)\,. \tag{75.16}$$

and

$$\mathcal{R}_L(\xi) = \min_{a^2>0} \mathcal{R}_l(a^2, \xi)\,. \tag{75.17}$$

§ 76. Taylor Series in Two Parameters

Busse (1962, 1967) used a double power series to study problems of bifurcation of conduction into cellular convection. His method was used by Krishnamurti (1968A) to study convection in the presence of heat sources and by Liang, Vidal and Acrivos (1969) to study convection in fluids with temperature dependent viscosity coefficients (see Exercise 76.3 and Fig. 77.1). For the heat source problem, the Taylor series is written in powers of ε and ξ.

$$\begin{bmatrix} \mathbf{Q}(\mathbf{x}, \varepsilon, \xi) \\ P(\mathbf{x}, \varepsilon, \xi) \\ \mathcal{R}(\varepsilon, \xi) \end{bmatrix} = \varepsilon \sum_{m=0} \sum_{n=0} \begin{bmatrix} \mathbf{q}_{mn}(\mathbf{x}) \\ p_{mn}(\mathbf{x}) \\ \mathcal{R}_{mn}/\varepsilon \end{bmatrix} \varepsilon^m \xi^n \tag{76.1}$$

If $\mathcal{R}_{00}$ is a simple isolated eigenvalue in a suitably defined class of solutions then the series converges (Fife and Joseph, 1969). This method has the undesirable effect of restricting analysis to small values of $|\xi|$, as well as to small values $|\varepsilon|$. An advantage of the two-parameter expansion is that the lowest order $(m, n) = (0, 0)$ problem reduces to the Bénard problem. The Bénard problem has constant coefficients and can be solved in terms of elementary functions in special cases.

To study bifurcation into convection of an internally heated fluid with two-parameter expansions we insert the series (76.1) into

$$(\mathbf{u}\cdot\nabla)\mathbf{M}\cdot\mathbf{Q} + \mathbf{L}\cdot\mathbf{Q} + \mathcal{R}\xi\mathbf{F}\cdot\mathbf{Q} = -\mathbf{d}P, \qquad \varepsilon = \langle \mathbf{Q}\cdot\mathbf{q}_{00}\rangle, \qquad \mathbf{Q}\in B \tag{76.2}$$

where B is a set of vectors $\mathbf{Q}$ satisfying $\mathbf{d}\cdot\mathbf{Q} = 0$ and appropriate boundary conditions. Then

(m, n)

$(0,0)$ $\quad \mathbf{L}_{00}\cdot\mathbf{q}_{00} = -\mathbf{d}p_{00}, \qquad 1 = \langle|\mathbf{q}_{00}|^2\rangle, \qquad \mathbf{q}_{00}\in B\,;$ (76.3 a)

$$(0,1)\quad \mathbf{L}_{00}\cdot\mathbf{q}_{01}+\mathbf{L}_{01}\cdot\mathbf{q}_{00}+\mathscr{R}_{00}\mathbf{F}\cdot\mathbf{q}_{00}=-\mathbf{d}p_{01} \tag{76.3 b}$$

$$(1,0)\quad \mathbf{L}_{00}\cdot\mathbf{q}_{10}+\mathbf{L}_{10}\cdot\mathbf{q}_{00}+\mathbf{u}_{00}\cdot\nabla\mathbf{M}\cdot\mathbf{q}_{00}=-\mathbf{d}p_{10} \tag{76.3 c}$$

$$(2,0)\quad \mathbf{L}_{00}\cdot\mathbf{q}_{20}+\mathbf{L}_{10}\cdot\mathbf{q}_{10}+\mathbf{L}_{20}\cdot\mathbf{q}_{00}+\mathbf{u}_{10}\cdot\nabla\mathbf{M}\cdot\mathbf{q}_{00} +\mathbf{u}_{00}\cdot\nabla\mathbf{M}\cdot\mathbf{q}_{10}=-\mathbf{d}p_{20} \tag{76.3 d}$$

$$(1,1)\quad \mathbf{L}_{00}\cdot\mathbf{q}_{11}+\mathbf{L}_{10}\cdot\mathbf{q}_{01}+\mathbf{L}_{01}\cdot\mathbf{q}_{10}+\mathbf{L}_{11}\cdot\mathbf{q}_{00} +\mathbf{u}_{01}\cdot\nabla\mathbf{M}\cdot\mathbf{q}_{00}+\mathbf{u}_{00}\cdot\nabla\mathbf{M}\cdot\mathbf{q}_{01} +\mathscr{R}_{00}\mathbf{F}\cdot\mathbf{q}_{10}=-\mathbf{d}p_{11} \tag{76.3 e}$$

where $\langle\mathbf{q}_{mn}\cdot\mathbf{q}_{00}\rangle=0$, $m^2+n^2\neq 0$, $\mathbf{q}_{mn}\in B$ and

$$\mathbf{L}_{mn}=-\mathscr{R}_{mn}\begin{bmatrix}0&0&0&0\\0&0&0&0\\0&0&0&1\\0&0&1&0\end{bmatrix}\equiv-\mathscr{R}_{mn}\tilde{\mathbf{L}}\,. \tag{76.4}$$

The lowest order problem is self-adjoint, $\mathbf{L}_{00}=\mathbf{L}_{00}^T-\mathbf{1}\nabla^2-\mathscr{R}_{00}\tilde{\mathbf{L}}$. Hence solvability in the two-parameter expansion does not oblige one to compute a separate adjoint.

Eq. (76.3 c) is solvable if

$$-\mathscr{R}_{10}\langle\mathbf{q}_{00}\cdot\tilde{\mathbf{L}}\cdot\mathbf{q}_{00}\rangle+\langle\mathbf{q}_{00}\cdot(\mathbf{u}_{00}\cdot\nabla)\mathbf{M}\cdot\mathbf{q}_{00}\rangle=0\,.$$

The second term of this equation vanishes and implies that $\mathscr{R}_{10}=0$. Eq. (76.3d) is solvable if

$$-\mathscr{R}_{20}\langle\mathbf{q}_{00}\cdot\tilde{\mathbf{L}}\cdot\mathbf{q}_{00}\rangle+\langle\mathbf{q}_{00}\cdot(\mathbf{u}_{10}\cdot\nabla)\mathbf{M}\cdot\mathbf{q}_{00}\rangle+\langle\mathbf{q}_{00}\cdot(\mathbf{u}_{00}\cdot\nabla)\mathbf{M}\cdot\mathbf{q}_{10}\rangle=0\,.$$

It is possible at this level of generality to prove that $\mathscr{R}_{20}>0$ (see Exercise 76.1). It follows that to lowest order

$$\mathscr{R}(\varepsilon,\xi)-\mathscr{R}_L(\xi)\sim\mathscr{R}_{11}\xi\varepsilon+\mathscr{R}_{20}\varepsilon^2 \tag{76.5}$$

and the bifurcation is two-sided when $\mathscr{R}_{11}\neq 0$.

Many difficult problems can be simplified, at a cost in generality, by the two parameter expansions. Applications of this method are given in the guided exercises 76.2 and 76.3.

Exercise 76.1: Show that $\mathscr{R}_{10}=0$,

$$\mathscr{R}_{11}\langle\mathbf{q}_{00}\cdot\tilde{\mathbf{L}}\cdot\mathbf{q}_{00}\rangle=\mathscr{R}_{00}\langle g(\theta_{10}w_{00}-w_{10}\theta_{00})+\mathscr{R}_{01}\langle\theta_{10}w_{00}+\theta_{00}w_{10}\rangle$$

and

$$\mathscr{R}_{20}\langle\mathbf{q}_{00}\cdot\tilde{\mathbf{L}}\cdot\mathbf{q}_{00}\rangle=\langle\mathbf{q}_{00}\cdot(\mathbf{u}_{00}\cdot\nabla)\mathbf{M}\cdot\mathbf{q}_{10}\rangle=-\langle\mathbf{q}_{00}\cdot\mathbf{L}_{00}\cdot\mathbf{q}_{00}\rangle\,.$$

Prove that $\langle \mathbf{q}\cdot\mathbf{L}_{00}\cdot\mathbf{q}\rangle=0$ is a maximum when $\mathbf{q}=\mathbf{q}_{00}$ and deduce that $\mathscr{R}_{20}>0$ (*Hint:* Consult the inequality B4.3). Show that $\mathbf{u}_{00}=\boldsymbol{\delta}\chi_{00}$; that the vertical component of the curl of $\mathbf{u}_{00}\cdot\nabla\mathbf{u}_{00}$ vanishes and that $\mathbf{u}_{10}=\boldsymbol{\delta}\chi_{10}$.

Exercise 76.2: Consider the bifurcation problem stated in Eqs. (76.3a) through (76.3e). Suppose that at lowest order the solution is represented by superposition of eigenfunctions of $\nabla_2^2\phi+a^2\phi=0$ having the same eigenvalue $|\mathbf{k}_\nu|^2=a^2$:

$$\phi(x,y)=\sum_{\substack{\nu=-N\\ \nu\neq 0}}^{N} C_\nu e^{i\mathbf{k}_\nu\cdot\mathbf{r}},\qquad C_\nu=\bar{C}_{-\nu},\qquad \sum_{\substack{\nu=-N\\ \nu\neq 0}}^{N}|C_\nu|^2=1 \tag{76.6}$$

where $\mathbf{r}=\mathbf{e}_x x+\mathbf{e}_y y$.

Derive the results listed below: The smallest value of $\mathscr{R}_{00}$ is given by $\mathscr{R}_{00}^2=f^3/a^2$ where $f=\pi^2+a^2$. The eigenfunctions of (76.3a) are given by

$$\begin{bmatrix}\chi_{00}\\ \theta_{00}\end{bmatrix}=A\begin{bmatrix}1\\ f/\mathscr{R}_{00}\end{bmatrix}\phi(x,y)\cos\pi z\qquad(-\tfrac{1}{2}\leqslant z\leqslant\tfrac{1}{2})$$

where $A^2=2/a^2f(1+\mathscr{P})$. The solutions of (76.3b) are given by

$$\begin{bmatrix}\chi_{10}\\ \theta_{10}\end{bmatrix}=A^2\sum_{\nu=-N}^{N}\sum_{l=-N}^{N}C_\nu C_l\begin{bmatrix}\chi_{10}^{\nu l}\\ \theta_{10}^{\nu l}\end{bmatrix}e^{i(\mathbf{k}_\nu+\mathbf{k}_l)\cdot\mathbf{r}}\sin 2\pi z$$

where

$$\begin{bmatrix}\chi_{10}^{\nu l}\\ \theta_{10}^{\nu l}\end{bmatrix}=\left(\theta_{00}^{\nu l}\begin{bmatrix}\mathscr{R}_{00}\\ \gamma_{\nu l}^2\end{bmatrix}+\chi_{00}^{\nu l}\begin{bmatrix}\gamma_{\nu l}/\beta_{\nu l}\\ \mathscr{R}_{00}\end{bmatrix}\right)\Big/\Delta_{\nu l}\,,$$

$$\begin{bmatrix}\chi_{00}^{\nu l}\\ \theta_{00}^{\nu l}\end{bmatrix}=-\pi a^2 f^{1/2}\begin{bmatrix}af^{1/2}\sin^2 a_{\nu l}\\ \tfrac{1}{2}\mathscr{P}(1-\cos a_{\nu l})\end{bmatrix},$$

$$\mathbf{k}_\nu\cdot\mathbf{k}_l=a^2\cos a_{\nu l}\,,$$

$$\beta_{\nu l}=2a^2(1+\cos a_{\nu l})\,,$$

$$\Delta_{\nu l}=\beta_{\nu l}\mathscr{R}_{00}^2-\gamma_{\nu l}^3\,,$$

and

$$\gamma_{\nu l}=4\pi^2+\beta_{\nu l}\,.$$

The following relations hold

$$\langle\theta_{10}w_{00}-\theta_{00}w_{10}\rangle=0\,,$$

$$a^3f^{1/2}\mathscr{R}_{11}=AI(g)\mathscr{R}_{00}\sum\sum\sum_{l,\nu,n=-N}^{N}C_\nu C_l C_n(a^2\theta_{10}^{\nu l}-f^2 w_{10}^{\nu l}/\mathscr{R}_{00})\,\delta(\mathbf{k}_\nu+\mathbf{k}_l+\mathbf{k}_n)$$

where

$$w_{10}^{\nu l}=\beta_{\nu l}\chi_{10}^{\nu l}$$

and

$$I(g)=\int_{-1/2}^{1/2}g(z)\cos\pi z\sin 2\pi z\,dz$$

$\mathscr{R}_{11}=0$ whenever $\mathbf{k}_\nu+\mathbf{k}_l+\mathbf{k}_n\neq 0$ for all possible combinations of allowed $\mathbf{k}$ vectors. $\mathscr{R}_{11}=0$ when $N=2$ and $N=4$. For hexagonally symmetric functions with $N=3$, $\nu\neq l$, $C_1=C_2=C_3=1/\sqrt{6}$ and $a_{\nu l}=2\pi/3$,

$$\mathscr{R}_{11}=\sqrt{\tfrac{27}{8}}\,\pi aAf^{3/2}I(g)\frac{\mathscr{P}(5\pi^2+2a^2)-f}{(4\pi^2+a^2)^3-f^3}\,.$$

Two-sided hexagonal bifurcation is not possible when the heat sources are given by an odd function of z.

$$\mathscr{R}_{20}=\frac{A^2\pi^2 f^{1/2}a^3}{16}\left\{\frac{2\mathscr{P}^2}{\pi^2}+\frac{3[(\pi^2+a^2)(4\pi^2+a^2)+2(\pi^2+a^2)^2\mathscr{P}+\mathscr{P}^2(4\pi^2+a^2)^2]}{[(4\pi^2+a^2)^3-(\pi^2+a^2)^3]}\right.$$
$$\left.+\frac{[(4\pi^2+3a^2)^2\mathscr{P}^2\sqrt{3}+2\mathscr{P}(\pi^2+a^2)^2+(\pi^2+a^2)(4\pi^2+3a^2)]}{[(4\pi^2+3a^2)^3-3(\pi^2+a^2)^3]}\right\}$$

When $a^2=\pi^2/2$,

$$\mathscr{R}_{20}=\frac{\sqrt{3}}{64}\pi^4A^2\{0.2835+0.1826\mathscr{P}+2.757\mathscr{P}^2\}, \tag{76.7}$$

and

$$\mathscr{R}_{11}=\frac{3}{52\sqrt{3}}\pi^3AI(g)\{4\mathscr{P}-1\}. \tag{76.8}$$

On the subcritical branch of the two-sided bifurcation the fluid will rise in the center of the hexagonal cell when $\mathscr{P}>\frac{1}{4}$ and will sink when $\mathscr{P}<\frac{1}{4}$.

Exercise 76.3 (Liang, Vidal, Acrivos, 1969): A circular cylinder is filled with an OB fluid whose viscosity varies with temperature. The cylinder wall and the top and bottom of the cylinder are shear-stress free. The temperature $\theta=0$ at the top and bottom of the cylinder and $\partial\theta/\partial r=0$ on the vertical side-walls. Show that there is an axisymmetric bifurcating solution which bifurcates to both sides (*Hint:* show that $\mathscr{R}_{11}\neq 0$).

Exercise 76.4: The aim of this exercise and the next is to show how to study bifurcation in a container filled with a fluid saturated porous material by the two-parameter method. This method allows one to relax the condition used in § 73 that the sidewalls are insulated. Now we shall require that the sidewalls be an ideal conducting material so that $\theta|_S=0$.

Show that at lowest order in the two-parameter expansion

$$\nabla^4\chi_{00}+\mathscr{R}_{00}\nabla_2^2\chi_{00}=0, \tag{76.9a}$$

$$\chi_{00}=\partial^2\chi_{00}/\partial z^2=0 \quad \text{on} \quad z=0,1, \tag{76.9b}$$

$$\chi_{00}=\partial\chi_{00}/\partial n=0 \quad \text{on} \quad S. \tag{76.9c}$$

Show that planform eigenfunctions will not separate the variables in (76.9). Show that a different separation into a Fourier sin series

$$\chi_{00}=\sum\bar{\bar{\chi}}_n(x,y)\sin n\pi z$$

where

$$(\nabla_2^2-n^2\pi^2)^2\bar{\bar{\chi}}_n+\mathscr{R}_{00}\nabla_2^2\bar{\bar{\chi}}_n=0, \tag{76.10a}$$

$$\bar{\bar{\chi}}_n=\partial\bar{\bar{\chi}}/\partial n=0 \quad \text{on} \quad S \tag{76.10b}$$

may be justified from the form of Eqs. (76.9).

Exercise 76.5: Consider problem (76.10) in an infinitely long container when the bifurcating solution is independent of y, $\bar{\bar{\chi}}_n=\bar{\bar{\chi}}_n(x)$ and

$$\bar{\bar{\chi}}_n=0 \quad \text{and} \quad \bar{\bar{\chi}}_n'=0$$

on the two parallel sidewalls $x=0$ and $x=l$ of the container. Show that

$$\bar{\bar{\chi}}_n(x)=[\alpha_4\sin\alpha_3 l-\alpha_3\sin\alpha_4 l][\sin\alpha_1 l\sin\alpha_2 x-\sin\alpha_2 l\sin\alpha_1 x]$$
$$-[\alpha_2\sin\alpha_1 l-\alpha_1\sin\alpha_2 l][\sin\alpha_3 l\sin\alpha_4 x-\sin\alpha_4 l\sin\alpha_3 x]$$

where $(i=1,2,3,4)$

$$\alpha_i=\pm\left[\frac{\mathscr{R}_{00}}{2}-n^2\pi^2\pm(-\mathscr{R}_{00}n^2\pi^2+\mathscr{R}_{00}^2/4)^{1/2}\right]^{1/2}$$

and

$$\bar{\bar{\chi}}_n'(l)=0\,.$$

Find the smallest eigenvalue $\mathscr{R}_{00}$. Compute $\mathscr{R}_{11}$.

§ 77. Two-Sided Bifurcation in a Laterally Unbounded Layer of Fluid

All experiments on convection are in bounded domains. Most of the analysis of convection is for unbounded domains. What then is the relation of the mathematical formulation of the problem on the unbounded layer to the physical problem? Two answers to this question are of interest here: (1) The difference between convection in a shallow container and an infinite layer tends to zero as the layer is made more shallow. (2) The uncertainties about boundary conditions at the finite boundaries of the convecting fluid make the laterally infinite layer a better model of the problem of convection in bounded domains. It is of interest to elaborate answer (1) by way of explanation of spectral crowding in ever more shallow containers. The degeneracy of the spectrum in unbounded layers means that the problem is basically more complicated in unbounded domains than in bounded domains where the eigenvalues are discrete and of finite (at most) multiplicity. This complication in mathematics means that one can study more complicated physics. This fact is the basis for answer (2). In thinking about (2) it is useful to maintain a distinction between the colloquial and mathematical definition of a bounded domain. The latter definition requires that one pose definite boundary conditions on the boundaries of some precisely delimited region of space. Our analysis of convection in containers is a good example of analysis of problems in a domain bounded in the mathematical sense. Fluid layers of finite extent with unconfined edges are bounded in the colloquial sense. The conditions at the edges are chaotic and unknown. Analysis should account for the fact that some conditions are unknown; one way is to extend the physical problem onto an unbounded domain where there are a greater variety of disturbances which do not artificially suppress possibly chaotic motions at the edges.

In the problem of stability of two-sided bifurcation of conduction into convection there can be a substantial difference between closed containers and horizontally infinite fluid layers. Mathematically, this difference arises as a result of the degeneracy of the spectrum in the unbounded case. Similar differences exist between containers for which $\mathscr{R}_0(\tilde{a}^2)$ is a simple eigenvalue of the governing partial differential equations and those for which $\mathscr{R}_0(\tilde{a}^2)$ is a multiple eigenvalue. When $\mathscr{R}_0(\tilde{a}^2)$ is simple, the supercritical branch of the two-sided bifurcation is stable to small disturbances. This branch can be unstable in domains with multiple eigenvalues and is unstable in hexagonal containers and fluid layers. These differences may be most easily understood by careful study of Fig. 77.1.

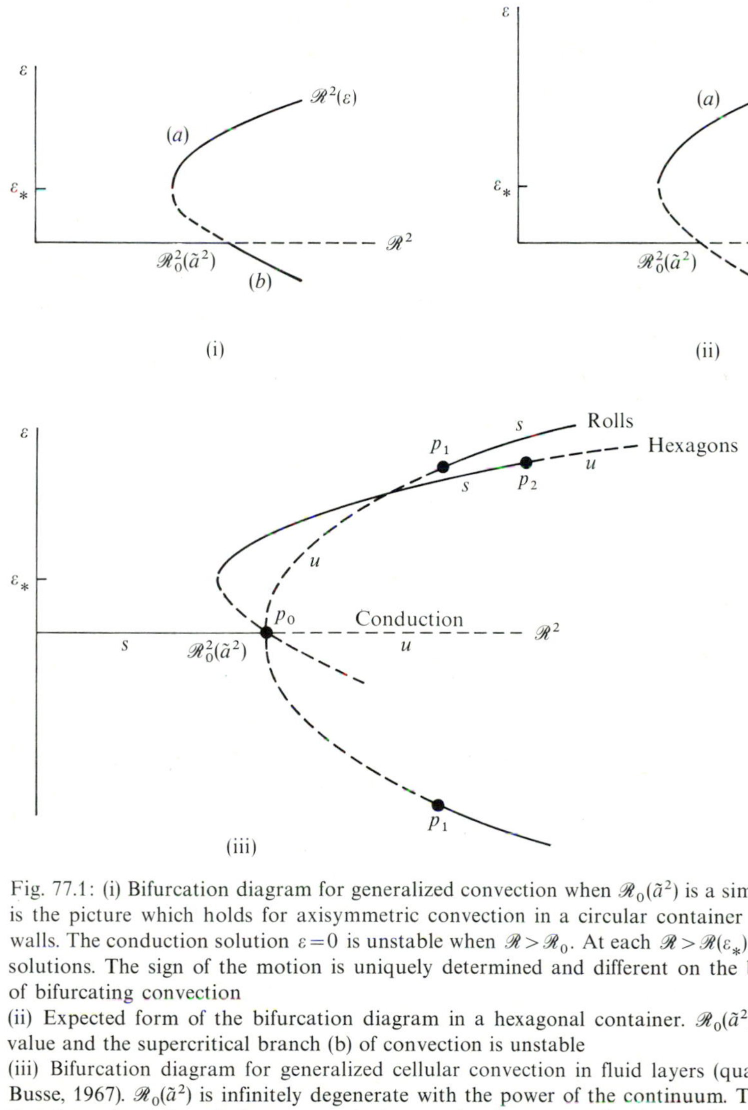

Fig. 77.1: (i) Bifurcation diagram for generalized convection when $\mathscr{R}_0(\tilde{a}^2)$ is a simple eigenvalue. This is the picture which holds for axisymmetric convection in a circular container with insulated sidewalls. The conduction solution $\varepsilon=0$ is unstable when $\mathscr{R}>\mathscr{R}_0$. At each $\mathscr{R}>\mathscr{R}(\varepsilon_*)$ there are two stable solutions. The sign of the motion is uniquely determined and different on the branches (a) and (b) of bifurcating convection
(ii) Expected form of the bifurcation diagram in a hexagonal container. $\mathscr{R}_0(\tilde{a}^2)$ is a double eigenvalue and the supercritical branch (b) of convection is unstable
(iii) Bifurcation diagram for generalized cellular convection in fluid layers (qualitative sketch after Busse, 1967). $\mathscr{R}_0(\tilde{a}^2)$ is infinitely degenerate with the power of the continuum. The analysis holds in the class of rotationally invariant solutions under the group of orthogonal rotations of the plane which divide 360° into equal parts. The bifurcation curve $\mathscr{R}(\varepsilon)$ for rolls is an even function of ε. In the Bénard problem $\mathscr{R}_1=0$ for all solutions; the points $p_1 \to p_0$, $p_2 \to p_0$. In the Bénard problem hexagons are unstable and rolls are stable

(a) Spectral Crowding

We will discuss spectral crowding in the context of the problems of convection in containers of porous materials with insulated side walls which was studied in § 73. That problem separates at lowest order and we can confine our considerations to an examination of the properties of the planform eigenfunction ϕ satisfying (73.7). This eigenfunction is unique and a^2 is simple in almost all bounded containers. In containers with symmetric planforms it is possible to have eigenvalues a^2 of higher multiplicity. In this case the eigenfunctions $\chi_0=\bar{\bar{\chi}}(z)\phi(x,y)$ and $\theta_0=\bar{\bar{\theta}}(z)\phi(x,y)$ are not uniquely determined by the eigenvalue $\mathscr{R}_0$ of (73.4) and $\mathscr{R}_0$ may be regarded as an eigenvalue of (73.2) of higher multiplicity. It is also possible for the eigenvalue $\mathscr{R}_0$ of (73.4) to have more than one eigenfunction pair $(\bar{\bar{\chi}},\bar{\bar{\theta}})$ when a^2 is a simple eigenvalue of (73.7). This case, however, seems not to be of great importance in cellular convection since the domain of the functions $\bar{\bar{\theta}}$ and $\bar{\bar{\chi}}$ is always bounded, $0\leqslant z\leqslant 1$.

The main interest in stability is in the smallest eigenvalue

$$\min_{a^2}\mathscr{R}_0(a^2)=\mathscr{R}_0(\tilde{a}^2).$$

The solutions which bifurcate from this eigenvalue are the only ones which can bifurcate from stable conduction. The conduction solution is unstable when $\mathscr{R}>\mathscr{R}_0(\tilde{a}^2)$.

When viewed in the context of the spectral problems (71.1) and (71.2), $\mathscr{R}_0(\tilde{a}^2)$ is a critical value and $\sigma(\mathscr{R}_0)=0$ is the smallest of the eigenvalues of (71.1) and (71.2). Consider the values of σ in a rectangular container with insulated sidewalls. For these $a^2=\pi^2\left(\frac{m^2}{l^2}+\frac{n^2}{d^2}\right)$. If, in addition, $\xi=0$ and $f=1$, then it is easily shown that

$$\sigma_{\hat{n}mn}(\mathscr{R})=-\frac{\mathscr{R}(m^2/l^2+n^2/d^2)}{(\hat{n}^2\pi^2+m^2/l^2+n^2/d^2)}+(\hat{n}^2\pi^2+m^2/l^2+n^2/d^2).$$

The smallest of these eigenvalues has $\hat{n}=1$. The smallest of the critical values $\mathscr{R}_0$, $\sigma_{mn}(\mathscr{R}_0)=0$, is given by

$$\mathscr{R}_0=\min_{m,n}\frac{(\pi^2+m^2/l^2+n^2/d^2)^2}{m^2/l^2+n^2/d^2}=\frac{(\pi^2+\bar{m}^2/l^2+\bar{n}^2/d^2)^2}{\bar{m}^2/l^2+\bar{n}^2/d^2}\geqslant 4\pi^2.$$

Now $\sigma_{1\bar{m}\bar{n}}(\mathscr{R}_0)=0$ is an isolated simple eigenvalue of (71.1). However, in shallow boxes, those with large values of l and d, the eigenvalues crowd together and in the limit,

$$\lim_{\substack{l\to\infty\\ d\to\infty}}[\sigma_{1mn}(\mathscr{R})-\sigma_{1\bar{m}\bar{n}}(\mathscr{R}_0)]\to 0.$$

Hence, by increasing the lateral dimensions of the container it is possible to inject an arbitrarily large number of eigenvalues $\sigma_{1mn}(\mathscr{R})$ into any neighborhood, no matter how small, of the principal eigenvalue $\sigma_{1\bar{m}\bar{n}}(\mathscr{R}_0)=0$. Perturbations in the spectrum of course must be expected because the spectrum in the real physical system is only modeled by the ideal mathematical. If the eigenvalues of the ideal mathematical system are sufficiently dense near their principal value, such deviations will cause the eigenvalues to coalesce and change their order. In this sense the crowding of the spectrum can lead to effective "loss of simplicity" even when $\sigma(\mathscr{R}_0)=0$ is ideally simple. Hence, in shallow containers it is probably better to abandon the concept of a bounded domain and to model the physical system with a laterally infinite layer. In the infinite layer, the degeneracy is infinitely great. There are, in fact, at least as many solutions, for a given a^2, as there are modes of vibration with frequency $\omega^2=a^2$ of a tight membrane stretched over the whole (x,y) plane.

(b) Cellular Convection

Convection in fluid layers is generally considered to be cellular convection. There is no universally accepted definition of cellular convection. Some mathematicians regard cellular convection as a manifestation of spatially periodic solutions of the OB equations or the DOB equations. Kirchgässner and Kielhofer (1972) define stationary solutions of the nonlinear Bénard problem as the doubly periodic (in x and y) solutions which leave the nonlinear equations invariant under the group of orthogonal rotations of the plane (frame-indifferent solutions). They show that the planforms of allowed solutions are rolls, hexagons, rectangles or triangles and no others. These planforms are discussed by Chandrasekhar (1961) and, from a different point of view, by Stuart (1964). Some mathematical studies consider only those sub-classes of these invariant forms which allow one to study bifurcation at a simple eigenvalue. For example, one can demonstrate simplicity among doubly periodic rectangular solutions for almost all periods (Yudovich 1966; Rabinowitz, 1968) and among solutions possessing full hexagonal symmetry (Yudovich, 1966). The demonstration of simplicity in these cases is very much like the one required for the study of the spectrum for convection in the containers considered in § 73. The planform is restricted by symmetry requirements; it is first necessary to show that the planform is uniquely determined once a^2 is given. For example, a^2 is a double eigenvalue in the class of solutions of partial hexagonal symmetry and a simple eigenvalue among solutions of full hexagonal symmetry. It is then necessary to show that when a^2 is given, $\mathscr{R}_0(a^2)$ is a simple eigenvalue of the governing partial differential equations. This will be true if $\mathscr{R}_0(a^2)$ is a simple eigenvalue of the ordinary differential equations, like (73.8), which arise after separating the variables. By confining the solutions to planforms in the group which leave the nonlinear equations invariant and which in addition make the planform unique, it is possible to study special cases of doubly periodic solutions with the theory of bifurcation at a simple eigenvalue.

It is possible to study the stability of bifurcating solutions by artificially restricting allowed disturbances to those in which $\mathscr{R}_0(a^2)$ is a simple eigenvalue.

We saw in § 73 (c) that this kind of restriction is not good; it leads to the conclusion that the supercritical branch of solutions is stable; this conclusion is true in the restricted class but is false when other allowed disturbances not satisfying the artificial symmetry requirement are considered.

When the allowed eigenfunctions for doubly periodic convection are unrestricted the spectrum degenerates, and analysis of bifurcation and stability becomes more complicated. The problem of unrestricted doubly-periodic bifurcation into convection has been studied by Schwiderski (1972). The stability of convection was not considered in his study.

The most comprehensive results about stability and bifurcation into convection so far are due to Busse (1962, 1967) and to Schlüter, Lortz and Busse (1965). These authors do not require periodicity and at first they do not require invariance of the nonlinear equations to any group of transformations. They start with all the solutions of the form (76.6) which are also eigenfunctions of the horizontal Laplacian with eigenvalue a^2. The form (76.6) defines a quasi-periodic function and the number of eigenfunctions belonging to a given value a^2 can be infinite; different sets of eigenfunctions with $|\mathbf{k}_\nu|^2 = a^2$ correspond to different N, c_ν and $\mathbf{k}_\nu$. They attempt to determine the linear combinations of eigenfunctions which the nonlinear equations will allow by solvability conditions at higher orders. They are not able to carry out their analysis in full generality but are forced to confine their attention to a sub-class of "regular" solutions. These solutions can be described as solutions which are invariant to rotations through angles α which divide the circle into equal sectors ($\alpha = 45°$ corresponds to octagons, etc.). Since the regular solutions are defined by proper orthogonal rotations in the plane, and since the nonlinear equations of convection are invariant to orthogonal rotations, the regular solutions are "frame indifferent". The regular solutions may be represented by wave vectors $\mathbf{k}_\nu$ spaced evenly on the circle of radius $a = |\mathbf{k}_\nu|$. Within this class they show that the solvability requirements will allow only certain linear combinations to perturb. Their analysis is not carried beyond 2nd order in ε; convergence proofs have not been given as yet and it has not yet been demonstrated that the problem of small divisors which frequently arises when quasiperiodic representations are used is avoided in the class of "regular" solutions. Schlüter, Lortz and Busse find that in the Bénard problem rolls are the only possible stable form of "regular" convection. The three-dimensional forms of "regular" convection are all unstable.

The problem of two-sided bifurcation into generalized convection was treated by Busse (1962) and again by Busse (1967) using the method of Taylor series in two parameters. Among the many results achieved in these studies we wish here to emphasize one. Of all the many different forms of two-sided convection which can bifurcate, the only possible stable forms near criticality are those for which three $\mathbf{k}$ vectors close to form an equilateral triangle, $\mathbf{k}_1 + \mathbf{k}_2 + \mathbf{k}_3 = 0$. The only possibility in this class which was studied by Busse are the hexagons (73.23) and of these the only stable form is the hexagon (73.25) with full symmetry. Moreover, the supercritical hexagon is also unstable and the subcritical hexagon is unstable near the point of bifurcation. Using group theoretic methods. The subcritical hexagon with full symmetry is stable to small disturbances but not near criticality. The unstable subcritical hexagon recovers stability when the bifurcating curve turns around the bend (see Fig. 77.1)[5].

Stuart (1964) has considered the problem of cellular convection from a different point of view. He points out that certain cellular solutions, like some of those given by Chandrasekhar (1961) and in other early works, do not correspond to cellular motions which are observed in experiments. He notes that in the cells which are observed in experiments the cell walls are impermeable and the vertical velocity is one-signed there. He suggests that these two properties define a cell in the observational sense. In Fig. 77.2 two systems of cells are outlined by solid lines and dotted lines. The cell center for the cell outlined by the solid line is on a point of triple intersection of dotted lines. The cell center for cell outlined by the dotted line is on a point of triple intersection of the solid line. Hence, there are two centers, one on solid lines and one on dotted lines; but only one of these, the solid cell, is a cell in Stuart's observational sense. The solid cell and dotted cell are both eigenfunctions of partial hexagonal symmetry but only the solid cell has full hexagonal symmetry. Theoretically, it is not proper to discard solutions which seem unphysical. We should expect that all cells which exist *and are stable* can be observed.

(c) Stability and the Sign of the Motion in Cellular Convection

One of the most interesting differences between bounded and unbounded domains can be framed as a selection criterion for the sign of the motion in cellular convection. This criterion follows from the fact that only a part of one branch of convection is stable in problems of two-sided bifurcation into cellular con-

[5] The first attempt to explain the tendency for hexagonal cells to form in nonlinear cellular convection influenced by variations in viscosity is due to Palm (1960). Similar, but more complete results were given by Segel and Stuart (1962), Palm and Øiann (1964) and Segel (1966). These authors considered the stability problem in a more restricted class of disturbances than Busse (1962). All these authors seem to agree that only one branch of hexagonal convection is stable and that for larger values of $\mathscr{R} > \mathscr{R}_L$ both hexagons and rolls and then rolls alone are stable. Busse (1962, 1967) derived an interesting variational principle for a functional of the participation factors C_v which arise when a^2 is an eigenvalue of higher multiplicity. His variational principle gives the forms of convection allowed by the nonlinear equation (through $O(\varepsilon^2)$) and a criterion for the stability of the allowed forms. Palm (1975) has given an interesting and readable alternative formulation of Busse's variational principle. The Palm (1975) paper and the review by Segel (1966) are good sources for references to applied mathematical literature on this problem.

Graham's (1933) orginal explanation of hexagonal convection uses the packing property of hexagons. He report that "Hot water was poured into a cigarette tin to the depth of about 1 cm; when the disturbances had died away a little cold milk was carefully poured into the water. The cold milk sank to the bottom forming a white layer below and leaving clear water above. If now the free surface of the water was cooled by blowing upon it, a number of round holes appeared in the milk layer ... clearly demonstrating the descending columns.

"The increase in the number of centres affords a reason for the formation of hexagons. With a perfectly uniform layer, from consideration of symmetry, the cells will at first be round, of equal diameters and uniformly distributed. As the number of centres increases a state will be reached when the cells touch, each ring then having six equal rings symmetrically disposed around it. Any closer packing can now only take place by the elimination of the vacant space between the rings; there will be a further squeezing together and a uniform hexagonal pattern will be formed."

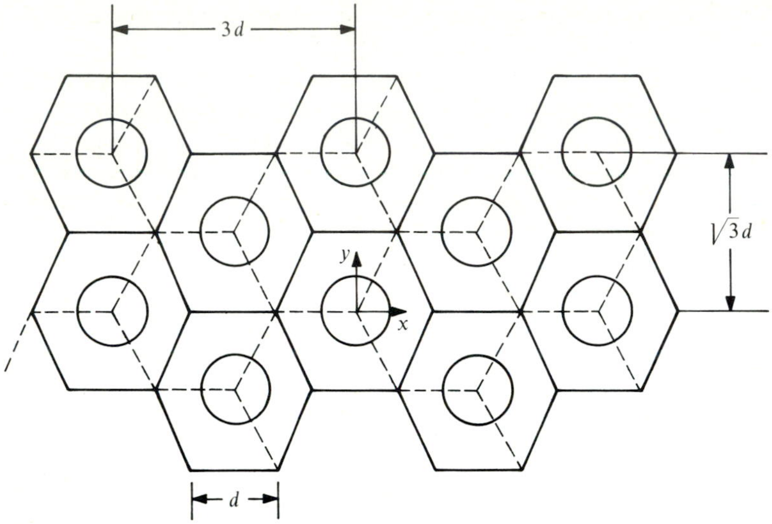

Fig. 77.2: Hexagonal convection. Let $d=4\pi/3a$ be the side of a hexagonal cell with overall wave number a. Functions of partial hexagonal symmetry are invariant to translations of $d\sqrt{3}$ in y and $3d$ in x and to rotations of $2\pi/3$. The solid lines and the dotted lines both fill the plane with hexagons. The circled regions represent sinking fluid and the solid hexagons represent cells with one-signed vertical velocity on cell walls in agreement with experiment (after Stuart (1964), Fig. 3 and Segel (1966), Fig. 7). Both sets of cells have partial hexagonal symmetry. The center of the solid hexagon is on dotted lines; the center of dotted hexagon is on the solid line. The solid hexagon is an even function of x and y and has full hexagonal symmetry

vection. The criterion does not apply to bounded domains where two branches of convection, associated with different signs for the motion, can be stable. It will be convenient to frame our discussion of this point in terms of the influence of a temperature-dependent viscosity on the bifurcating motion.

It has been suggested by Graham (1933) that the direction of flow in cellular convection is determined by the variation of the viscosity with temperature, and is such that the motion at the center of the cell is in the direction of increasing kinematic viscosity. In liquids the viscosity decreases as the temperature increases, $f'<0$; in gases $f'>0$. This suggestion was strengthened by Tippelskirch's (1956) experiment using liquid sulphur (the viscosity of which increases sharply with temperature in the range 153°—190 °C while decreasing everywhere else) in which flow reversal was noted when the temperature exceeded 153°.

We shall now show that this flow reversal would also be expected in a bounded domain but that the direction of flow need not be uniquely determined by stability.

Consider the problem of two-sided bifurcation of conduction of a DOB fluid into convection when the fluid saturated porous material is enclosed by a container with insulated side-walls. Suppose $\mathscr{R}_0(\tilde{a}^2)=\min_{a^2}\mathscr{R}_0(a^2)$ is a simple eigenvalue of (73.2) if $\tilde{a}^2$ is a simple eigenvalue of (73.7). We expect that $\tilde{a}^2$ will usually be simple when the cross-section A lacks symmetry. In symmetric containers, like the hexagonal container, or the ring container or the rectangular container, $\tilde{a}^2$ may have a multiplicity greater than one. Assuming simplicity we have, from (73.11) with $G'=1$, $\bar{\bar{\theta}}^*=\bar{\bar{\theta}}$, $\bar{\bar{\chi}}^*=\bar{\bar{\chi}}$, that

$$\mathscr{R}_1 = \frac{\int_0^1 f'\bar{\bar{\theta}}(\frac{1}{2}(D\bar{\bar{\chi}})^2 + a^2\bar{\bar{\chi}}^2)\,dz}{2\mathscr{R}_0\int_0^1 \bar{\bar{\chi}}\bar{\bar{\theta}}\,dz}\;\frac{\iint_{\mathscr{A}}\phi^3\,dx\,dy}{\iint_{\mathscr{A}}\phi^2\,dx\,dy}, \tag{77.1}$$

where

$$f' = f'\left(\frac{\Delta T}{T_0} + (1-z)\right).$$

The functions $\bar{\bar{\chi}}$ and $\bar{\bar{\theta}}$ are normalized by the last of conditions (72.2):

$$\langle|\mathbf{q}_0|^2\rangle = \langle((D\bar{\bar{\chi}})^2\,|\nabla_2\phi|^2 + (a^4\bar{\bar{\chi}}^2 + \bar{\bar{\theta}}^2)\phi^2\rangle = 1\,. \tag{77.2}$$

One consequence of (73.8 b) is that

$$\int_0^1 \bar{\bar{\chi}}\bar{\bar{\theta}}\,dz = \frac{1}{\mathscr{R}_0 a^2}\int_0^1((D\bar{\bar{\theta}})^2 + a^2\bar{\bar{\theta}}^2)\,dz\,. \tag{77.3}$$

The normalization (77.2) and the differential equations (73.7) and (73.8) do not fix the sign of the motion. If $(\bar{\bar{\chi}}(z)\,\phi(x,y), \bar{\bar{\theta}}(z)\,\phi(x,y))$ are solutions, then $(-\bar{\bar{\chi}}(z)\,\phi(x,y), -\bar{\bar{\theta}}(z)\,\phi(x,y))$ are also solutions. The choice of sign does not effect the final result since the sign of the motion at lowest order is determined by $(\varepsilon\bar{\bar{\chi}}\phi, \varepsilon\bar{\bar{\theta}}\phi)$ and the direction of bifurcation by the sign of $\varepsilon\mathscr{R}_1$. We can, without loss of generality, choose the sign by requiring that $\mathscr{R}_1>0$ (provided that $\mathscr{R}_1\neq 0$). Then the solutions $(\bar{\bar{\chi}}\phi, \bar{\bar{\theta}}\phi)$ are determined uniquely and the sign of the motion at the point (x_0, y_0, z_0) is given, when ε is small, by $W(x_0, y_0, z_0;\varepsilon)\simeq\varepsilon w_0(x_0, y_0, z_0)$.

As an example of the foregoing, consider axisymmetric convection. Many different forms of axisymmetric convection can occur in cylindrical containers with different $b=$radius/height aspect ratios. Given $\tilde{a}$ we will get axisymmetric convection for every aspect ratio $b_j=$radius/height for which $J_1(\tilde{a}b_j)=0$. We have normalized $\phi(r)=J_0(\tilde{a}r)$ so that $\phi(0)=1$ and it follows that $\iint_{\mathscr{A}}\phi^3/\iint_{\mathscr{A}}\phi^2>0$ (see Table 73.1), and $\mathscr{R}_1>0$ if the sign of $(\bar{\bar{\theta}}(z), \bar{\bar{\chi}}(z))$ is chosen so that

$$\int_0^1 f'\bar{\bar{\theta}}[\tfrac{1}{2}(D\bar{\bar{\chi}})^2 + a^2\bar{\bar{\chi}}^2]\,dz\Big/\int_0^1\bar{\bar{\chi}}\bar{\bar{\theta}}\,dz>0$$

For purposes of the discussion let us suppose that $\bar{\bar{\chi}}(z)$ and $\bar{\bar{\theta}}(z)$ are very nearly of one sign. When $f'=0$, our problem reduces to the Bénard problem for a constant viscosity DOB fluid, $\bar{\bar{\chi}}(z)$ and $\bar{\bar{\theta}}(z)$ are both proportional to $\sin\pi z$, and from (73.8 b) with $G'=1$, $\mathscr{R}_0\tilde{a}^2\bar{\bar{\chi}}(z)=(\pi^2+\tilde{a}^2)\bar{\bar{\theta}}(z)$. Since $\bar{\bar{w}}(z)=a^2\bar{\bar{\chi}}(z)$, $\bar{\bar{w}}(z)$ and $\bar{\bar{\theta}}(z)$ have the same sign. For liquids, $f'<0$. Hence, $\mathscr{R}_1>0$ if $\bar{\bar{\theta}}(z)$ is negative. Then $\bar{\bar{w}}(z)$ is negative, $w_0(r,z)=\bar{\bar{w}}(z)\,\phi(r)$ is negative when $r=0$ and

$$W(0,z;\varepsilon)\simeq\varepsilon w_0(0,z)<0 \quad \text{when} \quad \varepsilon\mathscr{R}_1>0 \quad \text{(supercritical bifurcation)}$$

and

$$W(0,z;\varepsilon)\simeq\varepsilon w_0(0,z)>0 \quad \text{when} \quad \varepsilon\mathscr{R}_1<0 \quad \text{(subcritical bifurcation)}\,.$$

It follows that liquids rise in the center for subcritical solutions and sink in the center on for supercritical solutions. The motions are reversed in gases where $f'>0$.

We now combine these observations with the results of stability analysis. Referring to Fig. 77.1 we see that the subcritical axisymmetric motion is unstable when $\varepsilon<\varepsilon_*$ but that this motion regains stability when the bifurcating curve turns around the bend $\varepsilon>\varepsilon_*$. Axisymmetric motions which rise in the center alone are stable when $\mathscr{R}<\mathscr{R}_0(\tilde{a}^2)$. When $\mathscr{R}>\mathscr{R}_0(\tilde{a}^2)$ there are two stable forms of motion; the rising motion with $\varepsilon>\varepsilon^*$ just discussed, and the sinking motion on the supercritical branch of bifurcation. Both motions are stable and both are observed (see Fig. 77.3).

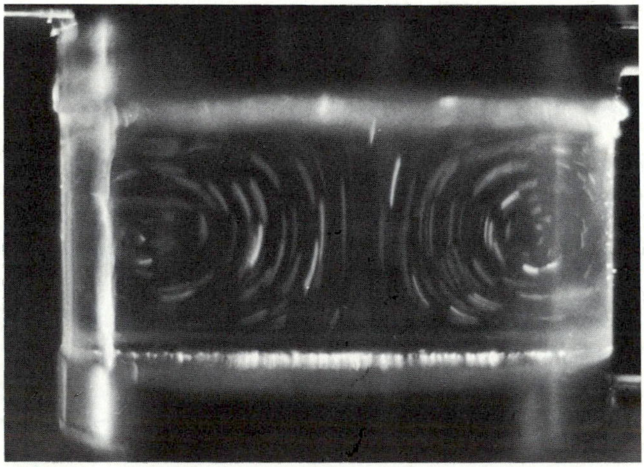

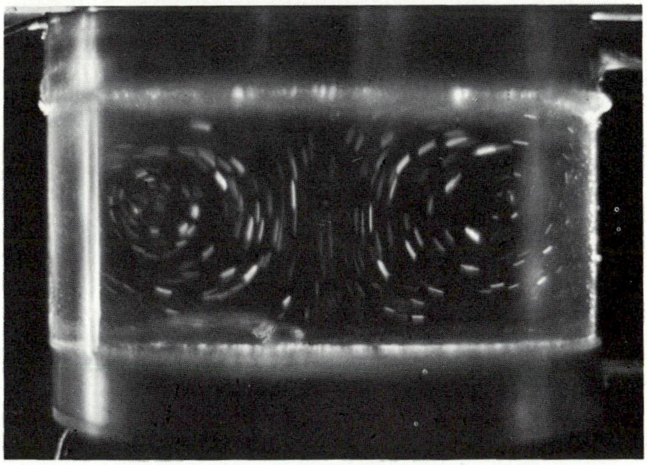

Fig. 77.3 (Liang, Vidal and Acrivos (1969)): Convection in a circular cylinder (diameter = 5.7 cm, height = 3 cm) of white oil with a temperature-dependent viscosity. Side walls are insulated and the upper surface is isothermal. (a) Downflow at the center, $\mathscr{R}^2=8000$ ($\simeq 2\mathscr{R}_L^2$); (b) Upflow at the center, $\mathscr{R}^2=8200$

I would have guessed that the same situation holds in containers of hexagonal shape. But this is evidently not the case. Hexagonal convection arises as a bifurcation from an eigenvalue of higher multiplicity and, as we showed in § 73 (c), is unstable on both sides of criticality; the only possible stable form of hexagonal convection is on the upper branch of convection with $\varepsilon > \varepsilon_*$ and rising motions in the center. The problem of cellular convection is very much like the problem of convection in a hexagonal container. The fluid rises in the center of the hexagonal cell in liquids and sinks in gases.

Similar considerations apply to the other problems of generalized convection mentioned in the introduction to this chapter (Joseph, 1971).

Exercise 77.1: Consider the set of planform eigenfunctions in the doubly periodic hexagonal lattice (periods $3d$ in x and $\sqrt{3}d$ in y). Suppose further that the allowed planforms are invariant to reflections in the planes $x=0$ and $y=0$. Show that when ε is small, fully symmetric hexagonal convection in a container is stable on the supercritical branch and is unstable on the subcritical branch. Show further that fully symmetric hexagonal convection in a laterally unbounded layer is unstable on both sides of criticality even though the eigenfunctions of partial hexagonal symmetry have been disallowed.

Addendum to Chapter X: Bifurcation Theory for Multiple Eigenvalues

In this addendum I shall give a general theory of bifurcation at an eigenvalue of multiplicity two for a simple example. I think that all of the main principles are involved in this example and that most of the extensions which are required for typical problems which arise in bifurcation analysis are straight-forward[6]. Bifurcation problems in hydrodynamics can usually be framed as operator equations of the form

$$\mathbf{L}\cdot\mathbf{U}+\mathbf{N}[\mathbf{U}]=0, \qquad \varepsilon=\langle\mathbf{U}\cdot\boldsymbol{\phi}_0\rangle$$

where $\mathbf{L}$ is a linear operator, $\mathbf{N}[\mathbf{U}]$ is an analytic operator beginning with quadratic terms, $\mathbf{N}[0]=0$ and $\boldsymbol{\phi}_0$ is the function $\mathbf{U}/\varepsilon$ in the limit $\varepsilon\to 0$. The problem

$$\mathbf{L}\cdot\boldsymbol{\phi}+\mathbf{N}[\boldsymbol{\phi},\varepsilon]=0, \qquad 1=\langle\boldsymbol{\phi}\cdot\boldsymbol{\phi}_0\rangle$$

arises from substituting $\varepsilon\boldsymbol{\phi}$ for $\mathbf{U}$. When $\mathbf{N}[\mathbf{U}]$ is a quadratic operator $\mathbf{N}[\boldsymbol{\phi},\varepsilon]=\mathbf{N}[\boldsymbol{\phi}]$ does not depend explicitly on ε; in any case the dependence of $\mathbf{N}[\mathbf{u},\varepsilon]$ on ε is not important and does not enter the bifurcation problem at the lowest significant order.

[6] We are assuming that the algebraic and geometric multiplicity of the eigenvalues are equal. If they are not equal, it is necessary to introduce generalized eigenvectors as in the case of systems of ordinary differential equations whose. Jacobian matrix reduces to block-diagonal Jordan form. If the algebraic and geometric multiplicities are different then the analysis of this addendum does not apply.

There are situations in which the appearance of multiple eigenvalues can be correlated with the appearance of secondary bifurcations (see Bauer, Keller and Reiss, 1975). The problem of secondary and repeated bifurcation is an important topic of current research in fluid dynamics.

(a) Membrane Eigenvalues Perturbed by a Nonlinear Term

We will study how the eigenfunctions $\phi(x,y;\varepsilon)$ and eigenvalues $\lambda(\varepsilon)$ of the problem

$$\nabla^2\phi+\lambda\phi+\varepsilon N[\phi]=0, \quad \phi|_{\partial\mathscr{A}}=0, \quad \langle\phi\phi_0\rangle=1 \tag{AX.1}$$

vary with ε. Here $\mathscr{A}$ is a square of side π, ϕ_0 is an eigenfunction belonging to λ_0 of the problem

$$\nabla^2\phi_0+\lambda_0\phi_0=0, \quad \phi_0|_{\partial\mathscr{A}}=0, \quad \langle\phi_0^2\rangle=1 \tag{AX.2}$$

and $N[\phi]$ is an analytic operator which may be developed into a power series around the function $\phi_0(x,y)$ and, in addition, $N[0]=0$.

In many applications $N[\phi]$ will be a power series in ϕ, $\partial_x\phi$ and $\partial_y\phi$ whose leading terms are quadratic, cubic or of still higher order. In subsection (f) we study the perturbation of (AX.2) with a linear operator $L\phi=N[\phi]$. In subsection (g) we study (AX.1) when

$$N[\phi]=(\alpha\sin 2x+\beta\sin 2y+\gamma)\phi^2\,. \tag{AX.3}$$

The eigenvalues and eigenfunctions of (AX.2) are given by

$$\phi_0=c\sin nx\sin my, \quad \lambda_0=n^2+m^2\,.$$

The first two eigenvalues and eigenfunctions are

$$\lambda_0=\lambda_{11}=2, \quad \phi_0=\phi_{11}=\pm\frac{2}{\pi}\sin x\sin y$$

and

$$\lambda_0=\lambda_{12}=\lambda_{21}=5\,, \quad \phi_0=A_0h_1+B_0h_2 \tag{AX.4}$$

where

$$h_1=\frac{2}{\pi}\sin x\sin 2y\,, \quad h_2=\frac{2}{\pi}\sin 2x\sin y\,.$$

Any linear combination $A_0h_1+B_0h_2$ is an eigenfunction of $\lambda_0=5$, and the eigenfunctions h_1 and h_2 are said to be a basis for the null space of the operator ∇^2+5. The normalizing condition of (AX.2) requires that

$$\sqrt{A_0^2+B_0^2}=\pm 1\,. \tag{AX.5}$$

Without loss of generality we may take the plus sign in (AX.5) (see Exercise AX.1).

The eigenvalue $\lambda_0=2$ is simple. The eigenvalue $\lambda_0=5$ is double. We are going to compare the problems of bifurcation arising from (AX.1) at $\lambda_0=2$ and $\lambda_0=5$. The number of solutions which bifurcate from $\phi=0$ at $\lambda=\lambda_0$ depends on the multiplicity M of λ_0 and the properties of the operator N. When

λ_0 is simple, a single one-parameter (ε) family of solutions bifurcates. The number of branching solutions from a simple eigenvalue does not depend strongly on $N[\phi]$. Bifurcation from an eigenvalue of multiplicity $M>1$ is much more complicated, even when $N[\phi]$ is a linear operator (see subsection f and Exercise (AX.3)).

We construct the bifurcating solutions of (AX.1) as a power series

$$\begin{bmatrix} \phi(x,y;\varepsilon) \\ \lambda(\varepsilon) \end{bmatrix} = \sum \varepsilon^n \begin{bmatrix} \phi_n(x,y) \\ \lambda_n \end{bmatrix}. \tag{AX.6}$$

If M' branches, $\lambda(\varepsilon)$ and $\phi(x,y;\varepsilon)$, bifurcate from $\phi=0$ and $\lambda_0=5$, then there will be M' different series solutions (AX.6), each with the same value $\lambda_0=5$.

Equations for the coefficients of the terms in (AX.6) are obtained by inserting (AX.6) into (AX.1) and setting the coefficients of the independent powers of ε to zero. Independent of the multiplicity of λ_0, we have

$$(\nabla^2+\lambda_0)\,\phi_1+\lambda_1\phi_0+N[\phi_0]=0 \tag{AX.7}$$

and for $n>1$

$$(\nabla^2+\lambda_0)\,\phi_n+\lambda_n\phi_0+(\lambda_1+N_\phi[\phi_0])\,\phi_{n-1}+P_n+\sum_{\substack{l+v=n\\ l\neq 0,1,n}} \lambda_l\phi_v=0 \tag{AX.8}$$

where $N_\phi[\phi_0]$ is the Frechet derivative of $N[\phi]$ evaluated at $\phi=\phi_0$ and P_n is a polynomial in the derivatives of ϕ which, apart from factorials, arises from repeated differentiation of the operator $N[\phi]$ with respect to ε. P_n is independent of ϕ_v for $v>n-2$;

$$P_n=P[\phi_0,\phi_1,\ldots,\phi_{n-2}]\,. \tag{AX.9}$$

When $n>0$ we must also require that

$$\phi_n|_{\partial\mathscr{A}}=0 \quad \text{and} \quad \langle\phi_0\phi_n\rangle=0\,. \tag{AX.10}$$

(b) Bifurcation from a Simple Eigenvalue

If λ_0 is simple the problems (AX.8) are solvable if and only if

$$\lambda_1+\langle\phi_0 N[\phi_0]\rangle=0 \tag{AX.11}$$

and, for $n>1$,

$$\lambda_n+\langle\phi_0 N_\phi[\phi_0]\,\phi_{n-1}\rangle+\langle\phi_0 P_n\rangle=0\,.$$

It follows that $\phi_n(x,y)$ and λ_n may be determined sequentially and uniquely and the series (AX.6) is convergent.

(c) Bifurcation from a Multiple Eigenvalue

If λ_0 is a multiple eigenvalue, we cannot expect the bifurcating branches $\lambda(\varepsilon)$, $\phi(x,y,\varepsilon)$ to vary continuously with ε unless we have selected the system of unperturbed eigenfunctions for the multiple eigenvalue in a suitable way. For the double eigenvalue $\lambda_0=5$ we have a null space of dimension two and this space is spanned by the orthogonal vectors h_1 and h_2 ($\langle h_1 h_2\rangle=0$). To solve (AX.8) when $\lambda_0=5$, the inhomogeneous terms must be orthogonal to both h_1 and h_2. In general, we shall need to choose *two* parameters (not just λ_n) at each order to guarantee solvability. In addition we must determine the number M' of analytic branches $\phi^{(\nu)}(x,y;\varepsilon)$, $\lambda^{(\nu)}(\varepsilon)$; $\nu=1,2,\ldots,M'$ which bifurcate from the double eigenvalue $\lambda_0=5$. The number M' is determined by a bifurcation equation which arises from the solvability conditions at the lowest nontrivial order. (AX.7) has solutions if and only if

$$\lambda_1\langle\phi_0 h_1\rangle+\langle h_1 N[\phi_0]\rangle=0\,, \tag{AX.12}$$

$$\lambda_1\langle\phi_0 h_2\rangle+\langle h_2 N[\phi_0]\rangle=0\,. \tag{AX.13}$$

If $\lambda_1\neq 0$ (AX.12) and (AX.13) define algebraic equations in A_0 and B_0 which determine the number M' of branches which may be initiated with starting values

$$A_0^{(\nu)},\ B_0^{(\nu)};\qquad \nu=1,2,\ldots,M' \tag{AX.14}$$

If $\lambda_1=0$ the bifurcation equation which gives M' occurs at $n=2$; for example, when $N[\phi]$ is given by (AX.3) and $\alpha=\beta=0$, then $\lambda_1=0$ but $\lambda_2\neq 0$ on two branches (see subsection h). If $\lambda_1=\lambda_2=0$ then the bifurcation equation arises at $n=3$ and so on.

(d) The Orthogonal Decomposition

We treat the two problems of determining M' and satisfying the two solvability conditions simultaneously by the orthogonal decomposition

$$\phi(x,y;\varepsilon)=A(\varepsilon)h_1+B(\varepsilon)h_2+\psi(x,y;\varepsilon) \tag{AX.15}$$

where $A(\varepsilon)$ and $B(\varepsilon)$ are analytic functions giving the projections of $\phi(x,y;\varepsilon)$ into the two-dimensional null space of ∇^2+5

$$\begin{aligned} A(\varepsilon)&=\langle h_1\phi(x,y;\varepsilon)\rangle\,,\\ B(\varepsilon)&=\langle h_2\phi(x,y;\varepsilon)\rangle \end{aligned} \tag{AX.16}$$

and $A(0)=A_0$, $B(0)=B_0$. It follows from (AX.15) and (AX.16) that $\langle h_1\psi\rangle=\langle h_2\psi\rangle=0$ [7] and

$$\phi_n(x,y)=A_nh_1+B_nh_2+\psi_n(x,y) \tag{AX.17}$$

with $\psi_0(x,y)\equiv 0$ and for $n\geqslant 1$,

$$\langle h_1\psi_n\rangle=\langle h_2\psi_n\rangle=0\,. \tag{AX.18}$$

Moreover, from (AX.1, 5, 15, 16, 17) we have

$$1=\langle\phi\phi_0\rangle=A(\varepsilon)A_0+B(\varepsilon)B_0=A_0^2+B_0^2$$

and for $n\geqslant 1$

$$A_nA_0+B_nB_0=0\,. \tag{AX.19}$$

Then we define

$$\mu=B_0/A_0 \quad (\mu\to\infty \text{ corresponds to } A_0\to 0) \tag{AX.20}$$

and from (AX.19) we find that

$$A_n=-\mu B_n\,. \tag{AX.21}$$

Using (AX.20, 21) we may reduce (AX.17) to the following:

$$\phi_0=(1+\mu^2)^{-1/2}(h_1+\mu h_2)\,, \tag{AX.22a}$$

$$\phi_n=B_n(h_2-\mu h_1)+\psi_n(x,y) \tag{AX.22b}$$

where $\psi_n(x,y)$ satisfies the orthogonality conditions (AX.18), the boundary conditions $\psi_n|_{\partial\mathscr{A}}=0$,

$$(\nabla^2+\lambda_0)\psi_1+\lambda_1\phi_0+N[\phi_0]=0 \tag{AX.22c}$$

and, for $n>1$,

$$(\nabla^2+\lambda_0)\psi_n+\lambda_n\phi_0+B_{n-1}(\lambda_1+N_\phi[\phi_0])(h_2-\mu h_1)+G_n=0 \tag{AX.22d}$$

---

[7] When $\mathscr{A}$ is a square of side π,

$$\psi(x,y;\varepsilon)=\sum_{v=1}\sum_{l=1}C_{vl}(\varepsilon)\sin vx\sin ly$$

where $C_{12}=C_{21}=0$. The same analysis holds when A is arbitrary but then $\psi(x,y;\varepsilon)$ must be expressed as a series of eigenfunctions of ∇^2 in $\mathscr{A}$. In more general problems, in which the unperturbed operator is not self-adjoint, it will be necessary to introduce an adjoint.

where

$$G_n = G_{n1} + G_{n2}\,,$$

$$G_{n1} = (\lambda_1 + N_\phi[\phi_0])\psi_{n-1}\,,$$

$$G_{n2} = P_n + \sum_{\substack{v+l=n \\ l\neq 0,1,n}} \lambda_l[B_v(h_2-\mu h_1)+\psi_v] \equiv G[\phi_0,\phi_1,\ldots,\phi_{n-2}]$$

and thus G_{n2} is independent of B_v and $\psi_v(x,y)$ for $v>n-2$.

(e) Solvability Conditions

Now we have to show how (AX.22) determines the number M' of branches $\phi^{(v)}(x,y;\varepsilon)$, $\lambda^{(v)}(\varepsilon)$, $v=1,2,\ldots,M'$ which bifurcate from the double eigenvalue λ_0. We show this by constructing the series solution (AX.6) for each of the M' branches. Our construction requires that we determine M' from an initiating bifurcation equation giving M' real roots $\mu^{(v)}$ which determine M' starting functions $\phi_0^{(v)}(x,y)$. Then given $\phi_0^{(v)}$ we must determine the unknown functions $\psi_n^{(v)}(x,y)$ and the solvability parameters $\lambda_n^{(v)}$ and $B_n^{(v)}$.

The solvability conditions (AX.12, 13) which arise from the relation

$$\langle h_1(\nabla^2+\lambda_0)\psi_n\rangle = \langle h_2(\nabla^2+\lambda_0)\psi_n\rangle = 0 \tag{AX.23}$$

are basic in the construction. When $n=1$ these equations may be expressed as

$$\lambda_1(1+\mu^2)^{-1/2} + \langle h_1 N[\phi_0]\rangle = 0\,, \tag{AX.24a}$$

$$\lambda_1\mu(1+\mu^2)^{-1/2} + \langle h_2 N[\phi_0]\rangle = 0 \tag{AX.24b}$$

where $\phi_0=(1+\mu^2)^{-1/2}(h_1+\mu h_2)$. When $n>1$ (AX.23) may be expressed as

$$\lambda_n(1+\mu^2)^{-1/2} + B_{n-1}\{\langle h_1 N_\phi[\phi_0](h_2-\mu h_1)\rangle - \mu\lambda_1\} + \langle h_1 G_n\rangle = 0\,, \tag{AX.25a}$$

$$\lambda_n\mu(1+\mu^2)^{-1/2} + B_{n-1}\{\langle h_2 N_\phi[\phi_0](h_2-\mu h_1)\rangle + \lambda_1\} + \langle h_2 G_n\rangle = 0\,. \tag{AX.25b}$$

The initiating solvability condition is obtained by eliminating λ_1 from (AX.24a,b)

$$\mu\langle h_1 N[(1+\mu^2)^{-1/2}(h_1+\mu h_2)]\rangle = \langle h_2 N[(1+\mu^2)^{-1/2}(h_1+\mu h_2)]\rangle\,. \tag{AX.26}$$

To each root $\mu^{(v)}$ of this equation there corresponds a value $\lambda_1^{(v)}$ giving the slope of the branch $\lambda^{(v)}(\varepsilon)$ of eigenvalues at $\varepsilon=0$. When $N[\phi]=\text{const}\,\phi^k$, (AX.26) is a polynomial of degree $k+1$ in μ and it has $M'\leqslant k+1$ real roots.

$\lambda_1=0$ if the integrals on the right and left of (AX.26) vanish and (AX.26) does not determine the roots $\mu^{(v)}$. Then, as in the example of subsection h, we must consider (AX.25a,b) at the first integer $l>1$ for which a nonvanishing derivative $\lambda_l\neq 0$ exists.

Suppose the initiating bifurcation equation occurs at order $l>1$, *that* μ, λ_v, B_{v-1} *and* $\psi_v(x,y)$ *are given when* $l<v<n$ *and that*

$$\langle(h_2-\mu h_1)N_\phi[\phi_0](h_2-\mu h_1)\rangle+\lambda_1(1+\mu^2)\neq 0. \tag{AX.27}$$

Then λ_n *and* B_{n-1} *are uniquely determined by* (AX.26) *and* $\psi_n(x,y)$, *orthogonal to* h_1 *and* h_2, *is uniquely determined by* (AX.22d).

Given (AX.27) the statements about λ_n and B_{n-1} are obvious. The statement about invertibility of (AX.22d) follows from the Fredholm alternative for the inverse of the operator $\nabla^2+\lambda_0$ on the Hilbert space

$$\{u:u|_{\partial A}=0,\ \langle|\nabla u|^2\rangle<\infty,\ \langle h_1u\rangle=\langle h_2u\rangle=0\}.$$

When $\mathscr{A}$ is the square with side π and $\lambda_0=5$,

$$\psi_n(x,y)=\sum_{v=1}\sum_{l=1}\frac{K^n_{vl}}{\lambda_0-(v^2+l^2)}\sin vx\sin ly \tag{AX.28}$$

where $K^n_{12}=K^n_{21}=0$ and for other v, l

$$\frac{\pi^2}{4}K^n_{vl}+\langle\sin vx\sin ly[\lambda_n\phi_0+B_{n-1}(\lambda_1+N_\phi[\phi_0])(h_2-\mu h_1)+G_n]\rangle=0.$$

(f) Perturbation of a Linear Problem at a Double Eigenvalue

When $N[\phi]=L_0\phi$ is a linear operator the initiating solvability equation corresponding to $l=1$ is

$$\mu^2\langle h_1L_0h_2\rangle+\mu[\langle h_1L_0h_1\rangle-\langle h_2L_0h_2\rangle]=\langle h_2L_0h_1\rangle. \tag{AX.29}$$

If L_0 is a self-adjoint operator such that

$$\langle h_2L_0h_1\rangle=\langle h_1L_0h_2\rangle\neq 0, \tag{AX.30}$$

then (AX.29) has two real roots and, using (AX.24) and (AX.29), it can be shown that (AX.27) holds. Hence, under the condition (AX.30) all the terms in the perturbation series may be computed. If L_0 is an operator of the form

$$a(x,y)\frac{\partial}{\partial x}+b(x,y)\frac{\partial}{\partial y}+c(x,y)$$

where a, b and c are real analytic functions, then the theory of analytic operators of type A applies (Kato, 1966; Chapter 7) and the series representations of the two branches converge. In general, for linear operators of type A, M' real-valued analytic branches split off from an eigenvalue of multiplicity M' (see Courant-Hilbert, Vol. 1; §13.2 for a simple example). Real-valued branches do not always arise from perturbing real eigenvalues and eigenvectors of multiplicity $M'>1$ (see Exercise AX.3). The perturbing linear operators must satisfy certain conditions (like those of type A).

(g) Bifurcation from a Double Eigenvalue: An Example where the Initiating Solvability Condition Occurs at Order $l=1$ ($\lambda_1 \neq 0$)

If

$$N[\phi]=g(x,y)\phi^2$$

where

$$g(x,y)=\alpha\sin 2x+\beta\sin 2y+\gamma$$

and α, β and γ are real constants, then the solvability equations (AX.24) become

$$\lambda_1\sqrt{1+\mu^2}+\langle gh_1^3\rangle+2\mu\langle gh_1^2h_2\rangle+\mu^2\langle gh_1h_2^2\rangle=0\,, \tag{AX.32a}$$

$$\lambda_1\mu\sqrt{1+\mu^2}+\langle gh_1^2h_2\rangle+2\mu\langle gh_1h_2^2\rangle+\mu^2\langle gh_2^3\rangle=0\,; \tag{AX.32b}$$

the bifurcation equation (AX.26) is then

$$\begin{aligned}\mu^3\langle gh_1h_2^2\rangle+\mu^2(2\langle gh_1^2h_2\rangle-\langle gh_2^3\rangle)\\ +\mu(\langle gh_1^3\rangle-2\langle gh_1h_2^2\rangle)-\langle gh_1^2h_2\rangle=0\end{aligned} \tag{AX.32c}$$

and the higher-order solvability condition (AX.27) becomes

$$\begin{aligned}\mu^2\langle gh_1^3\rangle+(\mu^3-2\mu)\langle gh_1^2h_2\rangle+(1-2\mu^2)\langle gh_1h_2^2\rangle\\ +\mu\langle gh_2^3\rangle+\tfrac{1}{2}\lambda_1(1+\mu^2)^{3/2}\neq 0\end{aligned} \tag{AX.32d}$$

where

$$\langle gh_1h_2\rangle=0\,,$$

$$\langle gh_1^2\rangle=\langle gh_2^2\rangle=\gamma\,,$$

$$\langle gh_1^3\rangle=4\beta/\pi^2\,,$$

$$\langle gh_2^3\rangle=4\alpha/\pi^2\,,$$

$$\langle gh_1^2h_2\rangle=32\alpha/15\pi^2$$

and

$$\langle gh_2^2h_1\rangle=32\beta/15\pi^2\,.$$

Using these values we may rewrite (AX.32a, b, c, d) as

$$15\lambda_1\pi^2\sqrt{1+\mu^2}+60\beta+64\alpha\mu+32\beta\mu^2=0\,, \tag{AX.33a}$$

$$15\lambda_1\pi^2\mu\sqrt{1+\mu^2}+32\alpha+64\beta\mu+60\alpha\mu^2=0\,, \tag{AX.33b}$$

$$8\beta\mu^3+\alpha\mu^2-\beta\mu-8\alpha=0 \tag{AX.33c}$$

and

$$\beta\mu^2 + \frac{8\alpha}{15}(\mu^3 - 2\mu) + \frac{8\beta}{15}(1-2\mu^2) + \alpha\mu + \frac{\pi^2}{8}\lambda_1(1+\mu^2)^{3/2} \neq 0\,. \tag{AX.33 d}$$

To determine the number of real bifurcating branches we must determine the number of real roots of (AX.33 c). The discriminant of this cubic equation is

$$\Delta = 96\tau^4 - 109439\tau^2 + 96\,.$$

When $\Delta > 0$ there are three real roots corresponding to three bifurcating branches arising from the double eigenvalue $\lambda_0 = 5$. When $\Delta < 0$ only one solution branches. For example, when $\gamma = \alpha = \beta = 1$, $\Delta = -109247$ and (AX.33 c) becomes

$$(\mu - 1)(8\mu^2 + 9\mu + 8) = 0\,.$$

Hence,

$$\mu = 1, \qquad \lambda_1 = \frac{-26\sqrt{2}}{5\pi^2}, \quad \text{and (AX.33 d) is satisfied.}$$

When $\Delta = 0$, all roots are real and at least two are equal. Excluding this exceptional case, either one ($\Delta < 0$) or three ($\Delta > 0$) solutions bifurcate. It follows that the nonlinear terms need not reduce the degeneracy of the linear theory and may even increase this degeneracy; from two solutions which exist when $\lambda_0 = 5$ we can get three bifurcating solutions of the nonlinear problem.

(h) Bifurcation from a Double Eigenvalue: An Example where the Initiating Solvability Condition Occurs at Order $l = 2$ ($\lambda_1 = 0$, $\lambda_2 \neq 0$)

When $\alpha = \beta = 0$, then $\lambda_1 = 0$ and we must consider the solvability conditions which arise at the next order (two). Eq. (AX.22 c) now becomes

$$(\nabla^2 + \lambda_0)\psi_1 + \gamma\phi_0^2 = 0 \tag{AX.34}$$

and (AX.22 d) with $n = 2$ becomes

$$(\nabla^2 + \lambda_0)\psi_2 + \lambda_2\phi_0 + 2\gamma B_1\phi_0(h_2 - \mu h_1) + 2\gamma\phi_0\psi_1 = 0\,. \tag{AX.35}$$

Since $\langle h_1\phi_0^2\rangle = \langle h_2\phi_0^2\rangle = 0$ the solvability equations (AX.24) for (AX.34) are automatically satisfied. The solvability equations, (AX.25) with $n = 2$, for (AX.35) can be written as

$$\begin{aligned} &\lambda_2 + 2\gamma\langle\psi_1(h_1^2 + \mu h_1 h_2)\rangle = 0\,,\\ &\lambda_2\mu + 2\gamma\langle\psi_1(h_1 h_2 + \mu h_2^2)\rangle = 0\,. \end{aligned} \tag{AX.36}$$

The solution of (AX.34) is

$$\psi_1 = \sum_{l=0} \sum_{v=0} \psi_{2v+1,2l+1} \sin(2l+1)x \sin(2v+1)y + \sum_{l=1} \sum_{v=1} \psi_{2v,2l} \sin 2lx \sin 2vy \tag{AX.37}$$

where

$$\begin{aligned} & -\pi^4(1+\mu^2)[\lambda_0 - v^2 - l^2]\psi_{v,l}/16\gamma \\ & = \langle(\sin^2 x \sin^2 2y + \mu^2 \sin^2 y \sin^2 2x) \sin lx \sin vy\rangle \\ & \quad + 2\mu\langle \sin x \sin 2y \sin y \sin 2x \sin lx \sin vy\rangle \\ & = \frac{64}{lv} \begin{cases} \dfrac{1}{(4-l^2)(16-v^2)} + \dfrac{\mu^2}{(4-v^2)(16-l^2)} & l \text{ and } v \text{ are odd integers} \\ \dfrac{2\mu l^2 v^2}{(l^2-1)(v^2-1)(9-l^2)(9-v^2)} & l \text{ and } v \text{ are even integers} \\ 0 & \text{other } l \text{ and } v. \end{cases} \end{aligned}$$

Substitution of (AX.37) into (AX.36) yields

$$\begin{aligned} & \lambda_2 T - [u^2 + \mu^2(v^2 - \omega^2)] = 0, \\ & \lambda_2 \mu T - [\mu(v^2 - \omega^2) + \mu^3 u^2] = 0 \end{aligned} \tag{AX.38}$$

where $\lambda_0 = 5$,

$$\begin{aligned} T &= \pi^6(1+\mu^2)/2(64)^3\gamma^2, \\ u^2 &= \sum\sum_{\substack{l=1,3,\dots \\ v=1,3,\dots}} 1/[\lambda_0 - v^2 - l^2]\, l^2 v^2 (4-l^2)^2 (16-v^2)^2, \\ \omega^2 &= -\sum\sum_{\substack{l=2,4,\dots \\ v=2,4,\dots}} 2l^2v^2/[\lambda_0 - v^2 - l^2](l^2-1)^2(v^2-1)^2(9-l^2)^2(9-v^2)^2, \\ v^2 &= \sum\sum_{\substack{l=1,3,\dots \\ v=1,3,\dots}} 1/[\lambda_0 - v^2 - l^2]\, l^2 v^2 (4-l^2)(16-v^2)(4-v^2)(16-l^2). \end{aligned}$$

Since u^2, v^2 and ω^2 are positive, (AX.38) yields three roots

$$\mu = 0, 1, -1$$

corresponding to three bifurcating solutions with

$$\lambda_2 T = u^2, \quad u^2 + v^2 - \omega^2, \quad u^2 + v^2 - \omega^2.$$

Exercise AX.1: Show that $\Phi(\varepsilon, \phi_0) = -\Phi(-\varepsilon, \phi_0)$, where $\Phi(\varepsilon, \phi_0) \equiv \phi(x, y; \varepsilon)$ is the bifurcating solution (AX.6), and $\phi_0 \equiv \phi(x, y; 0)$. *Hint:* Decompose $\Phi = \Phi_S + \Phi_A$ where Φ_S is an even function of ε and Φ_A is an odd function. Show, by induction, that Φ_S is an odd function of ϕ_0 and Φ_A an even function of ϕ_0.

Exercise AX.2: Consider the problem of stability of the bifurcating solution (AX.6) when $\phi(x, y, \varepsilon)$ is a stationary solution of the parabolic problem

$$\frac{\partial \psi}{\partial t} = \nabla^2 \psi + \lambda \psi + \varepsilon N[\psi], \qquad \psi|_{\partial A} = 0 .$$

Show that subcritical solutions which branch from simple eigenvalues are unstable. McCleod and Sattinger (1973) have considered the problem of bifurcation and stability at a double eigenvalue. For their problem, subcritical solutions which bifurcate at a double eigenvalue need not be unstable.

Exercise AX.3 (branching at a double eigenvalue of a 2×2 matrix):

$$\begin{bmatrix} a & \varepsilon \\ -\varepsilon & b \end{bmatrix} \begin{bmatrix} x_1 \\ x_2 \end{bmatrix} = \lambda \begin{bmatrix} x_1 \\ x_2 \end{bmatrix}$$

has two eigenvalues $\lambda^{(1)}$ and $\lambda^{(2)}$ corresponding to the roots of

$$\lambda = \frac{b+a}{2} \pm \sqrt{\frac{(b-a)^2}{4} - \varepsilon^2} .$$

Two linearly independent eigenvectors are

$$\mathbf{x}^{(1)} = \begin{bmatrix} 1 \\ \alpha_1 \end{bmatrix} x_1^{(1)}, \qquad \alpha_1 = -\varepsilon/(\lambda^{(1)} - b) ;$$

$$\mathbf{x}^{(2)} = \begin{bmatrix} \alpha_2 \\ 1 \end{bmatrix} x_2^{(2)}, \qquad \alpha_2 = \varepsilon/(\lambda^{(2)} - a)$$

where $x_1^{(1)}$ and $x_2^{(2)}$ are arbitrary. We may normalize these eigenvectors $\mathbf{x}^{(1)} \cdot \mathbf{x}^{(1)} = \mathbf{x}^{(2)} \cdot \mathbf{x}^{(2)} = 1$. This determines the eigenvectors uniquely except for sign.

Construct the eigenvalues $\lambda(\varepsilon)$ and normalized eigenvectors $\mathbf{x}(\varepsilon)$ as a power series in ε when $b \neq a$. Show that when $\varepsilon = 0$ and $b = a$, $\lambda = a$ is a double eigenvalue with normalized eigenvectors

$$\mathbf{x}^{(1)} = \begin{bmatrix} 1 \\ 0 \end{bmatrix},$$

$$\mathbf{x}^{(2)} = \begin{bmatrix} 0 \\ 1 \end{bmatrix}$$

and that there are no real-valued branches $\lambda(\varepsilon)$, $\mathbf{x}(\varepsilon)$.

Chapter XI

Stability of Supercritical Convection-Wave Number Selection Through Stability

In this chapter we are going to study the stability of cellular convection in a layer of porous material filled with fluid and heated from below. This type of cellular convection could be described as Bénard convection in a DOB fluid. As in the Bénard problem $\mathscr{R}_1=0$, $\mathscr{R}_2>0$ and the bifurcation of conduction into convection is supercritical rather than two-sided. The analysis of Schlüter, Lortz and Busse (1965) for the stability of Bénard convection in an OB fluid applies to the DOB fluid. Their analysis shows that the common three-dimensional plan forms which are included in (76.6) are all unstable and only two-dimensional plan forms can be stable. This result is consistent with the stability picture for two-sided bifurcation into cellular convection which is shown in Fig. 77.1 (iii) in the limit $p_1 \to p_0$, $p_2 \to p_0$, $\varepsilon_* \to 0$.

We are going to study the stability of roll convection with a two-dimensional planform. The goal is to find the values of the parameters $(R, a, \varepsilon)=$(Rayleigh number, wave number of the bifurcating solution, amplitude of the bifurcating solution) for which the bifurcating solution is unstable. In general, the parameters for which bifurcating rolls exist lie on a surface $R=R(a,\varepsilon)$ in the three-dimensional parameter space. The intersection of the plane $\varepsilon=0$ and this surface, $R_0(a)=R(a,0)$, is called a neutral curve for the basic flow. The conduction solution is the unique solution of the DOB equations which exists when R is small. The application of linear stability theory to this flow leads one to an eigenvalue problem for $\sigma(R)$ where σ enters through disturbances proportional to $e^{-\sigma t}$. Conduction loses stability as R is increased past $R_0(a)$ and is replaced by a bifurcating solution which is periodic in x with period $2\pi/a$. The curve $R_0(a)$ is called the neutral curve because it gives the values of R for which $\sigma(R)=0$.

In the same way, we may seek expressions for the nonlinear neutral curve. This curve is also associated with an eigenvalue, γ, which arises in the study of the linear theory of stability of the $2\pi/a$-periodic bifurcating solution of amplitude ε to disturbances proportional to $e^{-\gamma t}$. If there is an eigenvalue $\gamma(a,\beta,\varepsilon)<0$ then the bifurcating solution with amplitude ε and the wavenumber a is unstable to disturbances with wavenumber β. The nonlinear neutral curve is given by the implicit function

$$0=\min_\beta \gamma(a,\beta,\varepsilon)=\gamma(a,\beta(a,\varepsilon),\varepsilon)\,.$$

This curve has three projections: $a=a(\varepsilon)$, $R=R(\varepsilon)$, $R=R(a)$. It is possible to find explicit mathematical expressions giving the projections locally, in the neighborhood of the critical point $(R, a, \varepsilon) = (R_c, \tilde{a}, 0)$ where $R_c = \min_{a^2} R_0(a^2) = R_0(\tilde{a}^2)$: the critical point is the only point shared by the neutral curve for the basic flow and the nonlinear neutral curve.

The analysis given in § 79—80 is based on joint work of D. D. Joseph and D.A. Nield. This work, which was performed while Nield was on sabbatical leave from the Department of Mathematics, University of Aukland, New Zealand, will not be published elsewhere. A related analysis for a similar problem has been given by F. Busse (1971). The analysis given in § 82 is an application of a method introduced by Newell and Whitehead (1969).

Notation: The reduced time called τ in Chapter X is t in this chapter. a, β and τ^2 and $k = (\pi^2 + a^2)^{1/2}$ are wave numbers and $R = \mathscr{R}^2$. The critical Rayleigh number $R_c = R_0(\tilde{a}^2) = \min_{a^2} R_0(a^2)$. We sometimes use the symbol $a_0 = \tilde{a}$ for the minimizing wave number. For the DOB convection considered in this chapter $R_c = 4\pi^2$, $\tilde{a}^2 = a_0^2 = \pi^2$.

§ 78. Statistically Stationary Convection and Steady Convection

The configuration to be considered is a fluid-filled porous layer heated from below. The layer has height l and the lower surface is ΔT degrees hotter than the upper surface. In the dimensionless variables the temperature excess and the height of the layer are both one. It will be assumed that the top and bottom surface of the layer are heat conducting rigid plates. Then the normal component of the velocity vanishes at the top and bottom plate and the temperature at the top and bottom is prescribed at the values given by the outside ambient temperatures.

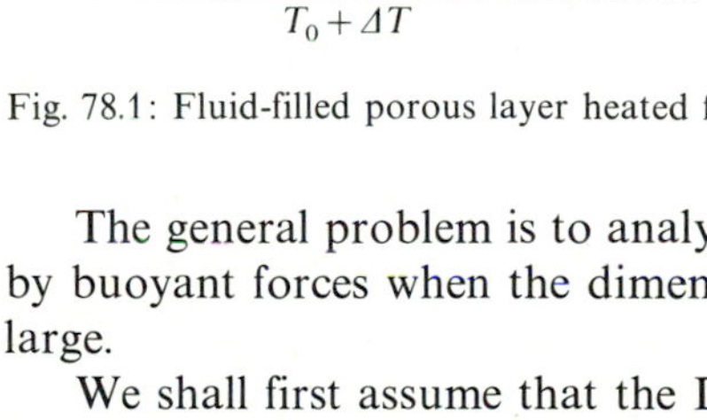

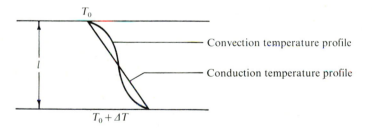

Fig. 78.1: Fluid-filled porous layer heated from below

The general problem is to analyze the convective motion which will be driven by buoyant forces when the dimensionless temperature contrast R is sufficiently large.

We shall first assume that the DOB system (70.10a, c, d), with $B \neq 0$, governs the convection[1]. The motion will be resolved into mean and fluctuating parts,

[1] I have argued that the assumption $B = 0$ is realistic for problems of thermal convection in DOB fluids. The assumption $B \neq 0$ is perhaps not so interesting for the physics of porous convection, but it allows greater mathematical generality and makes the analogy with convection in OB fluids more complete.

$$\mathbf{U}_m(\mathbf{x},t)=\overline{\mathbf{U}}_m(z,t)+\mathbf{u}(\mathbf{x},t)\,,$$

$$T(\mathbf{x},t)=\overline{T}(z,t)+\theta(\mathbf{x},t)\,,$$

and (78.1)

$$P(\mathbf{x},t)=\overline{p}(z,t)+p(\mathbf{x},t)$$

where the overbar designates horizontal averaging; $\overline{C}=C$ for any function $C(z,t)$ independent of (x,y) (see Eqs. following (28.3)).

The mean values of statistically stationary solutions are time independent. We also assume, consistent with the isotropy of the externally given conditions, that there is no forced convection,

$$\overline{\mathbf{U}}_m=0\,. \tag{78.2}$$

From now on, by statistically stationary we understand time-independent *mean* values and (78.2).

Using (78.1) and (78.2) one may deduce the equations for the mean motion

$$D\left\{\frac{B}{\hat{\phi}^2}\,\overline{w\mathbf{u}}+\overline{p}\,\mathbf{e}_z\right\}=R\left(\overline{T}-\frac{T_0}{\Delta T}\right)\mathbf{e}_z$$

where $D=\dfrac{d}{dz}$ and

$$D\overline{w\theta}=D^2\overline{T} \tag{78.3a}$$

and the equations governing the fluctuations

$$\nabla\cdot\mathbf{u}=0,\qquad \frac{B}{\hat{\phi}}\left[\frac{\partial\mathbf{u}}{\partial t}+\frac{1}{\hat{\phi}}\,\mathbf{u}\cdot\nabla\mathbf{u}-\frac{1}{\hat{\phi}}\,D\overline{w\mathbf{u}}\right]=-\mathbf{u}+\mathbf{e}_zR\theta-\nabla p \tag{78.3b}$$

and

$$C\frac{\partial\theta}{\partial\tau}+\nabla\cdot\{\mathbf{u}(\theta+\overline{T})-\overline{\mathbf{u}\theta}\}=\nabla^2\theta \tag{78.3c}$$

where $C=(C_0\rho_0)_f/(C\rho_0)_m$.

At the boundary $z=0$

$$\overline{T}=\frac{T_0}{\Delta T}+1,\qquad \theta=w=0\,,$$

and at $z=1$, (78.3d)

$$\overline{T}=\frac{T_0}{\Delta T},\qquad \theta=w=0\,.$$

Integrating (78.3a) we find that

$$D\overline{T}-\overline{w\theta}=\langle D\overline{T}\rangle-\langle w\theta\rangle=-1-\langle w\theta\rangle \tag{78.4}$$

where the angle bracket is the volume average which in dimensionless variables is

$$\langle A\rangle = \int_0^1 \bar{A}\,dz\,.$$

Eq. (78.4) shows that

$$D\bar{T} = -1 - \langle w\theta\rangle\,, \tag{78.5}$$

when $z=0$ and when $z=1$. The value of $D\bar{T}$ at the wall is a measure of the heat transported across the layer. Across a wall, since $w=0$, there is no heat transported by convection. The heat which enters the layer at the bottom by conduction is carried to the top by conduction and convection and this same quantity of heat leaves the system at the top by conduction.

The major physical relationship for the process which was just described is the relation between the driving temperature difference and the heat transport. In dimensionless variables this is the Rayleigh-Nusselt number curve. The Nusselt number is the ratio of the heat transported across any layer to the heat which would be transported by conduction alone.

We may conveniently form this ratio at $z=0$ as

$$Nu = \frac{D\bar{T}}{-1} = \frac{-1-\langle w\theta\rangle}{-1} = 1 + \langle w\theta\rangle\,.$$

Hence,

$$\langle w\theta\rangle = Nu - 1 \tag{78.6}$$

is a convenient measure of the difference between the heat transported and the heat which would be transported by conduction alone when the temperature difference is prescribed.

At a given temperature contrast, the amount of heat transported by statistically stationary convection exceeds the amount of heat which would be transported by conduction alone; i.e.,

$$Nu - 1 = \langle w\theta\rangle \geqslant 0$$

with equality if and only if $\mathbf{u}\equiv 0$.

The proof is an immediate consequence of the energy identity

$$\langle|\mathbf{u}|^2\rangle = R\langle w\theta\rangle \tag{78.7}$$

which follows from (78.3 b).

The heat transported by conduction ($D\bar{T}=-1$) alone is given by (78.5) when the fluctuations vanish. Then

$$T = T_c(z) = \frac{T_0}{\Delta T} + 1 - z\,. \tag{78.8}$$

The pressure stratification corresponding to the conduction temperature $T_c(z)$ is given by

$$\frac{dp_c}{dz} = R(1-z). \tag{78.9}$$

It is, of course, convenient to study the stability and bifurcation of the rest state by decomposing the motion into the rest state plus a disturbance:

$$\mathbf{U}_m(\mathbf{x},t)=\mathbf{u}(\mathbf{x},t),$$
$$T(\mathbf{x},t)=T_c(z)+\hat{\theta}(\mathbf{x},t),$$
$$p(\mathbf{x},t)=p_c(z)+\pi(\mathbf{x},t).$$

We think of this as a second decomposition of the motion (78.1). The fluctuations and disturbances are related by

$$\begin{aligned}\overline{T}(z)+\theta(\mathbf{x},t)&=T_c(z)+\hat{\theta}(\mathbf{x},t),\\ \overline{p}(z)+p(\mathbf{x},t)&=p_c(z)+\pi(\mathbf{x},t)\end{aligned} \tag{78.10}$$

where

$$\overline{T}(z)=T_c(z)+\overline{\hat{\theta}}(z)$$

and

$$\overline{p}(z)=p_c(z)+\overline{\pi}(z).$$

From (78.6) and (78.10) we find that

$$Nu-1=\langle w\theta\rangle=\langle w(T_c-T+\hat{\theta})\rangle=\langle w\hat{\theta}\rangle. \tag{78.11}$$

We turn next to the study of bifurcation of conduction into roll convection. The governing DOB equations are given by (70.10 b, c, d). It follows from (70.10 b) that $\mathbf{e}_z\cdot\operatorname{curl}\mathbf{U}_m=0$. Hence the velocity may be obtained from a poloidal potential

$$\mathbf{U}_m=\operatorname{curl}^2\hat{\chi}\equiv\boldsymbol{\delta}\hat{\chi}. \tag{78.12}$$

The equation

$$\nabla_2^2\{\nabla^2\hat{\chi}+R(T-T_0/\Delta T)\}=0, \qquad \nabla_2^2\equiv\frac{\partial^2}{\partial x^2}+\frac{\partial^2}{\partial y^2},$$

arises from the operation $\mathbf{e}_z\cdot\boldsymbol{\delta}$ applied to (2.1). We assume that $\mathbf{U}_m$ and T are periodic, or almost periodic, in x and y and $\mathbf{U}_m$ has a zero horizontal average. It follows that

$$\nabla^2\hat{\chi}+R(T-T_0/\Delta T)=f(z)$$

where $f(z)$ is an arbitrary function of z alone. Since $\hat{\chi}$ determines $\mathbf{U}_m$ to within an additive function $\tilde{g}(z)$ alone, we choose $\tilde{g}$ so that

$$\nabla^2\hat{\chi}=R(T_c(z)-T) \tag{78.13}$$

where

$$T_c(z)=1-z+T_0/\Delta T \tag{78.14}$$

is the temperature distribution of the conduction solution of the problem (70.10 b, c, d) and

$$\mathbf{U}_m=0\,,\qquad T=T_c(z)\,,\qquad p=p_c(z)\,. \tag{78.15}$$

Combining (70.10 b, c) and (78.12, 13) we find that

$$\left(\nabla^2-\frac{\partial}{\partial t}\right)\nabla^2\hat{\chi}+R\nabla_2^2\hat{\chi}=\boldsymbol{\delta}\hat{\chi}\cdot\nabla\nabla^2\hat{\chi} \tag{78.16}$$

where, at $z=0, 1$

$$\hat{\chi}=\nabla^2\hat{\chi}=0\,. \tag{78.17}$$

It is easy to verify that the conduction solution $\hat{\chi}=0$, of (78.16) and (78.17) is unstable when $R>4\pi^2$ and is globally stable when $R<4\pi^2$. The spectrum of the linearized equations corresponding to $\hat{\chi}$ is real-valued and the solutions which bifurcate from conduction take form as steady convection.

We consider steady, two-dimensional solutions of (78.16), (78.17) which are $2\pi/a$-periodic in x. The solutions of this problem may be constructed as a power series

$$\begin{aligned}\hat{\chi}&=\varepsilon\chi(x, z)=\varepsilon\chi_0(x, z)+\varepsilon^2\chi_1(x, z)+\varepsilon^3\chi_2(x, z)+\dots\,,\\ R(\varepsilon)&=R_0+\varepsilon^2R_2+\varepsilon^4R_4+\dots\,,\end{aligned} \tag{78.18}$$

in the Nusselt number discrepancy

$$\varepsilon^2=Nu-1=\langle w\hat{\theta}\rangle=\langle\nabla_2^2\hat{\chi}\nabla^2\hat{\chi}\rangle/R$$

where $\langle\;\rangle$ denotes an average over the whole fluid layer $-\infty<x<\infty$, $0<z<1$ and $\langle c\rangle=c$ for any constant c. The coefficients of the power series satisfy the following equations

$$\mathscr{L}_0\chi_0=0\,,\qquad\langle\nabla_2^2\chi_0\nabla^2\chi_0\rangle=R_0\,; \tag{78.19, 20}$$

$$\textstyle\sum_{k+l=n}\mathscr{L}_k\chi_l=\sum_{k+l+1=n}\boldsymbol{\delta}\chi_k\cdot\nabla\nabla^2\chi_l\,, \tag{78.21}$$

$$\textstyle\sum_{k+l=n}\langle\nabla_2^2\chi_k\nabla^2\chi_l\rangle=R_n\,,\qquad n\geq 1\,; \tag{78.22}$$

where

$$\mathscr{L}_0 = \nabla^4 + R_0(a)\nabla_2^2$$

$$\mathscr{L}_k = R_k(a)\nabla_2^2\,, \qquad k \geqslant 1\,.$$

In addition,

$$\chi_l = \nabla^2 \chi_l = 0 \quad \text{at} \quad z = 0, 1\,, \tag{78.23}$$

and

$$\chi_l(x, z) = \chi_l(x + 2\pi/a, z) \tag{78.24}$$

for $l \geqslant 0$. We find that

$$\begin{aligned}
R_0 &= (\pi^2 + a^2)^2/a^2\,,\\
\chi_0 &= \frac{2(\pi^2+a^2)^{1/2}}{a^2} \cos ax \sin \pi z\,,\\
\chi_1 &= -\frac{R_0}{8\pi^3} \sin 2\pi z\,,\\
R_2 &= R_0/2\,,\\
\chi_2 &= \frac{\chi_0}{4} + \frac{(\pi^2+a^2)^{1/2} R_0}{16\pi^2(5\pi^2+a^2)} \cos ax \sin 3\pi z\,,\\
\chi_3 &= \frac{\chi_1}{2} + \frac{(\pi^2+a^2)R_0}{128\pi^5}(\sin 2\pi z - \tfrac{1}{8}\sin 4\pi z)\\
&\quad + \frac{\pi(\pi^2+a^2)R_0}{24(5\pi^2+a^2)}\left[\frac{2\sin 2\pi z}{(\pi^2+a^2)^2} - \frac{\sin 4\pi z}{(3\pi^2+a^2)(7\pi^2+a^2)}\right]\cos 2ax\,,\\
R_4 &= \frac{1}{4} R_0 \left\{1 - \frac{(\pi^2+a^2)(11\pi^2+3a^2)}{16\pi^2 \quad (5\pi^2+a^2)}\right\}.
\end{aligned} \tag{78.25}$$

Exercise 78.1 (Horton and Rogers, 1945; Lapwood; 1948; Westbrook, 1969); Show that conduction is unstable to doubly-almost-periodic-disturbances when $R > 4\pi^2$. Show that conduction is globally and monotonically stable to doubly-almost periodic solutions when $R < 4\pi^2$. *Note:* An extension of the linearized stability analysis of conduction to include other boundary conditions and fluid mixtures was given by Nield (1968).

Exercise 78.2: Work out Eqs. (78.25) through 2nd order.

§ 79. Stability of Rolls to Noninteracting Three-Dimensional Disturbances

In this section we study stability of the steady spatially periodic two-dimensional solutions (78.25) of (78.16—17) to small disturbances. Substituting

$$\chi = \varepsilon\chi + \bar{\varepsilon}\Phi(x, y, z)e^{-\gamma t}$$

into (78.16) we linearize ($\bar{\varepsilon}\to 0$) and find that

$$(\nabla^2 + \gamma)\nabla^2\Phi + R\nabla_2^2\Phi = \varepsilon[\boldsymbol{\delta}\chi\cdot\nabla\nabla^2\Phi + \boldsymbol{\delta}\Phi\cdot\nabla\nabla^2\chi] \tag{79.1}$$

where Φ and $\nabla^2\Phi$ vanish at $z=0, 1$. The coefficients of this linear differential equation are independent of y and periodic in x with periodic $2\pi/a$. When $\varepsilon=0$, (79.1) may be solved by expressions of the form

$$e^{i(vx+\mu y)}\sin n\pi z \tag{79.2}$$

provided that

$$\gamma_0 = \frac{-\beta^2 R_0(a^2)}{n^2\pi^2+\beta^2} + (n^2\pi^2+\beta^2) \geqslant \frac{-\beta^2 R_0(a^2)}{\pi^2+\beta^2} + \pi^2 + \beta^2 = \frac{\beta^2}{\pi^2+\beta^2}[R_0(\beta^2) - R_0(a^2)]$$

where $\beta^2 = v^2 + \mu^2$. The smallest value of γ_0 is attained when $n=1$; moreover, γ_0 is negative whenever $R_0(\beta^2) < R_0(a^2)$. Since $R(a^2) > 4\pi^2$ if $a^2 \neq \pi^2$, it follows that if $a^2 \neq \pi^2$, then there is a β^2 (near π^2) such that $\gamma_0 < 0$.

The only size of the roll cells which could be stable when $\varepsilon\to 0$ *is the cell of period* $2 = 2\pi/a$; said in another way: *the points* $R(a, 0) = R_0(a^2)$ *on the bifurcating surface* $R(a, \varepsilon)$ *are all unstable except, possibly for the point* $R(\pi, 0) = 4\pi^2$.

The point $(R, a, \varepsilon) = (4\pi^2, \pi, 0)$ is simultaneously a point on the neutral curve R_0 for conduction and the nonlinear neutral curve $\min\gamma(a, \varepsilon) = 0$ for convection. We are going to find the nonlinear neutral curve among doubly-periodic noninteracting disturbances $\Phi(x, y, z) = \Phi(x + 2\pi/a, y + 2\pi/\mu, z)$. It is of interest to consider disturbances which are quasiperiodic in x and y. In fact, since the coefficients of (79.1) do not depend on y, the distinction between quasiperiodic and periodic functions of y is not important; in either case, we may assume a solution proportional to $e^{i\mu y}$. The quasiperiodic solutions we are going to consider are in a sort of Floquet form

$$\Phi(x, y, z) = e^{ivx} f(x, y, z) \tag{79.3}$$

where $f(x, y, z) = f(x + 2\pi/a, y, z)$. The function given by (79.2) is of this type. In this section we will ignore the possibility that solutions with different values of v^2 may be superposed. In fact, a superposition of solutions of this type is possible even when $\varepsilon=0$ and $\gamma_0 < 0$. For if $\gamma_0 < 0$ is prescribed, then there are two roots

β_1^2 and β_2^2 of the equation $(\pi^2+\beta^2)\gamma_0=R_0(\beta^2)-R_0(a^2)$ and four different solutions of the form

$$e^{\pm i(\nu_j x+\mu_j y)}\sin\pi z\,,\qquad \beta_j^2=\mu_j^2+\nu_j^2\qquad (j=1,2)\,.$$

From analysis carried out to $O(\varepsilon^2)$ we learn that these solutions do not interact unless $2a+\nu_m+\nu_l=0$. The interacting problem does not appear to give the most dangerous instabilities.

The construction of quasiperiodic solutions of the noninteracting type (79.3) as a series of powers in ε is formal, and more; not only is it formal but in some cases it is also incorrect (see § 81). The most interesting instabilities among the noninteracting disturbances (cross-roll and sinuous instabilities) fortunately occur among strictly periodic functions where the troublesome problem of small divisors (see § 81) does not arise.

Now we shall study the stability of rolls to three-dimensional disturbances

$$\Phi(x,y,z)=e^{iy\sqrt{\tau}}\psi(x,z)\,,\qquad \tau>0\,. \tag{79.4}$$

The boundary value problem governing ψ may be obtained by substituting (79.4) into (79.1). We find that

$$\hat{L}\psi=\varepsilon[\chi_{,xz}(\nabla^2-\tau)\psi_{,x}-\chi_{,xx}(\nabla^2-\tau)\psi_{,z}+\nabla^2\chi_{,x}\psi_{,xz}-\nabla^2\chi_{,z}(\nabla_2^2-\tau)\psi] \tag{79.5}$$

where

$$\hat{L}=(\nabla^2-\tau+\gamma)(\nabla^2-\tau)+R(\nabla_2^2-\tau^2)$$

and ψ and $\nabla^2\psi$ are to vanish on $z=0,1$.

Analysis of (79.5) proceeds along the usual lines. $R(a,\varepsilon)$ and $\chi(x,z;\varepsilon)$ are given analytic functions; we seek solutions of (79.5) in series

$$\psi(x,z;\varepsilon)=\psi_0(x,z)+\psi_1(x,z)\varepsilon+\psi_2(x,z)\varepsilon^2+\dots$$
$$\gamma=\gamma_0+\gamma_2\varepsilon^2+\gamma_4\varepsilon^4+\dots$$

When $\varepsilon=0$ we find that

$$\hat{L}_0\psi_0=0\,,\qquad \hat{L}_0=(\nabla^2-\tau+\gamma_0)(\nabla^2-\tau)+R_0(\nabla_2^2-\tau)\,. \tag{79.6}$$

Eq. (79.6), together with the boundary conditions, is satisfied by expressions of the form

$$e^{i\nu x}\sin n\pi z$$

if γ_0 obeys the equation

$$(n^2\pi^2+\nu^2+\tau)^2-\gamma_0(n^2\pi^2+\nu^2+\tau)-R_0(\nu^2+\tau)=0\,.$$

γ_0 is least when $n=1$; then

$$\gamma_0=\pi^2+v^2+\tau-\frac{R_0(v^2+\tau)}{\pi^2+v^2+\tau}. \tag{79.7}$$

The noninteracting eigensolutions of $\hat{L}_0\psi_0=0$ divide into two independent families, respectively, proportional to

$$\cos vx \sin \pi z$$

and

$$\sin vx \sin \pi z .$$

These two possibilities are represented by an expression

$$\psi_0=\sum_{l=-1}^{1} d_l e^{iv_l x} \sin \pi z \tag{79.8a}$$

where $v_{-l}=-v_l$ and $d_{-l}=\overline{d}_l$. For the even solutions ($\cos vx$) we put $\overline{d}_l=d_l$; for the odd solutions ($\sin vx$) we put $\overline{d}_l=-d_l$. The magnitude of d_1 is fixed by normalization. In the rest of the analysis it will be convenient to normalize ψ, satisfying (79.5), by requiring that

$$\langle\psi(\varepsilon)\psi_0\rangle=2C_1=2k/a^2, \qquad k^2=\pi^2+a^2. \tag{79.8b}$$

Returning now to the series solution of (79.5) we find that $\psi_n(x, z)$ and γ_n, $n>1$, satisfy

$$\begin{aligned}\sum_{m+l=n}\hat{L}_m\psi_l=\sum_{m+l+1=1}[\chi_{m,xz}(\nabla^2-\tau)\psi_{l,x}-\chi_{m,xx}(\nabla^2-\tau)\psi_{l,z}\\ +\nabla^2\chi_{l,x}\psi_{m,xz}-\nabla^2\chi_{m,z}(\nabla_2^2-\tau)\psi_l],\end{aligned} \tag{79.9a}$$

$$\begin{aligned}&\psi_n=\nabla^2\psi_n=0 \quad \text{on} \quad z=0, 1,\\ &\langle\psi_n\psi_0\rangle=0,\end{aligned} \tag{79.9b}$$

where $\hat{L}_0$ is given by (79.6) and

$$\hat{L}_m=\gamma_m(\nabla^2-\tau)+R_m(\nabla_2^2-\tau).$$

If problem (79.9) is solvable, the solutions are in the complement of the null space of the operator $\hat{L}_0$; that is,

$$\hat{L}_0\psi_n+f_n=0 \tag{79.10}$$

is solvable only if

$$\langle f_n \sin \pi z\, e^{-ivx}\rangle=0. \tag{79.11}$$

Proceeding now in the usual way, we find that

$$\psi_1=\sum_{l=-1}^{1}\sum_{m=-1}^{1}\frac{C_l d_m D_{lm}}{A(2,\, a_l+v_m,\, \tau)}\sin 2\pi z e^{i(a_l+v_m)x}\,, \tag{79.12}$$

where

$$a_1=-a_{-1}=a\,,$$
$$C_1=C_{-1}=k/a^2\,,\qquad k^2=\pi^2+a^2\,,$$
$$A(p,\, v,\, \tau)=(p^2\pi^2+v^2+\tau)^2-\gamma_0(p^2\pi^2+v^2+\tau)-R_0(v^2+\tau)$$

and

$$D_{lm}=\frac{\pi}{2}\{(a_l-v_m)^2(a_l v_m-\pi^2)-\tau(\pi^2+2a_l^2-a_l v_m)\}\,.$$

At second order, we find, using the expressions (79.8) and (79.12) that

$$\begin{aligned}\hat{L}_0\psi_2=&\sum_{m=-1}^{1}d_m\{\gamma_2(\pi^2+v_m^2+\tau)\sin\pi z+\tfrac{1}{2}R_0(v_m^2+\tau)\sin 3\pi z\}\,e^{iv_m x}\\&+\sum_k\sum_l\sum_m C_k C_l d_m\{P_{klm}^{+}\sin\pi z+P_{klm}^{-}\sin 3\pi z\}\,e^{i(a_l+a_k+v_m)x}\end{aligned} \tag{79.13}$$

where

$$\begin{bmatrix}P_{klm}^{+}\\ P_{klm}^{-}\end{bmatrix}=\pi D_{lm}\{a_k[4\pi^2+(a_l+v_m)^2+\tau](\pm 2a_k+a_l+v_m)-(\pi^2+a_k^2)[(a_l+v_m)(\pm 2a_k+a_l+v_m)+\tau]\}/2A(2,\, a_l+v_m,\, \tau)\,.$$

Applying (79.11) to (79.13) we find that

$$d_1\gamma_2(\pi^2+v^2+\tau)+\sum_k\sum_l\sum_m C_k C_l d_m P_{klm}^{+}\delta(a_l+a_k+v_m-v)=0 \tag{79.14}$$

where $\delta(x)$ is Dirac's delta function.

There are three separate cases to consider:

(I) $\quad v\neq a\,,\quad \gamma_2=\dfrac{-C_1^2(P_{1-11}^{+}+P_{-111}^{+})}{\pi^2+v^2+\tau}$ (79.15)

(II) $\quad v=a\,,\quad \gamma_2=\dfrac{-C_1^2\left(P_{1-11}^{+}+P_{-111}^{+}+\dfrac{d_{-1}}{d_1}P_{11-1}\right)}{k^2+\tau}$ (79.16)

(II a) $\quad \dfrac{d_{-1}}{d_1}=1$ (disturbances which are even functions of x) (79.16a)

(II b) $\quad \dfrac{d_{-1}}{d_1}=-1$ (disturbances which are odd functions of x). (79.16b)

It is apparent from (79.15) and (79.16) that there is a distinct difference between the cases $v \neq a$, $v \to a$, and $v = a$. This difference arises at 2nd order; at zeroth and first order there is no difference between the case $v \to a$ and $v = a$. In either case,

$$\gamma_0 = \frac{\tau(a^4 - \pi^4 + \tau a^2)}{a^2(k^2 + \tau)}$$

$$\psi_0 = \sum_{l=-1}^{1} d_l e^{i a_l x} \sin \pi z ,$$

and

$$\psi_1 = \sum_{l=-1}^{1} \sum_{m=-1}^{1} \frac{C_l d_m D_{lm}}{\Lambda(2, a_l + a_m, \tau)} e^{i(a_l + a_m)x} \sin 2\pi z .$$

With $d_1 = C_1$, we find that

$$\psi_0 = \chi_0 \quad \text{and} \quad \psi_1 = 2\chi_1 \quad \text{when} \quad d_{-1} = C_1$$

and

$$\psi_0 = -\frac{1}{a}\chi_{0,x}, \qquad \psi_1 = \chi_{1,x} = 0 \quad \text{when} \quad d_{-1} = -C_1 .$$

On the other hand, $v \to a$ and $v = a$ lead to different results at order two. From (79.15) we find that when $v \to a$

$$\gamma_2 = \frac{\pi^2 k^2}{4a^4(k^2 + \tau)}(G_1 - \tau^2 G_2) . \tag{79.17}$$

where

$$G_1 = \frac{\{8\pi^2 a^2 - \tau(\pi^2 - a^2)\}\{4a^2 k^2 + \tau(\pi^2 + 3a^2)\}}{(4\pi^2 + \tau)^2 - \gamma_0(4\pi^2 + \tau) - R_0 \tau}$$

and

$$G_2 = \frac{k^4}{(4k^2 + \tau)^2 - \gamma_0(4k^2 + \tau) - R_0(4a^2 + \tau)} .$$

But, when $v = a$, (79.16a) implies that

$$\gamma_2 = \frac{\pi^2 k^2}{4a^4(k^2 + \tau)}(2G_1 - \tau^2 G_2) \tag{79.18}$$

and (79.16b) implies that

$$\gamma_2 = \frac{-\pi^2 k^2 \tau^2 G_2}{4a^4(k^2+\tau)}.$$

Summarizing our results so far, we have found second order representations for the eigenvalues $\gamma = \gamma_0 + \varepsilon^2\gamma_2 + \ldots$ belonging to three different disturbances:

$$\gamma_{(1)}(a, v, \tau, \varepsilon) = \pi^2 + v^2 + \tau - \frac{R_0(v^2+\tau)}{\pi^2+v^2+\tau} - \frac{k^2(P^+_{1-11} + P^+_{-111})}{a^4(\pi^2+v^2+\tau)}\varepsilon^2 + O(\varepsilon^4) \tag{79.19}$$

for disturbances with $v \neq a$.

$$\gamma_{(2)}(a, \tau, \varepsilon) = \frac{\tau(a^4 - \pi^4 + \tau a^2)}{a^2(k^2+\tau)} + \frac{\pi^2 k^2(2G_1 - \tau^2 G_2)}{4a^4(k^2+\tau)}\varepsilon^2 + O(\varepsilon^4) \tag{79.20}$$

for disturbances with $v = a$ which are even functions of x (leading to varicose rolls; see Fig. 80.1).

$$\gamma_{(3)}(a, \tau; \varepsilon) = \frac{\tau(a^4 - \pi^4 + \tau a^2)}{a^2(k^2+\tau)} - \frac{\pi^2 k^2 \tau^2 G_2}{4a^4(k^2+\tau)}\varepsilon^2 + O(\varepsilon^4) \tag{79.21}$$

for disturbances with $v = a$ which are odd functions of x (leading to sinuous rolls; see Fig. 80.1). To compute the nonlinear neutral curve for the sinuous instability it will be necessary to obtain the term $\gamma_4\varepsilon^4$ in (79.21). The computation of γ_4 is straightforward but tedious. It is easiest to compute the function $F(x, y, z) = -F(-x, y, z)$ and the reduced eigenvalue $\hat{\gamma}$ where F and $\hat{\gamma}$ are defined by

$$\psi = \chi_{,x} + \tau F, \qquad \gamma = \tau\hat{\gamma}. \tag{79.22}$$

$\chi_{,x}$ is always an eigenfunction of (79.1) with $\gamma = 0$. First we substitute (79.22) into (79.1) to get the equation governing F. We then develop F and χ in Fourier series and equate coefficients (see § 81). The resulting equations among the Fourier coefficients are then developed into power series in ε. After a tedious but perfectly straightforward computation we find that

$$\begin{aligned} \hat{\gamma}_4 k^2 + R_4 = & -4\pi^3 a A_2 B_4/A_1 + (22\pi^3 a + 6\pi a^3) A_3 B_2/A_1 \\ & -4\pi^3 A_5 + \tfrac{5}{2}\pi k^2 A_7 + 4\pi^3 A_2 A_3/A_1 \end{aligned} \tag{79.23}$$

where the A_i are coefficients associated with χ and B_i are coefficients associated with F,

$$A_1 = 2k/a^2,$$

$$A_2 = -R_0/8\pi^3,$$

$$A_3 = kR_0/16\pi^2(5\pi^2 + a^2),$$

$$A_5 = R_0(a^2 - 7\pi^2)/128\pi^5,$$

$$A_7 = \pi k^2/12a^2(5\pi^2 + a^2),$$
$$B_2 = \pi/12a^3,$$
$$B_4 = -k^3(\pi^4 + 40\pi^2 a^2 + 55a^4)/96a^5(5\pi^2 + a^2)^2.$$

We may, therefore, rewrite (79.21) as

$$\gamma_{(3)} = \tau\hat{\gamma}_{(3)}(a, \tau; \varepsilon) = \gamma_0(a, \tau) + \varepsilon^2\gamma_2(a, \tau) + \varepsilon^4\gamma_4(a, \tau) + O(\varepsilon^6)$$

where

$$\hat{\gamma}_{(3)} = \frac{a^4 - \pi^4 + \tau a^2}{a^2(k^2 + \tau)} - \frac{\tau\pi^2 k^2 G_2}{4a^4(k^2 + \tau)}\varepsilon^2 + \hat{\gamma}_4\varepsilon^4 + O(\varepsilon^6) = \hat{\gamma}_0 + \tau\hat{\hat{\gamma}}_2\varepsilon^2 + \hat{\gamma}_4\varepsilon^4 + O(\varepsilon^6). \tag{79.24}$$

When $\tau = 0$, $\hat{\gamma}_3 = \hat{\gamma}_4\varepsilon^4 + O(\varepsilon^6)$. When $\tau = 0$ and $a = \pi$, then $\hat{\gamma}_4 = \frac{1}{288}$.

§ 80. Nonlinear Neutral Curves for Three-Dimensional Disturbances of Roll Convection

We want criteria separating the stable from the unstable cell sizes. These criteria take form in space curves $\tilde{\gamma}(a, \varepsilon) = 0$ on the surface $R(a, \varepsilon)$ of steady bifurcating roll convection. We shall call the space curve

$$\min_{v,\tau} \gamma(a, v, \tau, \varepsilon) = \tilde{\gamma}(a, \varepsilon) = 0 \tag{80.1}$$

a nonlinear neutral curve. The curve $a(\varepsilon)$ is the projection of (80.1) on the (a, ε) plane. The most dangerous disturbances are those which most narrowly restrict the band $a_1(\varepsilon) < a(\varepsilon) < a_2(\varepsilon)$ of stable wave numbers. To find the nonlinear neutral curve for roll convection we must compare the stable bands which are generated from the three equations $\tilde{\gamma}_{(l)} = 0$, $l = 1, 2, 3$. Nonlinear neutral curves for various disturbances are shown in Fig. 80.1.

(a) Oblique-Roll and Cross-Roll Instabilities

Oblique-roll instabilities arise from disturbances with $v \neq a$. For instability to disturbances $v \neq a$, it is enough that $\gamma_{(1)}(a, v, \tau, \varepsilon) < 0$. In studying these eigenvalues, it is convenient to introduce polar coordinates

$$\sqrt{\tau} = \beta \sin\phi,$$
$$v = \beta \cos\phi;$$

where $0 \leqslant \phi \leqslant \pi$. Then

$$\gamma_{(1)} = \gamma_0(a, \beta) + \gamma_2(a, \beta, \cos\phi)\varepsilon^2 \tag{80.2}$$

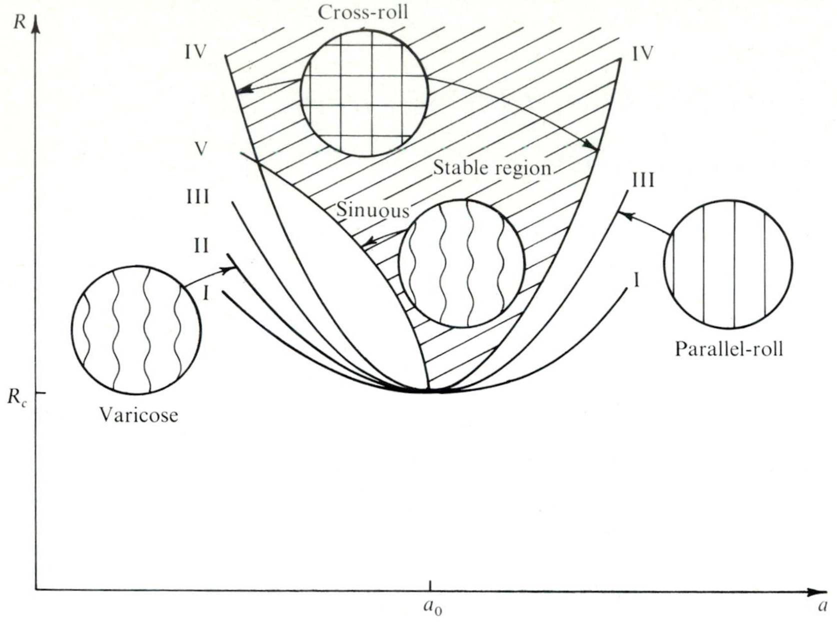

Fig. 80.1: Sketch of parabolic approximations of (a, R)-projections of nonlinear neutral curves for roll convection, valid in a neighborhood of (a_0, R_c). The neutral curve for the rest state with a constant temperature gradient is shown as I. Curves II, III, IV and V are nonlinear neutral curves for different disturbances of convection. Sinuous and varicose instabilities occur only when $a<a_0$. Curves I to IV are given by

$$\frac{R}{R_c}-1=n\eta\left(\frac{a}{a_0}-1\right)^2$$

where $n=1$ is the neutral curve for conduction and $n=(\frac{3}{2}; 3; \frac{10}{3} \text{ or } \frac{673}{200})$ corresponds to instabilities of convection to disturbances in the form of (varicose rolls; parallel rolls; cross rolls in a porous layer by or in a fluid with an infinite Prandtl number) $\sim$ (II; III; IV). Curve V is given by

$$\frac{R}{R_c}-1=\mu\left(1-\frac{a}{a_0}\right)^{1/2}$$

where

$$(a_0, R_c, \mu)=(\pi, 4\pi^2, 12/\sqrt{19}) \quad \text{(porous material)}$$

or

$$(a_0, R_c, \mu)=(\pi/\sqrt{2}, 27\pi^4/4, \sqrt{58345280/5264739}) \quad (P=\infty)$$

where

$$\gamma_0(a, \beta)=\pi^2+\beta^2-\frac{R_0(a^2)\beta^2}{\pi^2+\beta^2}, \tag{80.3}$$

$$\gamma_2(a, \beta, \cos\phi)=\frac{-k^2(P^+_{1-11}+P^+_{-111})}{a^4(\pi^2+\beta^2)}, \tag{80.4}$$

$$P^{+}_{1-11}=F(\alpha,\beta,\cos\phi)=\pi^2 N(\alpha,\beta,\cos\phi)/4D(\alpha,\beta,\cos\phi)$$

$$N(a,\beta,\cos\phi)=[(4\pi^2+a^2+\beta^2-2a\beta\cos\phi)(a^2+a\beta\cos\phi)\\ +k^2(a^2-\beta^2)][(a^2+2\pi^2+\beta^2)a\beta\cos\phi+\pi^2a^2+\pi^2\beta^2+2a^2\beta^2],$$

$$D(a,\beta,\cos\phi)=(4\pi^2+a^2+\beta^2-2a\beta\cos\phi)^2-\sigma_0(4\pi^2+a^2+\beta^2-2a\beta\cos\phi)\\ -R_0(a^2+\beta^2-2a\beta\cos\phi),$$

$$P^{+}_{-111}=F(\alpha,\beta,-\cos\phi).$$

To minimize $\gamma_{(1)}$, we first fix β and seek the angle ϕ for which $\gamma_{(1)}$ is a minimum:

$$\frac{\partial\gamma_{(1)}}{\partial\phi}=\frac{k^2}{a^4(\pi^2+\beta^2)}\{F_3(\alpha,\beta,\cos\phi)-F_3(\alpha,\beta,-\cos\phi)\}\sin\phi \tag{80.5}$$

where F_3 is the derivative of F with respect to the argument in the third place. Eq. (80.5) has two stationary points:

$$\phi=0,\quad v=\beta,\quad \tau=0;\quad \text{(parallel rolls)}$$

$$\phi=\frac{\pi}{2},\quad \tau=\beta^2,\quad v=0;\quad \text{(cross rolls)}.$$

The parallel roll solution is strictly formal; it involves small divisors and in the limit $\beta\to a$ it ceases to be a solution even in an asymptotic sense (see § 81). The cross-roll instability is doubly-periodic and convergence of the series giving this solution is easily established.

For the cross-roll instability

$$\gamma_{(1)}=\gamma_0(a,\tau)+\gamma_2(a,\tau,0)\varepsilon^2+\dots \tag{80.6}$$

This expression is now minimized with respect to τ. We find that

$$\tau(a,\varepsilon)=\tau_0+\varepsilon^2\tau_2+\dots$$

where

$$\tau_0=\frac{\pi}{a}(k^2-\pi a) \tag{80.7}$$

and

$$\gamma_0(a,\tau_0)=\frac{-k^2(\pi-a)^2}{a^2}.$$

Now we expand $\gamma_{(1)}(a,\tau(a,\varepsilon),\varepsilon)$ in powers of ε. After making use of the fact that $\partial\gamma_{(1)}/\partial\beta=0$ at each order in ε^2, we find that to lowest order

$$\tilde{\gamma}_{(1)}=\frac{-k^2(\pi-a)^2}{a^2}+\varepsilon^2\gamma_2(a,\tau_0,0)=0 \tag{80.8}$$

where $\gamma_2(a, \tau_0, 0)$ is equal to

$$\frac{\pi^2 k^2\{(4\pi^2+a^2+\tau_0)a^2+k^2(a^2-\tau_0)\}\{\pi^2 a^2+\pi^2\tau_0+2a^2\tau_0\}}{2a^4(\pi^2+\tau_0)[(4\pi^2+a^2+\tau_0)^2-\gamma_0(4\pi^2+a^2+\tau_0)-R_0(a^2+\tau_0)]}.$$

Eq. (80.8) defines an implicit function $\tilde{\gamma}_{(1)}(a, \varepsilon^2)=0$ with $\tilde{\gamma}_{(1)}(\pi, 0)=0$ and

$$\frac{\partial\gamma_{(1)}}{\partial(\varepsilon^2)}=\gamma_2(\pi, \pi^2, 0)=\tfrac{3}{7}\pi^2.$$

We may therefore solve (80.8) for $a(\varepsilon)$. We solve (80.8) by successive approximations: $a_0=\pi$, and then

$$\begin{aligned}
&-2(\pi-a)^2+\tfrac{3}{7}\varepsilon^2\pi^2=0,\\
&R-R_c=\tfrac{40}{3}(a-\pi)^2,\\
&R-R_c=\tfrac{20}{7}\varepsilon^2\pi^2.
\end{aligned} \tag{80.9}$$

The neutral curve for conduction is given to lowest approximation by

$$R_0-R_c=4(a-\pi)^2. \tag{80.10}$$

The ratio of the curvatures of the nonlinear and linear neutral curves is then

$$\frac{R(a, \varepsilon)-R_c}{R_0(a)-R_c}=\tfrac{10}{3}. \tag{80.11}$$

The parallel-roll instability gives $\gamma_{(1)}$ its maximum value and the cross-roll instability gives $\gamma_{(1)}$ its minimum value. The cross-roll instability is the most dangerous when $a>\pi$. For $a<\pi$ we must consider varicose ($\gamma_{(2)}$) and sinuous ($\gamma_{(3)}$) instabilities. As we see below, these cannot occur when $a>\pi$.

(b) Varicose Instabilities

We write $\gamma_{(2)}$ as given by (79.20) as $\gamma_{(2)}=\gamma_0+\varepsilon^2\gamma_2+\dots$ where

$$\gamma_0(a, \tau)=\frac{\tau(a^4-\pi^4+\tau a^2)}{a^2(k^2+\tau)}, \qquad \tau>0. \tag{80.12}$$

The nonlinear neutral curve is given by

$$\min_\tau \gamma_{(2)}(a, \tau; \varepsilon)=0. \tag{80.13}$$

When $\varepsilon=0$, and $a=\pi$, we find that $\tau=0$ solves (80.13); that is,

$$0=\gamma_{(2)}(\pi, 0; 0)=\gamma_0(\pi, 0).$$

Since $\gamma_2 > 0$ when τ is small, $\gamma_{(2)}$ cannot be negative when ε is small unless $\gamma_0 < 0$ and $\gamma_0 < 0$ only if $a < \pi$.

For $a < \pi$, we consider the minimum of $\gamma_{(2)}$

$$\frac{\partial \gamma_{(2)}}{\partial \tau} = \frac{\partial \gamma_0}{\partial \tau} + \varepsilon^2 \frac{\partial \gamma_2}{\partial \tau} + \cdots = 0 .$$

This gives the minimizing function

$$\tau(a, \varepsilon) = \tau_0 + \tau_2 \varepsilon^2 + \ldots$$

where

$$\tau_0 = \left(\frac{\pi}{a} - 1\right) k^2 .$$

Now we expand $\gamma_{(2)}(a, \tau(a, \varepsilon), \varepsilon)$ in powers of ε

$$\begin{aligned} 0 &= \gamma_0(a, \tau_0 + \tau_2 \varepsilon^2 + \ldots) + \varepsilon^2 \gamma_2(a, \tau_0 + \tau_2 \varepsilon^2 + \ldots) + \ldots \\ &= \gamma_0(a, \tau_0) + \varepsilon^2 \frac{\partial \gamma_0}{\partial \tau}(a, \tau_0) \tau_2 + \varepsilon^2 \gamma_2(a, \tau_0) + \ldots \\ &= \frac{-k^2(\pi - a)^2}{a^2} + \varepsilon^2 \gamma_2(a, \tau_0(a)) + \ldots . \end{aligned} \tag{80.14}$$

Eq. (80.14) defines a function $a(\varepsilon)$ with $a(0) = \pi$. In the next approximation

$$2(\pi - a)^2 = \varepsilon^2 \gamma_2(\pi, 0) = 2\pi^2 \varepsilon^2 \tag{80.15}$$

and

$$R - R_c = 6(a - \pi)^2 , \tag{80.16}$$

$$R - R_c = 6\pi^2 \varepsilon^2 , \tag{80.17}$$

where $R_c = 4\pi^2$. The curvature ratio is given by

$$\frac{R(a, \varepsilon) - R_c}{R_0(a) - R_c} = \tfrac{3}{2} . \tag{80.18}$$

(c) Sinuous Instabilities

The sinuous instability is governed by the eigenvalue $\gamma_{(3)}$ given by (79.24). We have already noted that $\gamma_0 = \tau \hat{\gamma}_0(a, \tau)$ is negative only when $a < \pi$; then its minimum value $-(\pi - a)^2 k^2 / a^2$ is attained when

$$\tau_0 = \left(\frac{\pi}{a} - 1\right) k^2 > 0 , \tag{80.19}$$

and then

$$\hat{\gamma}_0(a,\tau_0)=1-\frac{\pi}{a}<0\,. \tag{80.20}$$

Since $\gamma_2=\tau^2\hat{\hat{\gamma}}_2$ is negative, the nonlinear neutral curve

$$\min_\tau \gamma_{(3)}(a,\tau;\varepsilon)=0 \tag{80.21}$$

can be obtained, in the lowest approximation, only by including the terms of $O(\varepsilon^4)$.

We start by minimizing $\gamma_{(3)}$ when $a<\pi$. Then the function $\tau(a;\varepsilon)$ which minimizes appears as a root of the equation

$$0=\frac{\partial\gamma_{(3)}}{\partial\tau}=\frac{\partial\gamma_0}{\partial\tau}+\varepsilon^2\frac{\partial\gamma_2}{\partial\tau}+\varepsilon^4\frac{\partial\gamma_4}{\partial\tau}+\ldots\,. \tag{80.22}$$

Eq. (80.22) leads to the expression

$$\tau(a,\varepsilon)=\tau_0(a)+\varepsilon^2\tau_2(a)+\varepsilon^4\tau_4(a)+\ldots \tag{80.23}$$

for the minimizing τ. τ_0 is given by (80.19) and τ_2 and τ_4 are derived in the calculation which follows. We first expand (80.22) in powers of $\tau-\tau_0$ where τ is given by (80.23), getting

$$\begin{aligned}0&=\gamma_{0,\tau\tau}(\varepsilon^2\tau_2+\varepsilon^4\tau_4+\ldots)+\tfrac{1}{2}\gamma_{0,\tau\tau\tau}(\varepsilon^2\tau_2+\varepsilon^4\tau_4+\ldots)^2\\&\quad+\gamma_{2,\tau}\varepsilon^2+\gamma_{2,\tau\tau}\varepsilon^2(\varepsilon^2\tau_2+\ldots)+\gamma_{4,\tau}\varepsilon^4+\ldots\\&=(\tau_2\gamma_{0,\tau\tau}+\gamma_{2,\tau})\varepsilon^2+(\tau_4\gamma_{0,\tau\tau}+\tfrac{1}{2}\tau_2^2\gamma_{0,\tau\tau\tau}+\tau_2\gamma_{2,\tau\tau}+\gamma_{4,\tau})\varepsilon^4+\ldots\end{aligned} \tag{80.24}$$

where the subscript after comma means differentiate with respect to τ at $\tau=\tau_0$. The coefficients of different powers of ε^2 in (80.24) may be set to zero. After re-expressing the derivatives in terms of the roof functions defined in (79.24) we find that

$$\tau_2(2\hat{\gamma}_{0,\tau}+\tau_0\hat{\gamma}_{0,\tau\tau})+\tau_0(2\hat{\hat{\gamma}}_2+\tau_0\hat{\hat{\gamma}}_{2,\tau})=0 \tag{80.25}$$

and

$$\begin{aligned}&\tau_4(2\hat{\gamma}_{0,\tau}+\tau_0\hat{\gamma}_{0,\tau\tau})+\tfrac{1}{2}\tau_2^2(3\hat{\gamma}_{0,\tau\tau}+\tau_0\hat{\gamma}_{0,\tau\tau\tau})\\&\qquad+\tau_0\tau_2(2\hat{\hat{\gamma}}_{2,\tau}+\tau_0\hat{\hat{\gamma}}_{2,\tau\tau})+(\hat{\gamma}_4+\tau_0\hat{\gamma}_{4,\tau})=0\,.\end{aligned} \tag{80.26}$$

Hence,

$$\tau_2(a)=-\tau_0\hat{\hat{\gamma}}_2/\hat{\gamma}_{0,\tau}+O(\tau_0^2) \tag{80.27}$$

and

$$\tau_4(a) = -\hat{\gamma}_4/2\hat{\gamma}_{0,\tau} + O(\tau_0). \tag{80.28}$$

Eq. (80.19), (80.27) and (80.28) define the minimizing $\tau = \tau(a, \varepsilon)$ given by (80.23).

To find the nonlinear neutral curve we must study the function

$$\gamma_{(3)} = \tau(a; \varepsilon)\hat{\gamma}_{(3)}(a; \varepsilon) = 0.$$

Since $\tau > 0$ when $(\varepsilon, a) \neq (0, \pi)$ we have

$$\begin{aligned} 0 = \hat{\gamma}_{(3)} &= \hat{\gamma}_0(a, \tau_0 + \varepsilon^2\tau_2 + \varepsilon^4\tau_4) + \varepsilon^2[\tau_0 + \varepsilon^2\tau_2 + \varepsilon^4\tau_4]\hat{\hat{\gamma}}_2(a, \tau_0 + \varepsilon^2\tau_2 + \varepsilon^4\tau_4) \\ &\quad + \varepsilon^4\hat{\gamma}_4(a, \tau_0 + \varepsilon^2\tau_2 + \varepsilon^4\tau_4) + \dots \\ &= \hat{\gamma}_0(a, \tau_0) + \varepsilon^4\hat{\gamma}_4(a, \tau_0(a))/2 + O(\tau_0\varepsilon^2). \end{aligned} \tag{80.29}$$

Eq. (80.29) defines a function $a(\varepsilon) = a_0 + a_4\varepsilon^4 + \dots$ where $a_0 = \pi$. We find, using (79.25) that

$$a - \pi = -\frac{19\pi}{576}\varepsilon^4 + \dots. \tag{80.30}$$

Of course, $\tau_0 = ((\pi - a)/a)k^2 = O(\varepsilon^4)$; hence, all the neglected terms in (80.29) are of $O(\varepsilon^6)$.

Eq. (80.30) gives the projection of the nonlinear neutral curve on the (a, ε) plane. The projection on the (a, R) plane is determined as follows:

$$\begin{aligned} R &= R_0 + R_2\varepsilon^2 + \dots = R_0(1 + \varepsilon^2/2 + \dots) \\ &= R_0\left(1 + \tfrac{1}{2}\sqrt{\frac{576}{19\pi}(a_0 - a)}\right) + \dots = R_c\left[1 + \frac{12}{\sqrt{19}}\left(1 - \frac{a}{a_0}\right)^{1/2}\right] + \dots \end{aligned} \tag{80.31}$$

where $(a_0, R_c) = (\pi, 4\pi^2)$. The most dangerous disturbances when $a < \pi$ are the sinuous rolls.

The function $\tau(a(\varepsilon), \varepsilon)$, given by (80.23) and (80.30), gives the period, $2\pi/\sqrt{\tau}$, of the standing instability wave on the marginally stable bifurcating roll. This is a very long wave which tends to infinity with ε^{-3}. To show this we note that $\tau_0(a(\varepsilon)) = O(\varepsilon^4)$ and, using (80.23), (80.27) and (80.28),

$$\tau(a(\varepsilon), \varepsilon) = \tau_0(a(\varepsilon))[1 - \varepsilon^2\hat{\hat{\gamma}}_2/\hat{\gamma}_{0,\tau}] - \hat{\gamma}_4\varepsilon^4/2\hat{\gamma}_{0,\tau} + O(\varepsilon^6)$$

where, for example, terms of $O(\varepsilon^6)$ arise in τ_6 through γ_6. We did not calculate these $O(\varepsilon^6)$ terms. Since

$$\frac{\partial\gamma_0}{\partial\tau} = \tau\hat{\gamma}_{0,\tau} + \hat{\gamma}_0 = 0 \quad \text{at} \quad \tau = \tau_0 = \left(\frac{\pi}{a} - 1\right)k^2 \to \frac{19\pi^2}{288}\varepsilon^4 \tag{80.32}$$

and

$$\hat{\gamma}_{0,\tau} = -\hat{\gamma}_0/\tau_0 = k^{-2} \to \tfrac{1}{2}\pi^{-2}$$

we find, using (79.25), that the terms of $O(\varepsilon^4)$ cancel leaving

$$\tau(a(\varepsilon), \varepsilon) = O(\varepsilon^6)\,. \tag{80.33}$$

Exercise 80.1: Show

$$0 = \min_{\gamma} \gamma_{(1)}(a, v; \varepsilon)$$

where

$$\gamma_{(1)} = \gamma_0(a, v) + \gamma_2(a, v, 1)\varepsilon^2 + O(\varepsilon^4)$$

is given by

$$0 = \frac{-(\pi^2+a^2)(\pi-a)^2}{a^2} + \varepsilon^2 \gamma_2\left(a, \left[\frac{\pi}{a}(\pi^2+a^2-\pi a)\right]^{1/2}\right) + O(\varepsilon^4)\,.$$

Deduce that this parallel-roll instability leads formally to a curvature ratio of 2 but that (79.9) cannot be solved for ψ_2 when $n=2$, $\tau=0$ and $v=a$.

Exercise 80.2: Consider the Bénard problem for an OB fluid with an infinite Prandtl number between conducting and shear-stress free horizontal planes. Show that curvature ratio for the varicose instability is $\frac{3}{2}$, that the curvature ratio for the cross-roll instability is $\frac{673}{200}$ and that the nonlinear neutral curve for the sinuous instability is given by

$$\frac{R}{R_c} = 1 + \tfrac{1}{2}\varepsilon^2 = 1 + \sqrt{\frac{58345280}{5264739}}\left(1 - \frac{a}{a_0}\right)^{1/2}.$$

§ 81. Computation of Stability Boundaries by Numerical Methods

The results obtained in § 80 are valid only for small values of the amplitude ε. Busse (1967) introduced numerical methods to study bifurcation and stability which are not restricted to small values of ε. In his (1967) paper he considered the bifurcation and stability of roll solutions of the OB equations in an infinite Prandtl number fluid by a Galerkin method. He showed that rolls are stable to nonoscillatory disturbances of the type considered in § 79 for Rayleigh numbers up to about 23000. The results of his computations are displayed in Fig. 81.1. He found that rolls are unstable outside the closed nonlinear neutral curve shown in the Figure. For Rayleigh numbers above 23000 experiments indicate that stable, steady, three-dimensional solutions in the form of cross-rolls (bimodal convection) may bifurcate from roll convection. At still higher values of the Rayleigh number the motion becomes time-dependent and more complicated.

Busse's method has been applied to the problem of porous convection by Straus (1974) and to more complicated problems by Clever and Busse (1974). In these papers the bifurcating solutions are expanded into a Fourier series with

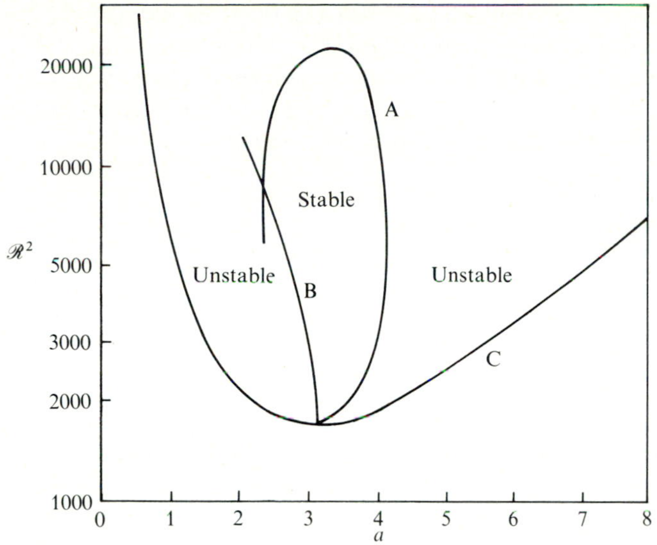

Fig. 81.1: Stability of roll convection in a infinite Prandtl number OB fluid between horizontally infinite rigid conducting plates: (A) Cross-roll instability, (B) Sinuous instability, (C) Neutral curve for conduction (Courtesy of F. Busse, 1967)

period $2\pi/a$ in x and the disturbances considered are restricted to the noninteracting type where

$$\Phi(x, y, z) = e^{ivx} e^{iy\sqrt{\tau}} f(x, z), \tag{81.1}$$

and

$$f(x, z) = f(x + 2\pi/a, z)$$

is represented as a Fourier series. In the numerical analysis it is not assumed that γ is real-valued and some cases of complex γ have been found by Clever and Busse (1974) for low Prandtl number convection. We shall consider the numerical methods for the problem of porous convection.

To study the stability of convection, the bifurcating roll and the disturbance are both represented as Fourier series which satisfy all the required boundary conditions:

$$\chi = \sum \chi_{lr} \sin l\pi z \, e^{irax}, \tag{81.2}$$

$$\Phi = e^{i\sqrt{\tau y}} \sum \psi_{nm} \sin n\pi z \, e^{iv_m x}, \qquad v_m = ma + v \tag{81.3}$$

$n, l = 1, 2, \ldots$
$r, m = 0, \pm 1, \pm 2, \ldots$

and ψ_{nm} is complex-valued. The coefficients χ_{lr} are obtained by solving the algebraic equations which arise when the representation (81.1) is inserted into (78.16). Some of these coefficients may be obtained from the solutions given by (78.25). The representations (81.2) and (81.3) are then inserted into (79.1) and after factoring e^{ivx} which is common to all terms, one obtains the following equation

$$\sum_{p,q} A_{pq}\psi_{pq} \sin p\pi z e^{iqax} = \frac{\pi\varepsilon}{2} \sum_{l,r} \sum_{n,m} \chi_{lr}\psi_{nm} e^{i(m+r)ax} \cdot [K_{nmlr} \sin(n+l)\pi z + H_{nmlr} \sin(n-l)\pi z]$$

where

$$A_{pq} = (p^2\pi^2 + \tau + v_q^2)^2 - \gamma(p^2\pi^2 + \tau + v_q^2) - R(\tau + v_q^2),$$
$$K_{nmlr} = ra(lv_m - rna)(n^2\pi^2 + \tau + v_m^2) + (l^2\pi^2 + r^2a^2)[v_m rna - l(v_m^2 + \tau)],$$
$$H_{nmlr} = ra(lv_m + rna)(n^2\pi^2 + \tau + v_m^2) - (l^2\pi^2 + r^2a^2)[v_m rna + l(v_m^2 + \tau)].$$

In the numerical method the system of algebraic equations $(n, l, p \geqslant 1)$

$$A_{pq}\psi_{pq} = \frac{\varepsilon\pi}{2} \sum_{m+r=q} \{\sum_{n+l=p} \chi_{lr}\psi_{nm} K_{nmlr} + \sum_{n-l=p} \chi_{lr}\psi_{nm} H_{nmlr}\}, \tag{81.4}$$

which arise from identifying Fourier coefficients, are truncated and the resulting set is solved numerically. The Eqs. (81.4) arise from a linear stability problem and they are linear in the coefficients ψ_{pq}. The scale of these functions may be selected arbitrarily. Truncated solutions are solutions ψ_{pq}^{NM} with the property that $\psi_{pq}^{NM} = 0$ if $|p| > N$ or $|q| > M$. Truncated solutions therefore satisfy a finite number of linear equations arising from (81.3). Convergence of the truncated solutions is checked by requiring that some measure of the difference of solutions computed for increasing values of (N, M) should tend to zero.

Solutions of (81.4) have a closedness property. Without loss of generality, χ may be taken to be an even function of x for which $\chi_{lr} = 0$ when $l + r = \text{odd}$. Then $p + q = m + n - l + r$ is even or odd according as $m + n$ is even or odd. Hence solutions of (81.4) separate into two families one for which $p + q$ is odd and one for which $p + q$ is even.

The perturbation analysis given in § 79 and § 80 and the numerical analysis under discussion both start from solutions of the form (81.1). To extract the perturbation results from (81.4) we need only to expand everything in sight in powers of ε:

$$[R, \gamma, A_{pq}, \chi_{pq}, \psi_{pq}] = \sum_{l=0} [R_l, \gamma_l, A_{pq}^{(l)}, \chi_{pq}^{(l)}, \psi_{pq}^{(l)}]\varepsilon^l.$$

Inserting these expressions in (81.4) we get

$$A_{pq}^{(0)}\psi_{pq}^{(0)} = 0 \tag{81.5}$$

and, for $k>0$,

$$A_{pq}^{(0)}\psi_{pq}^{(k)}+A_{pq}^{(k)}\psi_{pq}^{(0)}=-\sum_{\substack{v+\mu=k\\ \mu,v\neq k}}A_{pq}^{(v)}\psi_{pq}^{(\mu)}+\frac{\pi}{2}\sum_{m+r=q}\sum_{v+\mu=k-1} \cdot\left\{\sum_{n+l=p}\chi_{lr}^{(v)}\psi_{nm}^{(\mu)}K_{nmlr}+\sum_{n-l=p}\chi_{lr}^{(v)}\psi_{nm}^{(\mu)}H_{nmlr}\right\} \tag{81.6}$$

where

$$A_{pq}^{(0)}=[p^2\pi^2+\tau+v_q^2]^2-\gamma_0[p^2\pi^2+\tau+v_q^2]-R_0(v_q^2+\tau)$$

and

$$A_{pq}^{(v)}=-\gamma_v[p^2\pi^2+\tau+v_q^2]-R_v(v_q^2+\tau).$$

It will be noted that v appears in the expressions (81.5, 6) only in the combination $v_q=v+qa$. The form of (81.5) and (81.6) is invariant to a simultaneous change of the index q and the definition of v. To be explicit, set $q=q'+N$ and $m=m'+N$ and $\psi_{pq}=\psi_{p,q'+N}$ for an arbitrary given integer N. Then (81.5) and (81.6) hold with (q,m,v) replaced with $(q',m',v'=v-Na)$. We do not lose generality if we regard v as a small quantity and the disturbance (81.1) as a small deviation from a periodic solution of period $2\pi/a$.

To initiate the perturbation we must first solve (81.5) for all $p\geqslant 1$ and $q=0$, ±1, ±2, ... These equations are satisfied for preassigned values of a, v, τ when, for certain values of p and q, say, P and Q,

$$A_{PQ}^{(0)}\equiv\omega^2P^4\left[R_0(\omega^2)-\frac{R_0(a^2)}{P^2}-\frac{\gamma_0(\pi^2+\omega^2)}{P^2\omega^2}\right]=0\,,\qquad \omega^2=\frac{\tau+(Qa+v)^2}{P^2}\,, \tag{81.8}$$

whereas, for the other values of p and q, $\psi_{pq}^{(k)}=0$. Every equation $A_{PQ}^{(0)}=0$ must give the same value for γ_0. The smallest value of γ_0 corresponds to $p=1$ and is such that $\gamma_0=0$ when $a^2=\pi^2$. In this case $p^2=1$ and $R_0(\omega^2)=4\pi^2$ where

$$\omega^2=\pi^2=\tau+(\pm Q\pi+v)^2. \tag{81.9}$$

The two Eqs. (81.9) hold simultaneously if and only if $Qv=0$. There are then two sets of initiating values for (P,Q):

$$(P,Q)=(1,0)\,;\qquad A_{10}^{(0)}=0\,,\qquad \psi_{pq}^{(0)}=0\quad\text{when}\quad (p,q)\neq(1,0) \tag{81.10}$$

and

$$(P,Q)=(1,\pm1),\qquad v=0;\qquad A_{11}^{(0)}=A_{1-1}^{(0)}=0,\qquad \psi_{pq}^{(0)}=0\ \text{ when }\ (p,q)\neq(1,\pm1). \tag{81.11}$$

Case (81.10) corresponds to the oblique-roll instability which was discussed in § 79 and § 80. Case (81.11) corresponds to sinuous and varicose rolls. Cross-roll instabilities fall under (81.10) when $v=0$. The x-periodic disturbances have $v=0$. When $v=0$ we may put $A_{pq}^{(k)}=A_{p-q}^{(k)}$ and restrict our attention to values $q\geqslant 0$. Given initiating values (81.10) or (81.11), the higher order problems (81.6) may be solved sequentially for $\psi_{pq}^{(k)}$ provided that there are no values (p,q) which lie in an exceptional set. This set is defined as follows: We write (81.6) as

$$A_{pq}^{(0)}\psi_{pq}^{(k)}+A_{pq}^{(k)}\psi_{pq}^{(0)}=F_{pq}^{(k)}. \tag{81.12}$$

A pair $(p,q)\neq(P,Q)$ is said to belong to an exceptional set if for some $k>0$, $F_{pq}^{(k)}\neq 0$ and $A_{pq}^{(0)}=0$. If $(p,q)\neq(P,Q)$ then $\psi_{pq}^{(0)}=0$ and it is not possible to solve (81.12) for $\psi_{pq}^{(k)}$.

If the exceptional set is empty all of the Eqs. (81.6) are solvable. To illustrate, consider (81.10). Since $\psi_{10}^{(0)}\neq 0$ we may use the index separation (see paragraph above (81.5)) to conclude that the $\psi_{mn}^{(k)}=0$ when $m+n=$even. The solvability equation

$$A_{10}^{(k)}\psi_{10}^{(0)}=F_{10}^{(k)} \tag{81.13}$$

determines the value of γ_k uniquely. The remaining equations of (81.12) give

$$\psi_{pq}^{(k)}=F_{pq}^{(k)}/A_{pq}^{(0)} \tag{81.14}$$

determining $\psi_{pq}^{(k)}$ when $(p,q)\neq(1,0)$. It is necessary to note that the coefficients $\psi_{pq}^{(0)}$ are undetermined by the perturbation and may be selected to satisfy any normalizing condition which is imposed on $\Phi(x,y,z)$.

It is of interest to consider the values of parameters (a,v,τ) for which the exceptional set is not empty. For these parameters $A_{pq}^{(0)}=0$ and (81.14) cannot be solved. When (a,v,τ) are close to values for which $A_{pq}^{(0)}(a,v,\tau)=0$ then $A_{pq}^{(0)}=$ small and $\psi_{pq}^{(k)}=F_{pq}^{(k)}/$small is large. This troublesome problem is called the problem of "small divisors" or "near resonance".

Eq. (81.8) shows that $A_{pq}^{(0)}>0$ is bounded away from zero when $p^2>1$; γ_0 is not too much greater than zero and a^2 is not too much different from π^2. Hence, in the neighborhood of the nose of the nonlinear neutral curve the values (p,q) for which $p\neq 1$ cannot lie in the exceptional set. When $p=1$ we find that

$$\begin{aligned}A_{1Q}^{(0)}-A_{1m}^{(0)}=(Q-m)\{a^2(Q+m)+2va\}\{&2(\pi^2+\tau)\\&+(ma+v)^2+(Qa+v)^2-(\gamma_0+R_0)\}.\end{aligned} \tag{81.15}$$

Consider (81.15) for the case specified under (81.10). For this case, we may write (81.15) as

$$A_{10}^{(0)}-A_{1m}^{(0)}=-m\{m+2v/a\}\{-\pi^4+a^2[2\tau+(ma+v)^2+v^2-a^2-\gamma_0]\}. \tag{81.16}$$

The perturbation fails if $A_{10}^{(0)}-A_{1m}^{(0)}=0$ for any even value of $m\neq 0$. When $v=0$ (cross-rolls)

$$A_{10}^{(0)}-A_{1m}^{(0)}=-m^2\{(m^2-1)a^4-\pi^4+2a^2\tau-\gamma_0 a^2\},\quad m=\pm 2,\pm 4,\ldots$$

is not zero or close to zero near the nose of the nonlinear neutral curve. In this case the perturbation problems are solvable and possess bounded inverses. At the other extreme are the parallel-rolls; for these $\tau=0$ and, when $a=\pi$, $\gamma_0=0$ and

$$A_{10}^{(0)}-A_{1m}^{(0)}=-m\{m+2v/\pi\}\{v^2-2\pi^2+(m\pi+v)^2\}\pi^2 .$$

The perturbation fails when $v=\pi$. Then

$$A_{1-2}^{(0)}=A_{10}^{(0)}=0 . \tag{81.17}$$

$A_{1-2}^{(0)}$ is small when v is close to π so that the solutions of (81.6) do not have uniformly bounded inverses.

Exercise 81.1: Investigate the exceptional set corresponding to (81.11). Under what conditions do the Eqs. (81.12) posess bounded inverses?

Exercise 81.2: Derive (79.23) and (79.24) by the Fourier methods used in this section. (*Hint:* follow the recipe specified in 79.22.)

§ 82. The Amplitude Equation of Newell and Whitehead

A different approach to the problem of supercritical bifurcation and stability has been taken by Newell and Whitehead (1969). They study convection of an OB fluid between plane stress-free, conducting boundaries. They derive a partial differential equation for the amplitude of solutions which perturb the natural solution at the nose of the neutral curve. In the problem of convection of a DOB fluid considered here the neutral solutions at the node of the neutral curve are the solutions $\hat{\chi}\rightarrow\bar{\varepsilon}\hat{\chi}_0$

$$\mathscr{L}\hat{\chi}=\boldsymbol{\delta}\hat{\chi}\cdot\nabla\nabla^2\hat{\chi}\,, \qquad \hat{\chi}=\partial^2\hat{\chi}/\partial z^2=0 \quad \text{at} \quad z=0,1 \tag{82.1}$$

where

$$\mathscr{L}=\left[\nabla^2-\frac{\partial}{\partial t}\right]\nabla^2+R\nabla_2^2\,,$$

and

$$(a^2,\bar{\varepsilon},R)=(\pi^2,0,4\pi^2)\,.$$

The amplitude equation can be used to study the initial value problem for convection at slightly supercritical values of $R>4\pi^2$. Fig. 82.1 shows that there is a band of possible bifurcating solutions for each a^2 in an interval $[a_-^2, a_+^2]$ on the plane $R=\text{const}$, $R>4\pi^2$.

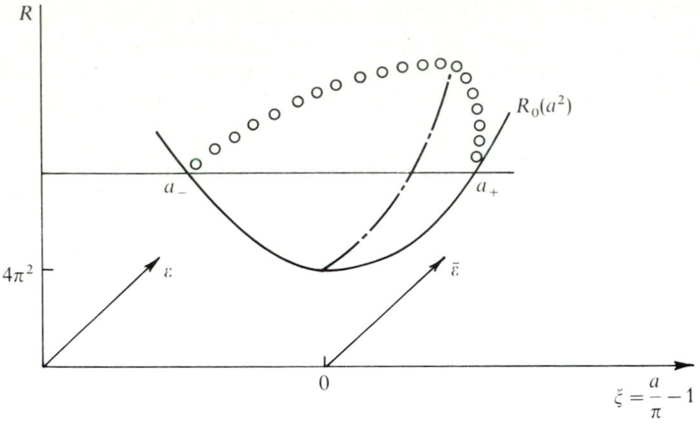

Fig. 82.1: Schematic sketch of the surface $R(a,\varepsilon)$ of values (R,a,ε) for which steady roll-convection exists. $R_0(a^2)$ is the neutral curve for conduction. Newell and Whitehead (1969) perturb the solution at the nose of the neutral curve. Their expansion parameter $\bar{\varepsilon}$ can be interpreted as the Nusselt number discrepancy $\bar{\varepsilon}^2 = \varepsilon^2$ in the plane $\xi = 0$. When ε and $\bar{\varepsilon}$ are small

$$R(a,\varepsilon) \sim R_0(a^2)(1+\tfrac{1}{2}\varepsilon^2), \qquad R(\pi,\bar{\varepsilon}) \sim 4\pi^2(1+\bar{\varepsilon}^2/2).$$

For a fixed value $R(a,\varepsilon) = R(\pi,\bar{\varepsilon})$ and small values of $\bar{\varepsilon}$, ε, ξ and $R - 4\pi^2$ the curve on the plane $R = \text{const}$ and the surface $R(a,\varepsilon)$ is given by

$$\bar{\varepsilon}^2 = \varepsilon^2 + 2\xi^2$$

At criticality, (82.1) reduces to

$$\nabla^4 \hat{\chi}_0 + 4\pi^2 \nabla_2^2 \hat{\chi}_0 = 0, \qquad \hat{\chi}_0 = \partial^2 \hat{\chi}_0 / \partial z^2 = 0 \quad \text{at} \quad z = 0, 1. \tag{82.2}$$

There are infinitely many quasiperiodic solutions of (82.2) of the form

$$\hat{\chi}_0 = \sum_{-N}^{N} \chi_j e^{i\mathbf{k}_j \cdot \mathbf{r}}, \qquad \bar{\chi}_j = \chi_{-j}, \qquad |\mathbf{k}_j|^2 = \pi^2, \qquad \mathbf{r} = \mathbf{e}_x x + \mathbf{e}_y y. \tag{82.3}$$

When $R - 4\pi^2 > 0$ is small, the solutions of (82.1) may be regarded as perturbations of (82.3). These perturbations stem from the nonlinearity associated with $\bar{\varepsilon}$ ($\bar{\varepsilon}^2$ is proportional to the discrepancy $R - 4\pi^2$ and will be defined more precisely by 82.19) and from small spatial and temporal deviations of the χ_j from constants. The spatial and temporal deviations are accommodated by allowing $\chi_j(X, Y, T)$ to depend on the "slow scales" X, Y, T. The relation of these scales to the discrepancy $\bar{\varepsilon}^2$ is determined by guesswork, based on experience, and checked *a-posteriori* for consistency.

Newell and Whitehead start with the case in which the slow scales are given by

$$X = \bar{\varepsilon} x, \qquad Y = \bar{\varepsilon} y, \qquad T = \bar{\varepsilon}^2 t. \tag{82.5}$$

At the end of this section we shall make a few remarks about the selection of slow spatial scales. For the present we may regard (82.5) as a hypothesis and regard X, Y and T as independent variables. Derivatives acting on functions of x, y, t, X, Y, T transform as follows

$$\begin{aligned}\frac{\partial}{\partial t}&\to\frac{\partial}{\partial t}+\bar{\varepsilon}^2\frac{\partial}{\partial T},\\ \frac{\partial}{\partial x}&\to\frac{\partial}{\partial x}+\bar{\varepsilon}\frac{\partial}{\partial X}\end{aligned}\tag{82.6}$$

and

$$\frac{\partial}{\partial y}\to\frac{\partial}{\partial y}+\bar{\varepsilon}\frac{\partial}{\partial Y}.$$

We next expand the functions $\hat{\chi}=\bar{\varepsilon}\hat{\chi}_0+\bar{\varepsilon}^2\hat{\chi}_1+\bar{\varepsilon}^3\hat{\chi}_2+\dots$ and the operators in (82.1), using (82.6), and retaining terms through $O(\bar{\varepsilon}^3)$:

$$\begin{aligned}&\mathscr{L}_0(\bar{\varepsilon}\hat{\chi}_0+\bar{\varepsilon}^2\hat{\chi}_1+\bar{\varepsilon}^3\hat{\chi}_2)+\bar{\varepsilon}\mathscr{L}_1(\bar{\varepsilon}\hat{\chi}_0+\bar{\varepsilon}^2\hat{\chi}_1)+\bar{\varepsilon}^2\mathscr{L}_2\bar{\varepsilon}\hat{\chi}_0\\ &\quad=\bar{\varepsilon}^2\boldsymbol{\delta}\hat{\chi}_0\cdot\nabla\nabla^2\hat{\chi}_0+\bar{\varepsilon}^3\left[\boldsymbol{\delta}\hat{\chi}_0\cdot\nabla\nabla^2\hat{\chi}_1+\boldsymbol{\delta}\hat{\chi}_1\cdot\nabla\nabla^2\hat{\chi}_0+M(\hat{\chi}_0,\hat{\chi}_0)\right]\end{aligned}\tag{82.7}$$

where

$$\mathscr{L}_0=\nabla^4-\frac{\partial}{\partial t}\nabla^2+R_0\nabla_2^2,$$

$$\mathscr{L}_1=2\hat{\mathscr{L}}\nabla_2\cdot\hat{\nabla}_2,$$

$$\mathscr{L}_2=\hat{\mathscr{L}}\hat{\nabla}_2^2+R_2\nabla_2^2+\left[4(\nabla_2\cdot\hat{\nabla}_2)^2-\frac{\partial}{\partial T}\nabla^2\right],$$

$$\nabla_2=\mathbf{e}_x\frac{\partial}{\partial x}+\mathbf{e}_y\frac{\partial}{\partial y},$$

$$\hat{\nabla}_2=\mathbf{e}_x\frac{\partial}{\partial X}+\mathbf{e}_y\frac{\partial}{\partial Y}$$

and

$$\hat{\mathscr{L}}=-\frac{\partial}{\partial t}+2\nabla^2+R_0.$$

A typical term of $M(\hat{\chi}_0,\hat{\chi}_0)$ is $\dfrac{\partial^2\hat{\chi}_0}{\partial X\partial z}\nabla^2\dfrac{\partial\hat{\chi}_0}{\partial x}$. Every term of $M(\hat{\chi}_0,\hat{\chi}_0)$ is differentiated with respect to a slow variable. At $z=0, 1$

$$\hat{\chi}_i=\partial^2\hat{\chi}_i/\partial z^2=0\qquad(i\geqslant 0).\tag{82.8}$$

Equating coefficients of different powers of $\bar{\varepsilon}$ in (82.7), we find that

$$\mathscr{L}_0\hat{\chi}_0=0, \tag{82.9}$$

$$\mathscr{L}_0\hat{\chi}_1+\mathscr{L}_1\hat{\chi}_0=\boldsymbol{\delta}\hat{\chi}_0\cdot\nabla\nabla^2\hat{\chi}_0 \tag{82.10}$$

and

$$\mathscr{L}_0\hat{\chi}_2+\mathscr{L}_1\hat{\chi}_1+\mathscr{L}_2\hat{\chi}_0=\boldsymbol{\delta}\hat{\chi}_0\cdot\nabla\nabla^2\hat{\chi}_1+\boldsymbol{\delta}\hat{\chi}_1\cdot\nabla\nabla^2\hat{\chi}_0+M(\hat{\chi}_0,\chi_0). \tag{82.11}$$

At zeroth order, superpositions of neutral solutions (82.3) are possible. For roll solutions

$$\hat{\chi}_0=[\chi(X,Y,T)e^{i\mathbf{k}\cdot\mathbf{r}}+\bar{\chi}(X,Y,T)e^{-i\mathbf{k}\cdot\mathbf{r}}]\sin\pi z \tag{82.12}$$

where $\mathbf{k}=\mathbf{e}_x\pi$ and the overbar designates complex conjugation. Of course, $\mathscr{L}_0\hat{\chi}_0=0$; in addition $\hat{\mathscr{L}}\hat{\chi}_0=0$ and

$$\boldsymbol{\delta}\hat{\chi}_0\cdot\nabla\nabla^2\hat{\chi}_0=-4\pi^5\chi\bar{\chi}\sin 2\pi z. \tag{82.13}$$

We find that

$$\hat{\chi}_1=-\frac{\pi}{4}\chi\bar{\chi}\sin 2\pi z. \tag{82.14}$$

Using (82.12) and (82.14), we find that the right side of (82.11) reduces to

$$-\nabla_2^2\hat{\chi}_0\frac{\partial^3\hat{\chi}_1}{\partial z^3}+M(\hat{\chi}_0,\hat{\chi}_0)=-\pi^6\chi\bar{\chi}(\chi e^{i\mathbf{k}\cdot\mathbf{r}}+\bar{\chi}e^{-i\mathbf{k}\cdot\mathbf{r}})(\sin\pi z-\sin 3\pi z)+M(\hat{\chi}_0,\hat{\chi}_0).$$

Moreover, $M(\hat{\chi}_0,\hat{\chi}_0)$ and $\mathscr{L}_1\hat{\chi}_1$ contain no term proportional to $\sin\pi z$. Hence,

$$\begin{aligned}\mathscr{L}_0\hat{\chi}_2+\mathscr{L}_2\hat{\chi}_0=&-\pi^6\chi\bar{\chi}(\chi e^{i\mathbf{k}\cdot\mathbf{r}}+\bar{\chi}e^{-i\mathbf{k}\cdot\mathbf{r}})\sin\pi z\\&+\text{terms proportional to }\sin n\pi z\quad(n\neq 1)\end{aligned} \tag{82.15a}$$

where on $z=0,1$

$$\hat{\chi}_2=\partial^2\hat{\chi}_2/\partial z^2=0. \tag{82.15b}$$

The problem (82.15) is solvable only if the inhomogeneous terms are orthogonal to the eigenfunctions of $\mathscr{L}_0$. Hence

$$\langle e^{-i\mathbf{k}\cdot\mathbf{r}}\sin\pi z\mathscr{L}_2\chi_0\rangle=-\pi^6\chi^2\bar{\chi}. \tag{82.16}$$

Recalling now the definition of $\mathscr{L}_2$, we find that

$$\frac{\partial \chi}{\partial T}-\frac{2}{\pi^2}(\mathbf{k}\cdot\hat{\nabla}_2)^2\chi=\tfrac{1}{2}(R_2-\pi^4\chi\bar{\chi})\chi\,. \tag{82.17}$$

Eq. (82.17) is the amplitude equation of Newell and Whitehead derived here for the problem of convection in a layer of porous material. When diffusion is neglected (82.17) reduces to an amplitude equation of the type studied by Segel (1966):

$$\frac{\partial \chi}{\partial T}=\tfrac{1}{2}(R_2-\pi^4\chi\bar{\chi})\chi\,. \tag{82.18}$$

There are two steady solutions of (82.18), $\chi=0$ and $\chi=\chi_e$ where $|\chi_e|^2=R_2/\pi^4$. Since $R_2>0$, bifurcating roll convection of period 2 in x is supercritical. $\chi=0$ is unstable when $R>4\pi^2$ $(R_2>0)$ and $\chi=\chi_e$ is stable and attracting for disturbances whose initial amplitudes satisfy the inequality $|\chi|^2_{T=0}<R_2/\pi^4$. The value of R_2 depends on the choice of $\bar{\varepsilon}$. We may regard $\bar{\varepsilon}^2$ as the Nusselt number discrepancy $Nu-1$: then (78.20) holds,

$$\langle\nabla_2^2\chi_e\nabla^2\chi_e\rangle=2\pi^4\chi_e^2=R_0=4\pi^2 \tag{82.19}$$

and $R_2=2\pi^2$.

The term

$$(\mathbf{k}\cdot\hat{\nabla}_2)^2\chi=\left(k_x\frac{\partial}{\partial X}+k_y\frac{\partial}{\partial Y}\right)^2\chi$$

of (82.17) introduces the effect of diffusion. Without loss of generality we can choose X to lie in the direction $\mathbf{k}$. Then (82.17) may be written as

$$\frac{\partial \chi}{\partial T}-2\frac{\partial^2\chi}{\partial X^2}=\tfrac{1}{2}(2\pi^2-\pi^4\chi\bar{\chi})\chi\,. \tag{82.20}$$

The relevant diffusion time is that of heat conduction in a fluid-saturated porous medium; spatial nonuniformities are propagated according to the law

$$x\sim 4\left[\frac{k_m t}{(\rho C_0)_m}\right]^{1/2}$$

where the subscript m stands for the mixture average defined by (70.8). In the more general problem treated by Newell and Whitehead the diffusion time depends on the Prandtl number and at low Prandtl numbers it is the viscous diffusion time which dominates. Newell and Whitehead highlight this difference by pointing out that the information that one roll is spinning is relayed to its neighbors by a thermal torque in a high Prandtl number fluid (and in a DOB fluid) and by a viscous torque in a low Prandtl number fluid.

(82.20) accommodates transient solutions

$$\chi = f(T)e^{iKX}, \qquad K^2 < \pi^2/2$$

which are all driven to the finite-amplitude steady state

$$\chi = \frac{2}{\pi}\left[\frac{1}{2} - \frac{K^2}{\pi^2}\right]^{1/2} e^{iKX}. \tag{82.21}$$

Returning to (82.12) with $\mathbf{k} = \mathbf{e}_x\pi$, (82.21), and $X = \bar{\varepsilon}x$, we find that, to lowest order

$$\hat{\chi} \sim \bar{\varepsilon}\hat{\chi}_0 = \frac{2\bar{\varepsilon}}{\pi}\left[\frac{1}{2} - \frac{K^2}{\pi^2}\right]^{1/2} \{e^{i(\pi + K\bar{\varepsilon})x} + e^{-i(\pi + K\bar{\varepsilon})x}\} \sin \pi z. \tag{82.22}$$

The expression (82.22) contains the slowly varying amplitude $\chi(X) = e^{iK\bar{\varepsilon}x}$. To compare (82.22) with neutral solutions for which $\varepsilon \to 0$ and $a^2 \neq \pi^2$ we set $\pi + K\bar{\varepsilon} = a$ and note, using Fig. 82.1, that

$$\varepsilon^2 = \bar{\varepsilon}^2 - 2\xi^2 = \bar{\varepsilon}^2 - 2\left(\frac{a}{\pi} - 1\right)^2 = \bar{\varepsilon}^2(1 - 2K^2/\pi^2).$$

Introducing these substitutions into (82.20) one finds that, in agreement with earlier results,

$$\hat{\chi} = \frac{\varepsilon\sqrt{2}}{\pi}(e^{iax} + e^{-iax}) \sin \pi z.$$

In the analysis leading to (82.17) we considered the amplitude equation corresponding to the scales (82.5) describing the effect of slow spatial variations which perturb bifurcating rolls. We turn now to the same problem when the bifurcating solution has a more complicated plan form, like hexagons and rectangles. This problem requires that one consider superpositions like (82.3) where the χ_j depend on X, Y, and T which, for the moment, are assumed to be given by (82.5). We therefore start by replacing (82.12) with such superpositions

$$\hat{\chi}_0 = \sum_{l=1}^{N} [\chi_l(X_l, Y_l, T)e^{i\mathbf{k}_l \cdot \mathbf{r}} + \bar{\chi}_l(X_l, Y_l, T)e^{-i\mathbf{k}_l \cdot \mathbf{r}}] \sin \pi z$$

where X_l is the coordinate along the vector $\mathbf{k}_l$ and Y_l is the coordinate perpendicular to this vector. Proceeding as in the derivation of (82.17) with computations which are essentially those leading to the expressions (79.15) for γ_2, we find that

$$\begin{aligned}\frac{\partial \chi_n}{\partial T} - \frac{2}{\pi^2}(k_n \cdot \hat{\nabla}_2)^2 \chi_n = \tfrac{1}{2}[R_2 - \pi^4 \textstyle\sum_{j=1}^{N} |\chi_j|^2]\chi_n \\ - \frac{\pi^4}{4} \sum_{\substack{j=1 \\ j \neq n}}^{N} \beta_{jn}|\chi_j|^2 \chi_n\end{aligned} \tag{82.23}$$

where

$$\beta_{jn}=\frac{(1-\cos\theta_{nj})^2(3+\cos\theta_{nj})}{(3+\cos\theta_{nj})^2-2(1+\cos\theta_{nj})}+\frac{(1+\cos\theta_{nj})^2(3-\cos\theta_{nj})}{(3-\cos\theta_{nj})^2-2(1-\cos\theta_{nj})}.$$

As an application of (82.23) consider the oblique-roll disturbances of bifurcating roll convection $(R_2=2\pi^2)$:

$$\chi_1=\frac{2}{\pi}\left[\frac{1}{2}-\frac{K^2}{\pi^2}\right]^{1/2}e^{iKX},$$

$$\chi_2=u_2(T), \qquad \chi_n=0 \quad \text{for} \quad n>2$$

where we understand that $u_2(T)$ is an infinitesimal disturbance of χ_1. We find that

$$\begin{aligned}\frac{du_2}{dT}&=\left\{\tfrac{1}{2}(2\pi^2-\pi^4|\chi_1|^2)-\frac{\pi^4}{4}\beta_{12}|\chi_1|^2\right\}u_2\\&=\left\{(2+\beta_{12})K^2-\frac{\pi^2}{2}\beta_{12}\right\}u_2.\end{aligned} \tag{82.24}$$

The growth rate $\lambda=(2+\beta_{12})K^2-\pi^2\beta_{12}/2$ of $u_2\propto\exp(\lambda T)$ is positive for wave numbers K^2 such that

$$K^2>\pi^2\beta_{12}/2(2+\beta_{12}) \tag{82.25}$$

and is maximum when the quantity on the right side of (82.25) is minimum; and this minimum is taken on when β_{12} is minimum. The maximum λ occurs for cross-rolls, $\theta_{12}=\pi^2/2$. Then, $\beta_{12}=\frac{6}{7}$ and $\lambda=(20K^2-3\pi^2)/7$. The minimum value of λ occurs when $\theta_{12}=0$.

The nonlinear neutral curve for cross-rolls is given by $\lambda=0$; that is, by (82.25) when the inequality is replaced by equality. For cross-rolls, $\beta_{12}=\frac{6}{7}$ and $K^2=3\pi^2/20$. Hence $(a-\pi)^2=K^2\bar{\varepsilon}^2=K^2\varepsilon^2/(1-2K^2/\pi^2)=3\pi^2\varepsilon^2/14$.

Newell and Whitehead note that the parallel-roll instability is not a true instability because it leads to an unbounded response. Though γ_2 exists when $\theta_{12}=0$, $\hat{\chi}_2$ does not exist. When θ_{12} is small $\hat{\chi}_2$ is large because of small divisors (see § 81). Hence the analysis just given requires that θ_{jn} should be strictly bounded away from zero.

Newell and Whitehead note that the scaling (82.5) is not correct when θ_{lj} is small. If one allows an $O(\varepsilon)$ band of modes in the direction X_l along $\mathbf{k}_l$ then an $O(\varepsilon^{1/2})$ band is possible in the direction Y_l (see Fig. 82.2). They conclude from this that (82.5) is not the correct scaling and should be replaced by

$$X=\bar{\varepsilon}x, \qquad Y=\sqrt{\bar{\varepsilon}}\,y, \qquad T=\varepsilon^2 t. \tag{82.26}$$

Retracing the steps of the deviation we arrive at (82.23), but with diffusion term $\frac{2}{\pi^2}(k_n \cdot \hat{\nabla}_2)^2 \chi_n$ replaced by $2\left(\frac{\partial}{\partial X_n} - \frac{i}{2\pi}\frac{\partial^2}{\partial Y_n^2}\right)^2 \chi_n$. This change in the slow-scale Y_n does not alter the results for the cross-roll instability but it does lead to a different type of parallel roll instability—the Eckhaus (1965) instability which is considered next.

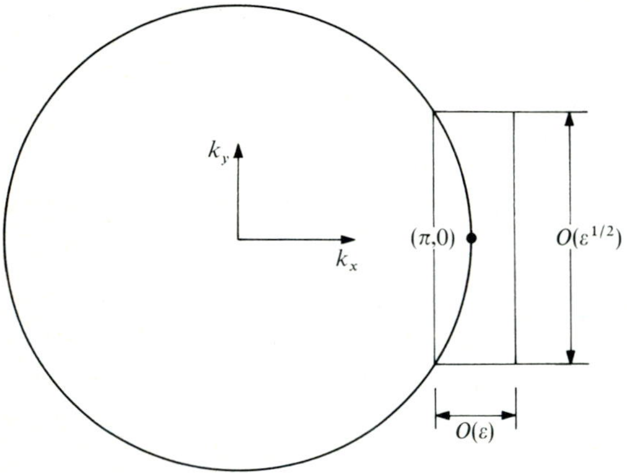

Fig. 82.2: The unperturbed neutral modes lie on the circle $|\mathbf{k}|^2 = \pi^2$. Disturbances of $O(\varepsilon)$ in k_x relative to the mode $\mathbf{k} = \mathbf{e}_x \pi$ are associated with $O(\varepsilon^{1/2})$ variations in k_y

The stability of roll-convection $(R_2 = 2\pi^2)$ is to be studied using the amplitude equation

$$\frac{\partial \chi}{\partial T} - 2\left[\frac{\partial}{\partial X} - \frac{i}{2\pi}\frac{\partial^2}{\partial Y^2}\right]^2 \chi = \left[\pi^2 - \frac{\pi^4}{2}|\chi|^2\right]\chi \tag{82.27}$$

which replaces (82.17) when the scales are given by (82.26).

A disturbance u of $\chi_1 = \frac{2}{\pi} \cdot \left(\frac{1}{2} - \frac{K^2}{\pi^2}\right)^{1/2} e^{iKX}$,

$$\chi = \chi_1(X) + u(X, Y, T),$$

satisfies

$$Au + (\pi^2 - 2K^2)\bar{u}e^{2iKX} \tag{82.28}$$

where

$$A = \frac{\partial}{\partial T} - 2\left[\frac{\partial}{\partial X} - \frac{i}{2\pi}\frac{\partial^2}{\partial Y^2}\right]^2 - (4K^2 - \pi^2).$$

The complex conjugate of (82.28) also holds. Using these two equations we find that

$$\bar{A}\{e^{-2iKX}Au\}=(\pi^2-2K^2)^2ue^{-2iKX}. \tag{82.29}$$

When the differentiations implied by the conjugate operator $\bar{A}$ are carried out, the exponential appears as a factor of all terms. The resulting equation for u has constant coefficients. The quasiperiodic solutions of (82.29) are in the form of exponentials

$$u=\exp\{\lambda T+i(\mu+K)X+i\sqrt{2}\pi M\} \tag{82.30}$$

where λ, μ, M are to be determined. Substituting (82.30) into (82.29) we find that

$$\begin{aligned}\{\lambda+2(K+\mu+M^2)^2-(4K^2-\pi^2)\}\{\lambda+2(K-\mu+M^2)^2\\ -(4K^2-\pi^2)\}=(\pi^2-2K^2)^2\end{aligned} \tag{82.31}$$

Eq. (82.31) is unchanged under the transformation $\mu\to-\mu$. Hence, there are two eigenfunctions for (82.29): (82.30) and (82.30) with μ replaced by $-\mu$.

The roots of (82.31) are given by

$$\lambda=-p^2+2K^2-U\pm\sqrt{p^4+V^2} \tag{82.32}$$

where

$$p^2=\pi^2-2K^2,$$

$$U=[K+\mu+M^2]^2+[K-\mu+M^2]^2$$

and

$$V=4\mu(K+M^2).$$

The maximum growth rate is associated with the plus sign in (82.32). There are two values (μ, M) at which λ is maximum:

$$\text{(I)}\ \mu=0, \quad M^2=-K, \quad \lambda=2K^2 \tag{82.33}$$

and

$$\begin{aligned}\text{(II)}\ M=0, \quad &\mu^2=(6K^2-\pi^2)(2K^2+\pi^2)/16K^2,\\ &\lambda=(6K^2-\pi^2)^2/8K^2.\end{aligned} \tag{82.34}$$

For purposes of comparison we recall that for cross-rolls:

$$\text{(III)}\ \lambda=(20K^2-3\pi^2)/7. \tag{82.35}$$

The nonlinear neutral curve for sinuous instability at 2nd order (I)[2], for the Eckhaus (1965) instability (II) and for the cross-roll instability (III) are shown in Fig. 80.1. The growth rates for the three instabilities (82.33, 34, 35) are sketched in Fig. 82.3.

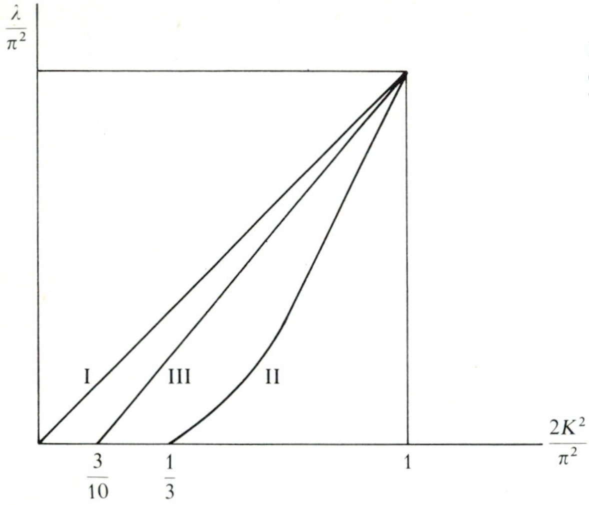

Fig. 82.3: Growth rates for three types of instability given by (82.33, 34, 35). The disturbances are proportional to $\exp\{(-\gamma_2\varepsilon^2+O(\overline{\varepsilon}^4))t\}$ where $\gamma_2=-\lambda$

Experiments suggest that the region of stable cellular flow may be considerably more circumscribed than that predicted on the basis of linearized theories (see Fig. 80.1). This is especially true if the preferred cell size is observed as a function of very slow changes in the Rayleigh number. This type of experiment appears to lead to something more akin to a curve than to a region. For convection, the experiments of Koschmeider (1969) and others show that when the Rayleigh number is slowly changed, the preferred wave number is a decreasing function of R. This curve lies closest to the nonlinear neutral curve for sinuous instability. It should be noted that the growth rate for this instability is greater than for the other forms of instability. Newell and Whitehead have used their amplitude equation to study the stability of bifurcating solutions of wave number $|\mathbf{k}|^2 \neq \pi^2$ as they evolve from a state of rest. They conclude that the only evolving solution which could be stable is the one starting at the nose of the neutral curve (in our case $|\mathbf{k}|^2=\pi^2$). If this is true, the nonlinearities at higher order might destabilize

[2] In § 80 we showed that the nonlinear neutral curve for the sinuous instability is given to lowest order by

$$\frac{R}{4\pi^2}-1=\frac{12}{\sqrt{19}}(a-\pi)^{1/2}.$$

This curve has a vertical tangent at the nose of the neutral curve (Fig. 80.1). At 2nd order this result reduces to (82.33). The more complete result requires that the perturbation be carried out to the fourth order.

all solutions in the shaded region of Fig. 80.1 lying to the right of some line in the vicinity of the nonlinear neutral curve for the sinuous instability.

We have seen that stability results which follow from standard bifurcation theory at 2nd order may also be obtained by the methods of multiple scales leading to an amplitude equation. In addition, the amplitude equation can be used to describe the spatial modulations which occur as the result of nonuniform forcing or from general initial and boundary conditions.

Further results, following from amplitude equations, can be found in the paper of Newell and Whitehead (1969) and in the references listed in the notes for Chapter II.

Chapter XII

The Variational Theory of Turbulence Applied to Convection in Porous Materials Heated from below

This chapter is about the variational theory of turbulence. The theory gives bounds on some of the average properties of statistically stationary, possibly turbulent, flow. Already in § 35 we applied this theory to find a bound on the mass flux which could be transported by turbulent Poiseuille flow at a fixed, externally given mean pressure gradient. Now we turn to a much simpler problem: we shall seek a bound on the maximum possible value of the heat transported which can be achieved by a statistically stationary motion of a fluid in a saturated porous material heated from below. We exploit the problem of porous convection to illustrate the aims, assumptions and methods of the variational theory of turbulence in its simplest mathematical context.

The basic idea of the variational theory of turbulence is due to L. N. Howard (1963). F. Busse (1969A) extended Howard's analysis to shear flows. In (1969B) Busse conjectured that Howard's problem could be solved by multi-α solutions and he constructed what is thought to be a boundary layer approximation to these solutions (see Notes for Chapter IV, § 31, 35). A mathematical characterization of the multi-α solutions was given by Busse and Joseph (1972). Joseph (1974A) gave a mathematically rigorous theory of bifurcation for the multi-α solutions. I will not present the bifurcation theory here. My treatment here follows the work of Busse and Joseph and Gupta and Joseph (1973).

§ 83. Bounds on the Heat Transported by Convection

A heat transport curve may be defined as a functional relation between the dimensionless temperature difference R and the Nusselt number *Nu*. A unique heat transport curve may be expected to associate with each solution of the DOB equations. Heat transport functionals are response functionals for fluid systems brought to motion by the application of temperature differences. These functionals are of foremost interest in the nonlinear theory of convection since, when evaluated on a solution, they lead to the response function, the heat transport curve.

The variational theory of turbulence takes form through the remark that the heat transport functional is a bounded functional in a class of functions which contains all of the statistically stationary solutions of the DOB equations.

Every statistically stationary solution satisfies (78.4), (78.7) and the energy identity

$$-\langle w\theta D\bar{T}\rangle = \langle|\nabla\theta|^2\rangle\,. \tag{83.1}$$

This identity follows from (78.3c) by multiplication by θ and integration using $\nabla\cdot\mathbf{u}=0$ and $w=\mathbf{u}\cdot\mathbf{n}=0$ at $z=0,1$. Elimination of $D\bar{T}$ from (83.1) using (78.4) leads to

$$\langle|\nabla\theta|^2\rangle = \langle w\theta\rangle - \langle\overline{w\theta}^2\rangle + \langle w\theta\rangle^2\,. \tag{83.2}$$

Eqs. (78.7) and (83.2) can be combined into the following expression:

$$\mathrm{R} = \frac{\langle|\nabla\theta|^2\rangle\langle|\mathbf{u}|^2\rangle}{\langle w\theta\rangle^2} + \frac{\langle(\overline{w\theta}-\langle w\theta\rangle)^2\rangle}{\langle w\theta\rangle^2}\langle|\mathbf{u}|^2\rangle\,. \tag{83.3}$$

It is convenient to introduce the Nusselt number into (83.3) as follows: Let

$$\mu = \langle|\mathbf{u}|^2\rangle\,. \tag{83.4}$$

Then, using (78.7) we find that

$$Nu - 1 = \mu/\mathrm{R}\,. \tag{83.5}$$

We call μ the "heat transport".

We next define the class of kinematically admissible fluctuation fields[1]:

$$\bar{\mathbf{H}} \equiv \{\mathbf{u},\theta\colon \operatorname{div}\mathbf{u}=0,\ \mathbf{u}=\theta=0|_{z=0,1},\ \bar{\mathbf{u}}=\bar{\theta}=0, \mathbf{u},\theta\in AP(x,y)\}\,. \tag{83.6a}$$

---

[1] The requirement that $\mathbf{u}$ and θ be doubly almost periodic functions of (x,y) allows one to formulate the variational problem in a Hilbert space of almost periodic functions. In this space the inner product is defined in terms of the horizontal average

$$\overline{fg} = \lim_{L\to\infty}\frac{1}{4L^2}\int_{-L}^{L}\int_{-L}^{L} fg\,dx\,dy$$

and the associated norm is Bohr's norm. On the other hand, the variational problem has a peculiar structure in which (x,y) derivatives enter only through the operator ∇_2^2. The problem is nonlinear, but since the nonlinearities depend on z alone, it is possible to superpose solutions which are eigenfunctions of the Laplace operator in the plane, as in (84.3). The intersection of these planform eigenfunctions and the set of almost periodic functions (discussed under (62.4a, b)) would therefore appear to be an appropriate setting for the variational problem. It is perhaps useful to note two points which may lead to an extension and clarification of the variational theory. First, our problem can be rigorously posed in bounded containers and if the side walls are insulated, as in the problem treated in Chapter X, then the problem is separable into functions of z and planform eigenfunctions. In the bounded domain the planform eigenfunctions are uniquely determined by the side wall boundary conditions. Second, the basic assumptions of statistical stationarity which are required to make the theory viable are speculative; I am not at all certain that mean values over the area of a bounded container are time-independent. It may be possible to relax the stringent requirements of statistical stationarity along the lines suggested in Exercise 30.1.

Among the elements $(\mathbf{u},\theta)\in\bar{\mathbf{H}}$ are those which satisfy normalizing conditions

$$\bar{\mathbf{N}}\equiv\{\mu=\langle|\mathbf{u}|^2\rangle,\ \mathrm{R}\langle w\theta\rangle=\langle|\mathbf{u}|^2\rangle\}. \tag{83.6b}$$

Every statistically stationary solution is simultaneously an element of $\bar{\mathbf{H}}\cap\bar{\mathbf{N}}$.

Consider next the variational problem

$$F(\mu)=\min_{\bar{\mathbf{H}}}\mathscr{F}[\mathbf{u},\theta;\mu], \tag{83.7}$$

where

$$\mathscr{F}[\mathbf{u},\theta;\mu]=\frac{\langle|\nabla\theta|^2\rangle\langle|\mathbf{u}|^2\rangle}{\langle w\theta\rangle^2}+\mu\frac{\langle(\overline{w\theta}-\langle w\theta\rangle)^2\rangle}{\langle w\theta\rangle^2} \tag{83.8}$$

and $\mu>0$ is preassigned. Since $\mathscr{F}$ is a homogeneous functional of degree zero, we may always renormalize the elements $(\tilde{\mathbf{u}},\tilde{\theta})$ which give $\mathscr{F}$ its minimum value F in $\bar{\mathbf{H}}$ so as to satisfy the conditions $\bar{\mathbf{N}}$. Hence, we may obtain a unique F among elements in $\bar{\mathbf{H}}\cap\bar{\mathbf{N}}$. Since statistically stationary solutions are also elements of $\bar{\mathbf{H}}\cap\bar{\mathbf{N}}$, we have, using (83.3) and (83.7), that $\mathrm{R}\geqslant F(\mu)$.

We may conclude that statistically stationary convection with heat transport μ cannot exist when

$$\mathrm{R}<F(\mu). \tag{83.9}$$

The function $F(\mu)$ is called the *bounding heat transport curve*.

Exercise 83.1: Show that the heat transported by statistically stationary convection (the quantity $Nu-1$), when the dimensionless temperature R is given, is bounded above by the maximum value of the functional

$$\langle w\theta\rangle^2\left[1-\frac{1}{\mathrm{R}}\frac{\langle|\nabla\theta|^2\rangle}{\langle w\theta\rangle}\right]\Big/\langle(\overline{w\theta}-\langle w\theta\rangle)^2\rangle \tag{83.10}$$

over functions in $\bar{\mathbf{H}}$.

Exercise 83.2 (Howard, 1963): Formulate the analysis and § 83 for statistically stationary solutions of the OB equations in a fluid layer heated from below.

Exercise 83.3 (Auchmuty, 1973): Consider the problem of Exercise 83.3 when the top and bottom are confined by conducting planes and

$$\oint\theta\frac{\partial\theta}{\partial n}ds=0 \tag{83.11}$$

where s is the rigid side wall. Show that (83.7) and (83.8) hold for the finite domain when the angle brackets are interpreted as volume averaged integrals.

Exercise 83.4: Consider the expanded class of admissibility $\bar{\mathbf{H}}_0$ which is the same as (83.6a) except that functions for which $\operatorname{div}\mathbf{u}\neq0$, $\bar{\mathbf{u}}\neq0$, $\bar{\theta}\neq0$ are also admitted. Let $F_0(\mu)=\min_{\mathbf{H}_0}\mathscr{F}[\mathbf{u},\theta;\mu]$.

(a) Show that $F_0(\mu)\leqslant F(\mu)$.
(b) Show that $F_0(0)=\pi^2$.
(c) Show that the maximizing fields are such that

$$\mathbf{u}=\mathbf{e}_z w(z) \quad \text{and} \quad \theta=\theta(z)$$

are functions of z alone.
(d) Find the Euler equations for F_0. Solve these Euler equations by quadrature in elliptic integrals.
(e) Prove that as $F_0\to\infty$

$$\mu\sim\left(\frac{e}{8}\right)(-1+\sqrt{F_0+1})\,e^{\sqrt{1+F_0}}.$$

The graph of $F_0(\mu)$ is given in Fig. 89.1.

§ 84. The Form of the Admissible Solenoidal Fluctuation Field Which Minimizes $\mathscr{F}[\mathbf{u},\theta;\mu]$

We need to characterize the minimizing solutions for $\mathscr{F}[\mathbf{u},\theta;\mu]$. Several simplifications of this problem are possible. Since $\mathbf{u}\in\overline{\mathbf{H}}$ is solenoidal (and $\overline{\mathbf{u}}=0$) we can decompose $\mathbf{u}$ into $\mathbf{u}_1+\mathbf{u}_2$ where $\mathbf{u}_1$ is a poloidal field and $\mathbf{u}_2$ is a toroidal field (see section B6 of Appendix B). These two fields are orthogonal in the sense that $0=\langle\mathbf{u}_1\cdot\mathbf{u}_2\rangle$. Moreover, since the toroidal field $\mathbf{u}_2$ has no vertical velocity we have $w=\mathbf{e}_z\cdot\mathbf{u}_1$, and the value

$$F(\mu)=\min_{\mathbf{H}}\frac{\langle|\mathbf{u}_1|^2+|\mathbf{u}_2|^2\rangle\langle|\nabla\theta|^2\rangle}{\langle w\theta\rangle^2}+\mu\frac{\langle(\overline{w\theta}-\langle w\theta\rangle)^2\rangle}{\langle w\theta\rangle^2}$$

must be taken for fields for which $\mathbf{u}_2=0$. The minimizing fields are therefore poloidal; $\mathbf{u}=\mathbf{u}_1=\boldsymbol{\delta}\chi$, $w=-\nabla_2^2\chi$

$$\langle\mathbf{u}_1^2\rangle=\langle\boldsymbol{\delta}\chi\cdot\boldsymbol{\delta}\chi\rangle=\langle\nabla_2^2\chi\nabla^2\chi\rangle$$

where the last equality follows after integrating by parts and

$$F=\min_{\chi,\theta}\left\{\frac{\langle\nabla_2^2\chi\nabla^2\chi\rangle\langle|\nabla\theta|^2\rangle}{\langle w\theta\rangle^2}+\mu\frac{\langle(\overline{w\theta}-\langle w\theta\rangle)^2\rangle}{\langle w\theta\rangle^2}\right\} \tag{84.1}$$

for almost periodic fluctuations χ and θ which vanish at $z=0,1$.

The Euler equations for (84.1) are

$$\langle\nabla_2^2\chi\nabla^2\chi\rangle\nabla^2\theta+\{(F+\mu)\langle w\theta\rangle-\mu\overline{w\theta}\}\,w=0 \tag{84.2a}$$

and

$$\langle|\nabla\theta|^2\rangle\nabla^2\nabla_2^2\chi+\{(F+\mu)\langle w\theta\rangle-\mu\overline{w\theta}\}\,\nabla_2^2\theta=0\,. \tag{84.2b}$$

These equations are nonlinear, but the nonlinear term $\overline{w\theta}$ is of a special kind; it is a function of z alone. This, coupled with the fact that x and y derivatives appear only in the combination ∇_2^2, makes it possible to superpose separable solutions of the form

$$\begin{bmatrix}\chi(x,y,z)\\ \theta(x,y,z)\end{bmatrix}=\begin{bmatrix}\chi^{(N)}\\ \theta^{(N)}\end{bmatrix}=\sum_{n=1}^{N}\begin{bmatrix}w_n(z)\alpha_n^{-3/2}\\ \theta_n(z)\alpha_n^{-1/2}\end{bmatrix}\phi_n(x,y) \tag{84.3a}$$

where

$$\nabla_2^2\phi_n=-\alpha_n^2\phi_n \tag{84.3b}$$

are eigenfunctions of the Laplacian in the plane which include all of the possible modes of vibration of a tight membrane stretched over the x,y plane. These eigenfunctions may be assumed to be orthogonal

$$\overline{\phi_n\phi_m}=\delta_{nm}$$

where δ_{nm} is Kronecker's delta. We note that the zero mean condition $\overline{w}=\overline{\theta}=0$ automatically requires that $\alpha_n^2\neq 0$.

The possibility that $N\to\infty$ in (84.3a) is allowed at the outset but analysis indicates that the minimizing solutions have finite values of N when μ is finite. For this reason problem (84.2) may be properly posed in the set (84.3) of multi-α solutions though this has not been proved yet[2].

The multi-α ansatz (84.3) could also be regarded as stemming from a Galerkin approximation scheme. However, we shall adopt the hypothesis that almost periodic solutions of the variational problem in the form (84.3) do exist; This hypothesis may be true and it allows us to exploit more fully the rigorous formulation of the variational problem to model subtle properties which occur in "true" turbulence.

Substitution of (84.2, 3) into (84.1) leads to the problem

$$\begin{aligned}
&F(\mu)=\inf_{N=1,2,\ldots,\infty}F_N(\mu)\,,\\
&F_N(\mu)=\min_{\alpha_l}\hat{F}_N(\alpha_l,\mu)\,,\\
&\hat{F}_N(\alpha_l,\mu)=\min_{w_l,\theta_l}\mathscr{F}_N[w_l,\theta_l;\alpha_l,\mu]\,,\\
&\mathscr{F}_N=\frac{I_1[w_l]I_1[\theta_l]}{I_2^2}+\mu\frac{I_3^2}{I_2^2}
\end{aligned} \tag{84.4a,b,c,d}$$

where

$$I_1[w_l]=\sum_{l=1}^{N}\langle w_l'^2/\alpha_l+\alpha_l w_l^2\rangle$$

$$I_2=\sum_{l=1}^{N}\langle w_l\theta_l\rangle$$

and

$$I_3^2=\langle(\sum_{l=1}^{N}[w_l\theta_l-\langle w_l\theta_l\rangle])^2\rangle\,.$$

---

[2] G. Auchmuty (1973) has proved that the variational problem (83.7), formulated in a bounded domain as in Exercise 83.3, may be solved in the Sobolev space which is a functional analytic interpretation of the requirements that **u** and θ be differentiable, that $\mathrm{div}\,\mathbf{u}=0$ and that **u** and θ satisfy the boundary conditions. He also gives an existence theory for the turbulent convection problem treated by Howard (1963).

We shall see that when $\mu=0$, then $N=1$. Present evidence suggests that $N=1$ as long as μ is below a first critical value $\mu=\mu_2$. At this critical value a new solution with $N=2$ also minimizes $F_N(\mu_{12})=F_1(\mu_{12})=F_2(\mu_{12})$. Thereafter, as μ is increased further, the two-wave number polynomial minimizes: $F_N(\mu)=F_2(\mu)$. There is second critical value μ_{23}, a third, and so on.

Since the range of $N(\mu)$ is integer-valued, changes in N can only occur as discrete jumps. The process by which the minimizing values $N(\mu)$ increase as μ is increased corresponds physically to successive instability of the boundary layer. Each new value of N allows for a new "eddy" size through which the larger scale motion in the interior can adjust to the smaller scales which are required near the boundary. Here these scales are chosen to minimize the ratio of the total dissipation to the heat transported by admissible fluctuations.

Exercise 84.1: Find the Euler equations for the minimum problem (84.1).

Exercise 84.2: Suppose $\phi(x,y)$ is an almost periodic eigenfunction of ∇_2^2 with eigenvalue α. Show that

$$\phi\sim\sum\phi_l e^{i(\gamma_l x+\beta_l y)}$$

where the summation is over all integers l such that $\gamma_l^2+\beta_l^2=\alpha^2$ and the constants ϕ_l are Fourier coefficients.

§ 85. Mathematical Properties of the Multi-α Solutions

We shall show that the functions which give $\mathscr{F}_N$ its minimum value $\hat{F}_N(\alpha_l,\mu)$ have $w_n=\theta_n$; then we shall prove that the minimizing functions $\theta_n(z)$ are symmetric or antisymmetric with respect to the channel midplane; then we shall prove that if the N functions θ_n are different then the corresponding N values of α_n are different; then we shall derive orthogonality conditions (85.22) satisfied by the functions $\theta_n(z)$ belonging to wave numbers α_n; then we give a formula (85.24) giving the optimizing value of the wave number; and finally we give an energy integral (85.25) which is satisfied by multi-α solutions.

To prove that

$$w_n=\theta_n \tag{85.1}$$

we use an adaption (Busse and Joseph, 1972) of a proof devised by Howard (1963) in his study of turbulent convection. The Euler equations for an extremum of the functional $\mathscr{F}_N$ can be written in the form

$$\begin{aligned}(D^2-\alpha_n^2)\,w_n+\alpha_n\phi\theta_n&=0\,,\\(D^2-\alpha_n^2)\,\theta_n+\alpha_n\phi w_n&=0\,,\end{aligned} \tag{85.2}$$

where

$$\phi=(\hat{F}_N+\mu)\textstyle\sum_\nu\langle w_\nu\theta_\nu\rangle-\sum_\nu w_\nu\theta_\nu \tag{85.3}$$

and $D \equiv d/dz$. Here and in the following, the summation over the index ν runs from 1 to N unless indicated otherwise.

For mathematical convenience we have replaced the normalization conditions (83.6b) by the conditions

$$\sum_\nu \langle w_\nu'^2/\alpha_\nu + \alpha_\nu w_\nu^2 \rangle = \sum_\nu \langle \theta_\nu'^2/\alpha_\nu + \alpha_\nu \theta_\nu^2 \rangle = 1 . \tag{85.4}$$

In addition, we can assume without loss of generality that

$$\sum_\nu \langle w_\nu \theta_\nu \rangle > 0 . \tag{85.5}$$

In preparation for the proof that $w_n = \theta_n$ we introduce the variables

$$\sigma_n \equiv \tfrac{1}{2}(w_n + \theta_n) \quad \text{and} \quad \tau_n = \tfrac{1}{2}(w_n - \theta_n)$$

and find from (85.2) that

$$(D^2 - \alpha_n^2)\sigma_n + \alpha_n \phi \sigma_n = 0, \qquad (D^2 - \alpha_n^2)\tau_n - \alpha_n \phi \tau_n = 0, \tag{85.6}$$

where ϕ is given in terms of σ_n and τ_n by

$$\phi = \hat{F}_N \sum_n \langle \sigma_n^2 - \tau_n^2 \rangle + \mu \sum_n \langle \sigma_n^2 - \tau_n^2 \rangle - \mu \sum_n (\sigma_n^2 - \tau_n^2), \tag{85.7}$$

Multiplication of (85.6a) by σ_n yields

$$\frac{1}{2} \frac{d^2}{dz^2} \sigma_n^2 = -\alpha_n \phi \sigma_n^2 + (D\sigma_n)^2 + (\alpha_n \sigma_n)^2 ,$$

and after summation over all n we obtain

$$\frac{1}{2} \frac{d^2}{dz^2} \sigma^2 = -\phi \sum_n \alpha_n \sigma_n^2 + \sum_n \{(D\sigma_n)^2 + (\alpha_n \sigma_n)^2\} \tag{85.8}$$

and, analogously,

$$\frac{1}{2} \frac{d^2}{dz^2} \tau^2 = \phi \sum_n \alpha_n \tau_n^2 + \sum_n \{(D\tau_n)^2 + (\alpha_n \tau_n)^2\} , \tag{85.9}$$

where the abbreviation $\sigma^2 = \sum_n \sigma_n^2$ has been used. Eq. (85.9) shows that τ^2 cannot have a maximum value in any interval on which $\phi > 0$. Since $\tau^2 = 0$ at $z = 0, 1$ and $\tau^2 \geqslant 0$ we may conclude that

$$\tau^2 = \sum_n \tau_n^2 \equiv 0 \quad \text{if} \quad \phi \geqslant 0 \quad \text{for} \quad -\tfrac{1}{2} \leqslant z \leqslant \tfrac{1}{2} . \tag{85.10}$$

We note that $\phi > 0$ at $z = 0, 1$. To prove $\phi \geqslant 0$ throughout $[-\frac{1}{2}, \frac{1}{2}]$ we show that it is not possible to have $\phi < 0$ on some interior interval $z_1 < z < z_2$ with

$\phi=0$ at z_1 and z_2. Assume that it is possible. Then we find

$$0 \geqslant \frac{d}{dz}\phi = \frac{d}{dz}(\tau^2-\sigma^2) \quad \text{at} \quad z=z_1$$

or

$$\left.\frac{d\sigma^2}{dz}\right|_{z=z_1} \geqslant \left.\frac{d\tau^2}{dz}\right|_{z=z_1} \tag{85.11 a}$$

and similarly

$$\left.\frac{d\tau^2}{dz}\right|_{z=z_2} \geqslant \left.\frac{d\sigma^2}{dz}\right|_{z=z_2}. \tag{85.11 b}$$

Since $\phi<0$ in the open interval (z_1, z_2), integration of (85.8) over the closed interval yields

$$\left.\frac{d\sigma^2}{dz}\right|_{z=z_2} > \left.\frac{d\sigma^2}{dz}\right|_{z=z_1}, \tag{85.12}$$

where the inequality is strict because σ^2 and α_n are positive. From (85.11 a, b) and (85.12) we conclude that

$$\left.\frac{d\tau^2}{dz}\right|_{z=z_2} > \left.\frac{d\tau^2}{dz}\right|_{z=z_1}. \tag{85.13}$$

Hence $d\tau^2/dz$ has a strictly positive increase over any interval with $\phi<0$. Since it cannot decrease over intervals with $\phi \geqslant 0$, as is evident from (85.9), $d\tau^2/dz$ shows a strictly positive increase from $z=0$ to $z=1$. Since $\tau=0$ at $z=0,1$ we cannot have $\phi<0$ and it follows that $\tau^2=0$, proving (85.1).

Using (85.1) we rewrite the variational functional (84.4d) for the multi-α solutions in the form

$$\mathscr{F}_N = (I_1^2 + \mu I_3^2)/I_2^2 \quad \text{where} \quad I_2 = \langle |\boldsymbol{\theta}|^2 \rangle \equiv \langle \textstyle\sum_1^N \theta_l^2 \rangle, \qquad I_3^2 = \langle (|\boldsymbol{\theta}|^2 - \langle |\boldsymbol{\theta}|^2 \rangle)^2 \rangle \tag{85.14}$$

and

$$I_1[\theta_l] \equiv I(\boldsymbol{\alpha}, \boldsymbol{\theta}) \equiv \left\langle \textstyle\sum_1^N \dfrac{\theta_l'^2}{\alpha_l} + \alpha_l \theta_l^2 \right\rangle.$$

To shorten the notation we have introduced the N-dimensional vector $\boldsymbol{\theta}^{(N)}$, which has the functions $\theta_n(z)$ as its components. Similarly, we have combined the wave-numbers α_n to form the vector $\boldsymbol{\alpha}^{(N)}$. The index (N) will be dropped as in (85.14) except when different multi-α solutions have to be distinguished. The Euler equations for the minimum $\hat{F}_N(\mu)$ of the functional (85.14) are

$$\theta_n'' - \alpha_n^2 \theta_n + (\alpha_n/I)\{(\hat{F}_N + \mu)\langle |\boldsymbol{\theta}|^2 \rangle - \mu |\boldsymbol{\theta}|^2\}\, \theta_n = 0 \tag{85.15}$$

and the corresponding boundary conditions are

$$\theta_n(0)=\theta_n(1)=0 \quad \text{for} \quad n=1,\dots,N. \tag{85.16}$$

The symmetry of the equations suggests that the solutions $\theta_n(z)$ are either symmetric or antisymmetric with respect to $z=\frac{1}{2}$. To prove this we separate the vector $\boldsymbol{\theta}$ into its symmetric and antisymmetric parts:

$$\boldsymbol{\theta}=\boldsymbol{\theta}_s+\boldsymbol{\theta}_a,$$

with $\boldsymbol{\theta}_s\equiv\frac{1}{2}\{\boldsymbol{\theta}(z)+\boldsymbol{\theta}(1-z)\}$ and $\boldsymbol{\theta}_a=\frac{1}{2}\{\boldsymbol{\theta}(z)-\boldsymbol{\theta}(1-z)\}$. The functional (85.14) obviously satisfies the relation

$$\mathscr{F}_N(\mu,\theta,\boldsymbol{\alpha})\geqslant\tilde{\mathscr{F}}_N(\mu,\boldsymbol{\theta}_s,\boldsymbol{\theta}_a,\boldsymbol{\alpha})$$

with

$$\tilde{\mathscr{F}}_N\equiv\{[I(\boldsymbol{\alpha},\boldsymbol{\theta}_s)+I(\boldsymbol{\alpha},\boldsymbol{\theta}_a)]^2+\mu\langle(|\boldsymbol{\theta}_s|^2+|\boldsymbol{\theta}_a|^2-\langle|\boldsymbol{\theta}_s|^2+|\boldsymbol{\theta}_a|^2\rangle)^2\rangle\}\times\langle|\boldsymbol{\theta}_s|^2+|\boldsymbol{\theta}_a|^2\rangle^{-2} \tag{85.17}$$

since $\tilde{\mathscr{F}}_N$ is the same as expression (85.14) except that the positive term

$$4\mu\langle(\boldsymbol{\theta}_s\cdot\boldsymbol{\theta}_a)^2\rangle\langle|\boldsymbol{\theta}_s|^2+|\boldsymbol{\theta}_a|^2\rangle^{-2} \tag{85.18}$$

has been neglected. We consider $\tilde{\mathscr{F}}$ as a functional of the symmetric vector $\boldsymbol{\theta}_s$ and the antisymmetric vector $\boldsymbol{\theta}_a$, and obtain as necessary conditions for the minimum $\tilde{F}_N(\mu)$ of $\tilde{\mathscr{F}}_N$ the Euler equations

$$\left.\begin{aligned}\theta''_{sn}-\alpha_n^2\theta_{sn}+\alpha_n\phi\theta_{sn}=0,\\ \theta''_{an}-\alpha_n^2\theta_{an}+\alpha_n\phi\theta_{an}=0,\end{aligned}\right\} \tag{85.19}$$

with

$$\phi\equiv\{(\tilde{F}_N+\mu)\langle|\boldsymbol{\theta}_s|^2+|\boldsymbol{\theta}_a|^2\rangle-\mu(|\boldsymbol{\theta}_s|^2+|\boldsymbol{\theta}_a|^2)\}\,[I(\boldsymbol{\alpha},\boldsymbol{\theta}_s)+I(\boldsymbol{\alpha},\boldsymbol{\theta}_a)]^{-1}.$$

From (85.19) we obtain

$$\theta''_{sn}\theta_{na}-\theta''_{an}\theta_{sn}=0$$

and by integration of this relation over the interval $0\leqslant z\leqslant\frac{1}{2}$ we find that

$$\theta'_{an}(\tfrac{1}{2})\,\theta_{sn}(\tfrac{1}{2})=0, \tag{85.20}$$

since θ_{sn} as well θ_{an} vanishes at the boundaries $z=0,1$. Relation (85.20) requires that either θ_{an} or θ_{sn} vanishes together with its derivative at $z=\frac{1}{2}$ since by definition $\theta_{an}(\frac{1}{2})=\theta'_{sn}(\frac{1}{2})=0$. Because ϕ is non-singular, however, any solution of (85.19) which vanishes together with its derivative at some point must vanish identically. Hence we can conclude that the vectors $\boldsymbol{\theta}_s$ and $\boldsymbol{\theta}_a$ minimizing $\tilde{\mathscr{F}}$ have the property

$$\boldsymbol{\theta}_s \cdot \boldsymbol{\theta}_a \equiv 0 . \tag{85.21}$$

Since the functionals $\mathscr{F}_N$ and $\tilde{\mathscr{F}}_N$ become identical for vectors $\boldsymbol{\theta}_s$ and $\boldsymbol{\theta}_a$ with the property (85.21), the vectors $\boldsymbol{\theta}_s$ and $\boldsymbol{\theta}_a$ minimizing $\tilde{\mathscr{F}}_N$ must also minimize $\mathscr{F}_N$. This proves the conjecture that the minimizing solutions θ_n of the Euler equations (85.15) are either symmetric or antisymmetric with respect to $z=\frac{1}{2}$.

It is worth pointing out some further properties of the multi-α solutions. Without loss of generality we can assume that all wavenumbers α_n are different. If two wavenumbers are equal the N-α solution can always be reduced to a $(N-1)$-α solution, as the following consideration shows. Suppose that $\alpha_1=\alpha_2$. In this case it is readily seen that the corresponding solutions θ_1 and θ_2 of (85.15) satisfy the relation

$$\theta_1''\theta_2-\theta_2''\theta_1=0 ,$$

which by integration and use of (85.16) yields $\theta_1'\theta_2-\theta_2'\theta_1=0$ or $(\theta_2/\theta_1)'=0$ in any interval with non-vanishing θ_1. Hence θ_2 must be a multiple of θ_1, say $\theta_2=\gamma\theta_1$. Then the vector $\boldsymbol{\theta}^{(N)}$ of N functions can be reduced to a vector of $N-1$ functions by neglecting θ_2 and α_2 and replacing θ_1 by $\theta_1(1+\gamma^2)^{1/2}$.

The solutions of (85.15) are characterized by two orthogonality relations. The first relation is obtained by multiplying the equation for θ_n by $\alpha_n^{-1}\theta_m$ and the equation for θ_m by $\alpha_m^{-1}\theta_n$. Averaging the equations and subtracting them yields after integration by parts

$$\frac{\alpha_m-\alpha_n}{\alpha_m\alpha_n}\{\langle\theta_m'\theta_n'\rangle-\alpha_m\alpha_n\langle\theta_m\theta_n\rangle\}=0 . \tag{85.22 a}$$

Since $\alpha_m\neq\alpha_n$ we have

$$\langle\theta_m'\theta_n'\rangle-\alpha_m\alpha_n\langle\theta_m\theta_n\rangle=0 . \tag{85.22 b}$$

The other relation is obtained by the same process without using the factors α_n^{-1} and α_m^{-1}, respectively:

$$\langle\{(\hat{F}_N+\mu)\langle|\boldsymbol{\theta}|^2\rangle-\mu|\boldsymbol{\theta}|^2-(\alpha_n+\alpha_m)I\}\,\theta_n\theta_m\rangle=0 . \tag{85.23}$$

We call (85.22) and (85.23) orthogonality relations although θ_m and θ_n are not necessarily orthogonal functions in the usual sense.

The wavenumbers α_n can be determined by minimizing the functional (85.14) as a function of α. The condition $\partial\mathscr{F}(\mu,\boldsymbol{\theta},\boldsymbol{\alpha})/\partial\alpha_m=0$ yields the wavenumber formula

$$\alpha_n^2=\langle\theta_n'^2\rangle/\langle\theta_n^2\rangle . \tag{85.24}$$

Formula (85.24) shows that relation (85.22b) includes the case $m=n$ if the minimizing value of the wavenumbers are used. The interpretation of (85.24) is that the dissipation $\langle|\theta_n'|^2\rangle$ associated with the vertical scale is equal, for each function θ_n, to the dissipation $\alpha_n^2\langle\theta_n^2\rangle$ associated with the horizontal scale.

Finally, we note that the multi-α solutions have an energy integral which may be written as

$$I\sum_{n=1}^{N}\left(\frac{\theta_n'^2}{\alpha_n}-\alpha_n\theta_n^2\right)+(\hat{F}_N+\mu)\langle|\boldsymbol{\theta}|^2\rangle(|\boldsymbol{\theta}|^2-\langle|\boldsymbol{\theta}|^2\rangle)-\tfrac{1}{2}\mu(|\boldsymbol{\theta}|^4-\langle|\boldsymbol{\theta}|^4\rangle)$$

$$=I\sum_{n=1}^{N}\left\langle\frac{\theta_n'^2}{\alpha_n}-\alpha_n\theta_n^2\right\rangle. \tag{85.25}$$

When the α_n have their optimal values, given by (85.24), the right-hand side of this equation vanishes.

Exercise 85.1: Derive Eqs. (85.24) (see Exercise B4.6), and (85.2). Derive Eq. (85.25).

Exercise 85.2: Show that

$$F_N(\mu)=\left[4\sum_{l=1}^{N}\langle\theta_l'^2\rangle/\langle\theta_l^2\rangle+\mu\langle(\sum_{l=1}^{N}\langle\theta_l^2\rangle-\theta_l^2)^2\rangle\right]/\langle\sum_{l=1}^{N}\theta_l^2\rangle^2. \tag{85.26}$$

§ 86. The Single-α Solution and the Situation for Small μ

When $\mu=0$ we must solve the linear problem

$$\theta_m''-\alpha_m^2\theta_m+\hat{F}_N\frac{\alpha_m\langle|\boldsymbol{\theta}|^2\rangle}{I}\theta_m=0, \tag{86.1}$$

with $\theta_m=0$ at $z=0,1$. From (86.1) we see that all even derivatives of $\theta_m(z)$ vanish at the boundary. Hence $\theta_m(z)$ can be developed as $\theta_m(z)=\sum_{n=1}^{\infty}A_{mn}\sin n\pi z$, which together with (86.1) implies that

$$\hat{F}_N=\frac{I}{\langle|\boldsymbol{\theta}|^2\rangle}\left\{\frac{n^2\pi^2}{\alpha_m}+\alpha_m\right\}. \tag{86.2}$$

On the other hand, on multiplying (86.1) by θ_m, integrating and summing over m we find that

$$\hat{F}_N=I^2/\langle|\boldsymbol{\theta}|^2\rangle^2. \tag{86.3}$$

Hence the minimum of $\hat{F}_N^{1/2}=(n^2\pi^2/\alpha_m)+\alpha_m$ over α_m is $\hat{F}_N^{1/2}(n)=2n\pi$ for $\alpha_m=n\pi$. The expression $\hat{F}_N(n)$ reaches its lowest value $\hat{F}_N=4\pi^2$ for $n=1$, $\alpha_m=\pi$ and $\theta_m(z)=A_m\sin\pi z$. The minimizing solutions $\theta_m(z)$ are therefore identical apart from a multiplicative constant. The number N of different horizontal scales α_m^{-1} required for minimizing F_N when $\mu=0$ is just one: $N(\mu)=1$ when $\mu=0$ and $\min_N F_N = F_1 = F(0)$.

Now we shall seek the minimum $\hat{F}_1(\mu)$ of $\mathscr{F}_1[\theta;\alpha,\mu]$ over functions with a single wavenumber α. More than one wavenumber is required to minimize $\mathscr{F}$ when μ is large. When $N=1$ we may write (85.25) as

$$I_1((D\theta)^2/\alpha - \alpha\theta^2) + (\hat{F}_1 + \mu)\langle\theta^2\rangle\theta^2 - \tfrac{1}{2}\mu\theta^4 = A = \text{constant}, \tag{86.4}$$

where

$$I_1 = \langle (D\theta)^2/\alpha + \alpha\theta^2\rangle$$

and, with θ normalized so that the maximum value of θ is one,

$$A = -\alpha I_1 + (\hat{F}_1 + \mu)\langle\theta^2\rangle - \tfrac{1}{2}\mu\,. \tag{86.5}$$

The wavenumber α is to have the value given by $\alpha^2 = \langle (D\theta)^2\rangle/\langle\theta^2\rangle$ from (85.24). This implies that

$$I_1/\alpha = 2\langle\theta^2\rangle, \qquad \alpha I_1 = 2\langle (D\theta)^2\rangle\,. \tag{86.6}$$

Elimination of α in (86.4) with the constants as given in (86.5) and (86.6) gives

$$(D\theta)^2 = \lambda(1-\theta^2)(1-k^2\theta^2)\,, \tag{86.7}$$

where $k^2 = \mu/\{2(F_1+\mu)\langle\theta^2\rangle - 4\langle(D\theta)^2\rangle - \mu\}$ and $\lambda = \mu/4k^2\langle\theta^2\rangle$.

To write the required result we employ the complete elliptic integrals $K(k^2)$ and $D(k^2)$. Here k lies in the range $[0,1]$. One finds by integration of (86.7) that

$$\lambda^{1/2} = 2nK(k^2)\,, \tag{86.8}$$

where n is the number of half-periods of θ on $0<z<1$,

$$\langle\theta^2\rangle = 2n\textstyle\int_0^1 W^2 \dfrac{dz}{d\theta}\,d\theta = \dfrac{2n}{\lambda^{1/2}}D(k^2) = D(k^2)/K(k^2) \tag{86.9}$$

and

$$\langle (D\theta)^2\rangle = 2n\lambda \textstyle\int_0^1 (1-\theta^2)(1-k^2\theta^2)\dfrac{dz}{d\theta}\,d\theta = 2n\lambda^{1/2}\textstyle\int_0^1 [(1-\theta^2)(1-k^2\theta^2)]^{1/2}\,d\theta\,.$$

The last expression may be reduced by a standard transformation for elliptic integrals (Hancock 1958, p. 63) to

$$\langle (D\theta)^2\rangle = 4n^2 K[\tfrac{2}{3}K - \tfrac{1}{3}(k^2+1)D]\,. \tag{86.10}$$

Using the expressions (86.8), (86.9) and (86.10) in the definition of k^2 and λ, we have

$$F_1 = \tfrac{16}{3}n^2\{(k^2+1)K^2 + K^3/D - 3k^2DK\} \tag{86.11}$$

and

$$\mu = 16n^2k^2DK\,. \tag{86.12}$$

Eqs. (86.11) and (86.12) give the value $F_1(\mu, n)$ parametrically. The smallest value $F_1(\mu)$ of $F_1(\mu, n)$ is taken on for $n=1$.

We have therefore shown that among functions with a single wavenumber (chosen optimally) the minimum value $F_1(\mu)$ of the functional $\mathscr{F}_1$ is given parametrically by (86.11) and (86.12) with $n=1$. Statistically stationary convection with heat transport μ and a single overall wavenumber cannot exist when $R < F_1(\mu)$. The curve $F_1(\mu)$ for $N=1$ is given in Fig. 89.1

When $\mu=0$, $k^2=0$ and one finds that $F_1=4\pi^2$. The slope of the heat transport curve at the point $\mu=0$ is

$$\frac{dF_1}{d\mu} = \frac{\langle(\theta^2 - \langle\theta^2\rangle)^2\rangle}{\langle\theta^2\rangle} = \tfrac{1}{2}\,. \tag{86.13}$$

We shall use this result to show the following. Of all the small statistically stationary solutions of the Darcy-Boussinesq equations in the case $B=0$, two-dimensional rolls maximize the heat transport. Moreover, the values

$$F_1(0) = 4\pi^2\,, \qquad dF_1(0)/d\mu = \tfrac{1}{2} \tag{86.14}$$

are the best possible.

Summarizing the results, we found, in § 78, that roll convection exists for small μ in the case $B=0$ with the following dependence of the Rayleigh number on the convective heat transport:

$$R(\mu) = 4\pi^2 + \tfrac{1}{2}\mu + O(\mu^2)\,.$$

We have also found, see (86.13), that

$$F_1(\mu) = 4\pi^2 + \tfrac{1}{2}\mu + O(\mu^2)\,.$$

Recall that statistically stationary convection cannot exist when

$$R < F(\mu) = \min_N F_N(\mu) \leqslant F_1(\mu)\,.$$

Hence, when μ is sufficiently small

$$F(\mu) = 4\pi^2 + \tfrac{1}{2}\mu + O(\mu^2)\,.$$

This proves that when μ is small the single-α bounding solution gives the same heat transport as an exact steady solution in the case $B=0$.

The perturbation analysis which was just given has been carried out to order ε^6 by Palm, Weber and Kvernvold (1972). The results of Palm, *et al.*, are compared with the results of the variational analysis of Gupta and Joseph (1973) in Fig. 92.1. The discussion of this comparison is best deferred to § 92.

§ 87. Boundary Layers of the Single-α Solution

For large μ it is of interest to construct a boundary layer analysis similar to that given by Howard (1963). The Euler equation for the single-α solution is

$$I\left(\frac{D^2\theta}{\alpha}-\alpha\theta\right)+\{(F_1+\mu)\langle\theta^2\rangle-\mu\theta^2\}\theta=0, \qquad \theta(0)=\theta(1)=0. \tag{87.1}$$

For the best values of α, (85.24) holds. Because of homogeneity, we may normalize θ, $\langle\theta^2\rangle=1$. Call $\langle(D\theta)^2\rangle=C^2$. Then we may rewrite (87.1) as

$$2D^2\theta-2C^2\theta+F_1\theta+\mu(\theta-\theta^3)=0. \tag{87.2}$$

The basis of the boundary layer analysis is the fact that when μ is large, the first three terms are of smaller order than μ except in narrow layers at $z=0$ and $z=1$. In the interior $\theta=\theta^3=1$, nearly.

For the boundary layer analysis we introduce the boundary layer coordinate $z=\zeta/\sqrt{\mu}$ and anticipate that $F_1/\mu\to 0$ as $\mu\to\infty$. Then, noting that as $\mu\to\infty$

$$\frac{C^2}{\mu}=\frac{1}{\mu}\int_0^1(D\theta)^2\,dz=\frac{2}{\mu^{1/2}}\int_0^\infty\left(\frac{d\theta}{d\zeta}\right)^2 d\zeta,$$

we find that in the boundary layer,

$$2\frac{d^2\theta}{d\zeta^2}+\theta-\theta^3=0 \tag{87.3 a}$$

and

$$\theta(0)=\frac{d\theta(\infty)}{d\zeta}=0, \qquad \theta(\infty)=1. \tag{87.3 b}$$

Multiply (87.3 a) by $d\theta/d\zeta$ and use (87.3 b) to form the energy integral

$$\left(\frac{d\theta}{d\zeta}\right)^2=\tfrac{1}{4}(1-\theta^2)^2.$$

Hence,

$$\frac{d\theta}{d\zeta}=\tfrac{1}{2}(1-\theta^2), \qquad C^2=\tfrac{1}{2}\sqrt{\mu}\int_0^1(1-\theta^2)^2\frac{d\zeta}{d\theta}d\theta=\tfrac{2}{3}\sqrt{\mu},$$

and, using the relations (85.24) and those just found,

$$\begin{aligned}F_1(\mu)&=4\langle(D\theta)^2\rangle+\mu\langle(\theta^2-1)^2\rangle\\&=\tfrac{8}{3}\mu^{1/2}+\mu\left\{\frac{2}{\mu^{1/2}}\int_0^1(\theta^2-1)^2\frac{d\zeta}{d\theta}d\theta\right\}=\tfrac{16}{3}\mu^{1/2}.\end{aligned}$$

In terms of the Nusselt number $(Nu \sim \mu/F_1)$,

$$Nu = (\tfrac{3}{16})^2 F_1 .$$

Exercise 87.1 (Howard, 1972): Use the inequality (C.11) to prove that

$$F_1 \leqslant \tfrac{16}{3}\mu^{1/2}, \qquad Nu - 1 \leqslant (3\mathrm{R}/64)^{1/2} \tag{87.4}$$

for all $\mu > 0$.

Exercise 87.2: Express the solution of (87.1) in terms of Jacobi elliptic functions. Show that as $\mu \to \infty$

$$F_1(\mu) \to \tfrac{16}{3}\mu^{1/2} \quad \text{and} \quad Nu - 1 \to (3\mathrm{R}/64)^{1/2} .$$

Exercise 87.3: Find the mean temperature profile $\overline{T}(z)$ (see (78.4)) from the solution which minimizes $\mathscr{F}_1$ when $\mu \to \infty$. Sketch the graph of this function and compare it with Fig. 78.1.

§ 88. The Two-α Solution

When $\mu = 0$ we have found that $F_N(0) = F_1(0) = 4\pi^2$ corresponding to a single wave number $N = 1$. We have also derived a solution which gives the functional $\mathscr{F}_1(\mu)$ its minimum value $F_1(\mu)$ for all μ. This solution, given in terms of elliptic functions and integrals in the previous section, is continuous (in μ) with the one which minimizes $\mathscr{F}$ when $\mu = 0$. Does a solution with more than one wave number give $\mathscr{F}(\mu)$ a value smaller than $F_1(\mu)$! Numerical work, using a Galerkin method, indicates that there are no solutions of (85.15,16) with $N > 1$ when $\mu < 318.506 = \mu_{12}$ and $F < F_1(\mu_{12}) = 113.115$. For larger values of μ it is possible to construct numerical solutions with two distinct values of α which give values

$$F_2(\mu) < F_1(\mu) . \tag{88.1}$$

The Euler equations for the two-α solution are given by (85.15) with $n = 1$ and $n = 2$. The symmetry properties of minimizing solutions and numerical work suggest that θ_1 is a symmetric function with respect to $z = \frac{1}{2}$ and θ_2 is an antisymmetric function; then, at the boundary

$$\theta_1(0) = \theta_1'(\tfrac{1}{2}) = 0 , \qquad \theta_2(0) = \theta_2(\tfrac{1}{2}) = 0 . \tag{88.2a, b}$$

The orthogonality conditions (85.22) are automatically satisfied by functions with the assumed symmetry properties and need not be considered further.

The exact values of $\mu = 318.506$, $\alpha = \alpha_{12}$ and $F(\mu_{12}) = 113.115$ at which a two-α solution becomes possible can be obtained by a "stability" analysis. This analysis shows that the two-α solution bifurcates from the single-α with the property that $\theta_2 \to 0$ when $\mu \downarrow \mu_{12}$. Further analysis (see Fig. 91.4) shows that when $\mu - \mu_{12}$ is small, $\theta = \sqrt{\mu - \mu_{12}}\,\hat{\theta}(\mu)$ where $\hat{\theta}(\mu_{12})$ is finite and not identically zero. To construct the stability analysis we first linearize the Euler problem for the two-α solution for small values of θ; then,

$$\theta_1'' - \alpha_1^2\theta_1 + (\alpha_1/I)\{(F_1+\mu)\langle\theta_1^2\rangle - \mu\theta_1^2\}\theta_1 = 0 \tag{88.3a}$$

and

$$\hat{\theta}_2'' - \alpha_2^2\hat{\theta}_2 + (\alpha_2/I)\{(F_1+\mu)\langle\theta_1^2\rangle - \mu\theta_1^2\}\theta_2 = 0 \tag{88.3b}$$

where

$$I = \langle\theta_1'^2/\alpha_1 + \alpha_1\theta_1^2\rangle, \qquad \alpha_1 = \alpha_1(\mu), \qquad F_1 = F_1(\mu)$$

and θ_1 and θ_2 satisfy (88.2). This problem is uncoupled in the sense the equation for θ_1 does not depend explicitly on α_2 or θ_2, and (88.2a, 3a) always accommodates the single-α solution which was derived in the previous section. We are maintaining that when $\mu = \mu_{12}$ and $F_1 = F_1(\mu_{12})$ there exists a value of α_2 such that (88.2a, b) and (88.3a, b) are satisfied.

The linear eigenvalue problem (88.2b, 3b) is of a standard kind. The coefficients I, F_1 and θ_1^2 for (88.3b) are given from the single-α solution of § 87, that is, the single-α solution of (88.2a, 3a). We seek the smallest values of

$$F_1(\mu(\alpha_2)) = F_2 \quad \text{and} \quad \mu(\alpha_2) = \mu_{12}$$

for which (88.2b, 3b) has a solution. The result of numerical studies of this problem are exhibited in Fig. 88.1a, b, c.

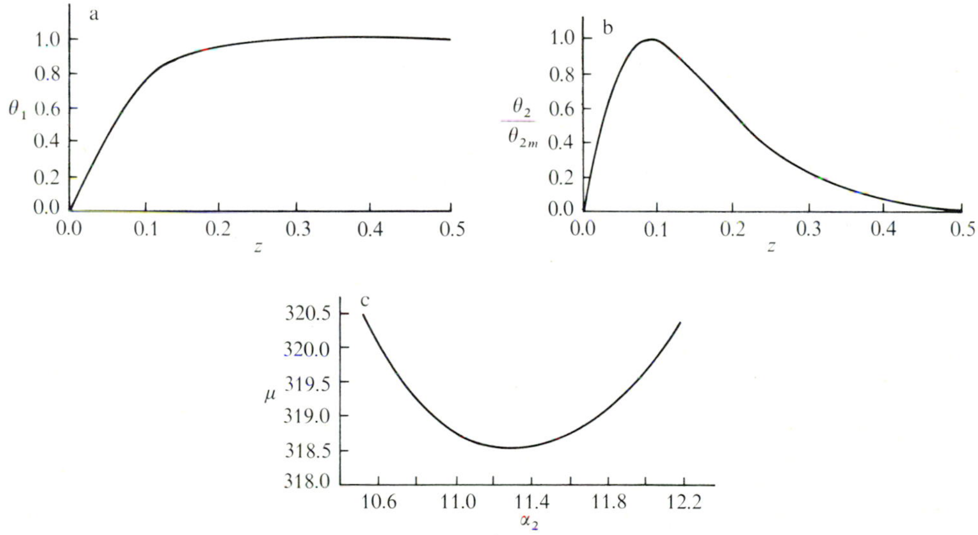

Fig. 88.1: (a) The single-α solution at the point of bifurcation. (b) The second component of the two-α solution at the point of bifurcation. Here θ_{2m} is the maximum value of $\theta_2(z)$. (c) Eigenvalue $\alpha_2(\mu)$ of (88.3b) (Joseph, 1974A).

Numerical computations of the two-α solution, with $\mu > \mu_{12}$ shows that $F_2(\mu) < F_1(\mu)$ (see Fig. 89.1). In Fig. 88.2 we have compared the numerical two-α solution when $\mu > \mu_{12}$ with a boundary layer approximation to be derived in the following section.

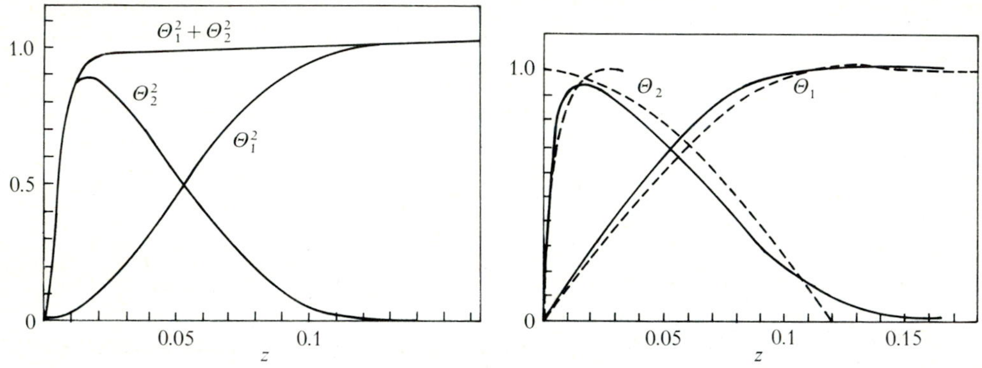

Fig. 88.2: (a) The two-α solution as given by numerical computation. The assumption in the boundary layer that $\Theta_1^2+\Theta_2^2 \approx 1$ in the boundary region where Θ_1^2 and Θ_2^2 are non-vanishing seems to be valid here even though $R=50\pi^2$ is not yet asymptotic. (b) The two-α solution. Graph of the functions $\Theta_1(z)$ and $\Theta_2(z)$ for $R=50\pi^2$. ——, numerical computation; ------ boundary-layer theory (Busse and Joseph, 1972)

Exercise 88.1: Assuming the bifurcation property $\theta_2 \to 0$ as $\mu \downarrow \mu_{12}$, show that $dF/d\mu$ is continuous at the point $\mu=\mu_{12}$.

§ 89. Boundary Layers of the Multi-α Solutions

From what has already been done we expect that the minimizing function

$$F(\mu)=\min_N F_N(\mu)$$

will lead to discrete bifurcations characterized by an increasing integer-valued function $N(\mu)$[3]. This is a very hard problem; to understand it better we seek an approximating solution which, like the single-α solution develops a boundary-layer structure as $\mu \to \infty$.

It is clear already from the form of $\mathscr{F}_N$ expressed in (86.14) that when $\mu \to \infty$ the quantity

$$\mu\langle(|\boldsymbol{\theta}|^2-\langle|\boldsymbol{\theta}|^2\rangle)^2\rangle$$

[3] See Joseph (1974A) for the rigorous theory of bifurcating multi-α solutions.

is proportional to μ unless $|\boldsymbol{\theta}|^2 \to \langle|\boldsymbol{\theta}|^2\rangle$ over most of the domain $0 \leqslant z \leqslant 1$ of $\boldsymbol{\theta}(z)$. By adopting, for convenience, the normalization

$$\langle|\boldsymbol{\theta}|^2\rangle = 1 \tag{89.1 a}$$

we see that $|\boldsymbol{\theta}(z)| \to 1$ everywhere except in boundary layers where $|\boldsymbol{\theta}(z)|$ drops to its zero value at $z = 0, 1$.

For the single-α solution the structure of the boundary layer is determined from the requirement that $\mathscr{F}_1$ be a minimum. This requirement can only be satisfied when the small boundary layer in which θ^2 drops from its interior unit value is large enough for the contributions of the large derivatives to I^2 to be of the same order $(\mu^{1/2})$ as $\mu\langle(1-\theta^2)^2\rangle$.

The same sort of minimizing balance can be anticipated in the multi-α case. Here again we must have

$$|\boldsymbol{\theta}|^2 \approx 1 \tag{89.1 b}$$

in the interior. To prevent the terms with derivatives in I^2 from growing more sharply than $\mu\langle(1-|\boldsymbol{\theta}|^2)^2\rangle$ the solution will develop a boundary layer for $|\theta|^2$ of order μ^{-r_N}, where r_N is a positive number to be determined.

The difference between the multi-α solution and the single-α solution is just this: the presence of the many different functions and wavenumbers allows for the development of boundary layers within boundary layers; indeed our formal calculation shows that the solution does develop a nested sequence of N boundary layers in which the sharp rise of θ_n^2 takes place in an interval of the same order (in μ) as the slow fall of the function θ_{n+1}^2. This process has the overall effect of allowing one to extend closer to the boundaries the region on which (89.1 b) holds while, at the same time, holding the dissipation integrals I^2 to within the same order of μ^{1-r_N} as the term $\mu\langle(1-|\boldsymbol{\theta}|^2)^2\rangle$. It is the presence of the different horizontal scales α_n^{-1} which moderates the increases of the dissipation I^2. Indeed inspection of (86.14) shows that I (and $\mathscr{F}_N$) is minimized when the horizontal scale α_n^{-1} equals the vertical scale of the functions θ_n. Hence we know that the functions θ_n^2 with the steepest boundary layers will also have the smallest horizontal scales α_n^{-1}.

In view of the foregoing discussion it seems plausible to postulate that each function $\theta_n^2(z)$ has its own boundary layer. The boundary layer for θ_N^2 is closest to the wall and has a steep rise of order μ^{-r_N}. The function θ_{N-1}^2 also has a boundary layer but here the rise of θ_{N-1}^2 is less sharp and is of $O(\mu^{-r_{N-1}})$, where $r_{N-1} < r_N$. Unlike the boundary-layer solution for the single-α case one cannot anticipate that θ_N with $N>1$ will tend to one in the interior. If this were the case one could not satisfy (89.1) with non-zero interior values for the other functions θ_n. Moreover, the contribution $\alpha_N\langle\theta_N^2\rangle$ has to be kept small because of the relatively large value of α_N. Hence it is plausible to suppose that θ_N^2 first rises and then falls to zero. It is clear from what we anticipate for the sizes μ^{-r_N} and $\mu^{-r_{N-1}}$ of the two layers closest to the wall that, as $\mu \to \infty$, θ_N^2 may experience a very rapid rise on an interval of order μ^{-r_N} on which θ_{N-1} is barely different from zero, and a relatively gentle fall to zero in an interval of order $\mu^{-r_{N-1}}$ on which θ_{N-1} is still rising rapidly relative to the rate of its own subsequent gentle decline in the layer of $O(\mu^{-r_{N-2}})$. The same sharp rise followed by gentle fall is anticipated of the sequence $\theta_N^2, \theta_{N-1}^2, \ldots, \theta_n^2, \theta_{n-1}^2, \ldots$ until the actually rapid but relatively (to θ_2^2) slow rise of θ_1^2 to its interior value $\theta_1^2 \simeq 1$ is complete.

We shall now show how the description of the multi-α solution for large μ just given does arise from formal analysis. The hypothesis whose consistency is being tested is that θ_n rises in the nth boundary layer and falls to zero in the $(n-1)$th boundary layer. Hence, each of the functions differs from zero essentially only in two consecutive boundary layers and it is convenient to separate the rising and falling parts. Thus we let

$$\theta_n(z) = \left\{ \begin{matrix} \hat{\theta}_n(\zeta_n) & \text{for} \quad z = O(\mu^{-r_n}), \\ \tilde{\theta}_n(\zeta_{n-1}) & \text{for} \quad z = O(\mu^{-r_{n-1}}) \end{matrix} \right\}, \tag{89.2}$$

where

$$n = 1, \ldots, N, \qquad r_n > r_{n-1} \quad \text{and} \quad \zeta_n = z\mu^{r_n} \tag{89.3}$$

is the nth boundary layer co-ordinate. In the limit $\mu \to \infty$ the relations

$$\hat{\theta}_n^2 + \tilde{\theta}_{n+1}^2 = 1 \quad \text{for} \quad z = O(\mu^{-r_n}) \qquad (0 \leqslant n < N) \tag{89.4}$$

and

$$\tilde{\theta}_1^2 = 1 \quad \text{for} \quad z = O(\mu^{-r_0}) \qquad r_0 = 0 \tag{89.5}$$

are implied by (89.1b) and (89.2). It is this feature which allows one to satisfy (89.1b) close to the wall. Only in the Nth layer is $1 - \theta_N^2 > 0$ when $\mu \to \infty$.

We note that when ζ_n is fixed and $\mu \to \infty$

$$\zeta_{n-1} = \zeta_n \mu^{r_{n-1} - r_n} \to 0 .$$

Hence we must have

$$\left. \begin{matrix} \hat{\theta}_n(0) = 0, & \hat{\theta}_n(\infty) = 1 & \text{for} \quad n = 1, \ldots, N, \\ \tilde{\theta}_n(0) = 1, & \tilde{\theta}_n(\infty) = 0 & \text{for} \quad n = 2, \ldots, N. \end{matrix} \right\} \tag{89.6}$$

Since the problem is symmetric with respect to the two boundaries, we assume that the same description holds for the boundary layer at $z = 1$ with $1 - z$ replacing z in the definitions (89.2)—(89.4).

Returning now to the expression (85.24) for the best α_n and (89.2) and (89.3) we find that

$$\alpha_n^2 = \mu^{r_n - r_{n-1}} b_n^2 \tag{89.7}$$

where

$$b_n^2 = \int_0^\infty \hat{\theta}_n'^2 \, d\zeta_n \Big/ \int_0^\infty (1 - \hat{\theta}_n^2) \, d\zeta_{n-1}, \qquad b_1^2 = 2 \int_0^\infty \hat{\theta}_1'^2 \, d\zeta_1 \tag{89.8}$$

are independent of μ at large μ. Here, the factor 2 arises from the fact that there are two boundary layers, one on each wall. We have eliminated

$$\tilde{\theta}_n^2(\zeta_{n-1}) = 1 - \hat{\theta}_{n-1}^2(\zeta_{n-1})$$

by using (89.4).

To determine the exponents we insert (89.3)—(89.7) into (85.14), let $\mu \to \infty$ and use (89.4) and (89.5) to eliminate $\tilde{\theta}_m^2 (m = 1, \ldots, N)$. We find that

$$\mathscr{F}_N[\theta; \alpha, \mu] = \left\{ \mu^{r_1/2} b_1 + \frac{2}{b_1} \int_0^\infty \hat{\theta}'^2 \, d\zeta_1 + 2 \sum_{m=2}^{N} \mu^{(r_m - r_{m-1})/2} \left(\frac{1}{b_m} \int_0^\infty \hat{\theta}_m'^2 \, d\zeta_m \right.\right.$$
$$\left.\left. + b_m \int_0^\infty (1 - \hat{\theta}_{m-1}^2) \, d\zeta_{m-1} \right) \right\}^2 + 2\mu^{1 - r_N} \int_0^\infty (1 - \hat{\theta}_N^2)^2 \, d\zeta_N \,. \qquad (89.9)$$

Differentiation of (89.9) with respect to the parameters r_m shows that a minimum can be reached only if all exponents of μ in the terms contributing additively to the right-hand side of (89.9) are equal:

$$1 - r_N = r_N - r_{N-1} = \ldots = r_2 - r_1 = r_1 \,.$$

This leads to the solution

$$r_m = m/(N+1) \quad \text{for} \quad m = 1, \ldots, N \,. \qquad (89.10)$$

Now, on using (89.10) in (89.9) we have

$$\mathscr{F}_N^\dagger \sim \mu^{1/(N+1)} \{ \hat{I}^2 + 2 \int_0^\infty (1 - \hat{\theta}_N^2)^2 \, d\zeta_N \} \equiv \mu^{1/(N+1)} \mathscr{F}_N^* \qquad (89.11)$$

where

$$\hat{I} = b_1 + \frac{2}{b_N} \int_0^\infty \hat{\theta}_N'^2 \, d\zeta_N + 2 \sum_{m=1}^{N-1} \int_0^\infty \left[\frac{\hat{\theta}_m'^2}{b_m} + b_{m+1}(1 - \hat{\theta}_m^2) \right] d\zeta_m \,,$$

$$F_N^* = \min_{\theta_l, b_l} \mathscr{F}_N^* [\hat{\theta}_l; b_l] \,.$$

$$F^*(\mu) = \inf_N F_N^*(\mu) \,, \qquad F_N^\dagger = \mu^{1/(N+1)} F^*(\mu) \,,$$

and $F^\dagger(\mu)$ is the boundary layer approximation to $F(\mu)$. The minimum of (89.11) is found for functions which satisfy the Euler equations

$$\hat{\theta}_m'' + b_m b_{m+1} \hat{\theta}_m = 0 \quad \text{when} \quad 0 \leqslant \zeta_m \leqslant \bar{\zeta}_m$$

and (89.12a)

$$\hat{\theta}_m(\zeta_m) = 1 \quad \text{when} \quad \zeta_m \geqslant \bar{\zeta}_m; \qquad m = 1, \ldots, N-1$$

and

$$(\hat{I}/b_N) \hat{\theta}_N'' + (1 - \hat{\theta}_N^2) \hat{\theta}_N = 0 \qquad (89.12\text{b})$$

subject to the conditions (89.6). The best values of b_m are given by (89.8).

A continuously differentiable solution of (89.12a, b) and (89.6) is given by

$$\hat{\theta}_n = \begin{cases} \pm\sin(b_n b_{n+1})^{1/2}\zeta_n & \text{for} \quad 0 \leqslant \zeta_n \leqslant \pi/2(b_n b_{n+1})^{1/2}, \\ \pm 1 & \text{for} \quad \pi/2(b_n b_{n+1})^{1/2} \leqslant \zeta_n, \end{cases} \tag{89.13}$$

$$\hat{\theta}_N = \pm\tanh\{(\tfrac{1}{2}b_N \hat{I}^{-1})^{1/2}\zeta_N\}. \tag{89.14}$$

Note that the signs of the functions $\hat{\theta}_n$ and $\tilde{\theta}_n$ remain undetermined. This corresponds to the property that $\theta_n(z)$ is either symmetric or antisymmetric. Hence the representations of $\theta_n(z)$ at the two boundaries may have opposite signs. Only when lower order terms, neglected in this approximation, are taken into account can the question of symmetry be decided.

It is of interest to calculate the thicknesses of the boundary layer of the minimizing N-α solution. We define the characteristic thickness $z=\delta_n$ by the property that the arguments of the sin-function and the tanh-function in expressions (89.13) and (89.14), respectively, assume the value 1;

$$\begin{aligned} \delta_n &= \mu^{-n/(N+1)}(b_n b_{n+1})^{-1/2} = \frac{2}{\pi}\left(\frac{64\mu}{9\pi^4 N}\right)^{-n/(N+1)} \quad \text{for} \quad n=1,\dots,N-1, \\ \delta_N &= \mu^{-N/(N+1)}(2b_N^{-1}\hat{I})^{1/2} = \frac{32}{3\pi^2}\left(\frac{64\mu}{9\pi^4 N}\right)^{-N/(N+1)}. \end{aligned} \tag{89.15}$$

The formulas (89.15, 16 and 17) are the main results of this section. They are, at worst, approximations to asymptotic results which hold as $\mu\to\infty$. Comparison of these results with the numerical results for the two-α solution are given in Figs. 89.1 and 88.2b. In comparing these results we must note that the highest value of R which was computed numerically, $\mathrm{R}=50\pi^2$ can barely be regarded as representative for asymptotic values. Ignoring, for the moment, that the asymptotic analysis is for $\mu\to\infty$, we note that for small μ (<2496)

(i) $N=1$ minimizes $\mathscr{F}_N^\dagger$ and

$$F_1^\dagger(\mu) \sim \tfrac{16}{3}\mu^{1/2}, \quad \alpha_1^2 \sim \tfrac{2}{3}\mu^{1/2} \quad \text{and} \quad \delta_1 \sim 2/\mu^{1/2},$$

for larger μ (>2496).

(ii) $N=2$ minimizes $\mathscr{F}_N^\dagger$ and

$$F_2^\dagger(\mu) \sim 6\pi^2(32\mu/9\pi^4)^{1/3}, \quad \alpha_2^2 \sim 8\mu/9\pi^2, \quad \alpha_1^2 \sim \frac{\pi^2}{4}(32\mu/9\pi^4)^{1/3},$$

$$\delta_1 \sim \frac{2}{\pi}(9\pi^4/32\mu)^{1/3} \quad \text{and} \quad \delta_2 \sim \frac{32}{3\pi^2}(9\pi^4/32\mu)^{2/3}.$$

We next calculate $\hat{I}$, b_n and F_N^*. Using (89.13) and (89.14) we find that

$$\hat{I} = 2\sum_{n=2}^{N-1}\frac{\pi}{2}\left(\frac{b_{n+1}}{b_n}\right)^{1/2} + b_1 + \tfrac{4}{3}(2b_N\hat{I})^{-1/2}.$$

The parameters b_n, $n=1,\dots,N$, can be determined either from relation (89.8) or by minimizing the functional (89.11) as a function of the b_n. In both cases we arrive at equations of the form

$$1-\frac{\pi}{2b_1}\left(\frac{b_2}{b_1}\right)^{1/2}=0,$$

$$\frac{\pi}{2b_n}\left\{\left(\frac{b_n}{b_{n-1}}\right)^{1/2}-\left(\frac{b_{n+1}}{b_n}\right)^{1/2}\right\}=0 \quad \text{for} \quad n=1,\ldots,N-1,$$

$$2\frac{\partial\hat{I}}{\partial b_N}(1+\tfrac{2}{3}(2b_N\hat{I})^{-1/2})-\tfrac{2}{3}(2b_N\hat{I})^{1/2}b_N^{-2}=0,$$

which yield the solution $(n=1,\ldots,N)$

$$b_n=\frac{\pi}{2}\left(\frac{8}{3\pi^2\sqrt{N}}\right)^{(2n-1)/(N+1)}, \qquad \hat{I}=2Nb_1,$$

and

$$\alpha_n^2=\frac{\pi^2}{4}\left(\frac{64\mu}{9\pi^4N}\right)^{(2n-1)/(N+1)}. \tag{89.16}$$

The final expression for $F_N^\dagger(\mu)$ is

$$F_N^\dagger(\mu)\sim\mu^{1/(N+1)}F_N^*=\mu^{1/(N+1)}N(N+1)4b_1^2=N(N+1)\pi^2(64\mu/9\pi^4N)^{1/(N+1)}. \tag{89.17}$$

$$F_2^\dagger(\mu)\sim6\pi^2(3\mu/9\mu^4)^{1/3}, \qquad \alpha_2^2\sim8\mu/9\pi^2,$$

$$\alpha_1^2\sim\frac{\pi^2}{4}(32\mu/9\pi^4)^{1/2}, \qquad \delta_2\sim\frac{32}{3\pi^2}\left(\frac{9\pi^4}{32\mu}\right)^{1/3}.$$

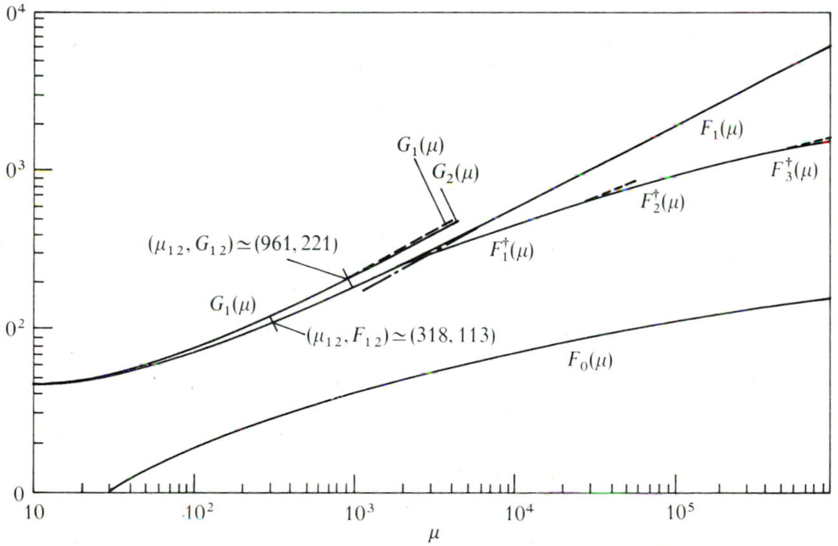

Fig. 89.1: Bounds for heat transport in a porous layer. The bounds $G_N(\mu)$ are defined by (90.10). The bounds $F_N(\mu)$ are defined by (84.4b). The points (μ_{12}, G_{12}) and (μ_{12}, F_{12}) are points of bifurcation. The functions $F_N^\dagger(\mu)$ are boundary layer approximations to $F_N(\mu)$. The curve $F_0(\mu)$ is defined in Ex. (83.4)

The value $\mu=2496$ is found by equating the expressions for $F_1^\dagger$ and $F_2^\dagger$. At still larger values of μ, $F_3^\dagger(\mu)$ minimizes $\mathscr{F}_N^\dagger$ and so on. The boundary layer approximation to

$$F(\mu)=\min_N F_N(\mu)$$

is a continuous curve with kinks. However, the bifurcation analysis of the multi-α problem shows that $F(\mu)$ is a continuously differentiable curve with discontinuous second derivatives (see Exercise 88.1). We also note that the boundary layer analysis overestimates the value of the first point of bifurcation $\mu_{12}=318.5$ by a factor of about eight (see Fig. 89.1).

The boundary layer approximation does appear to have many of the same qualitative features expected of bifurcating multi-α solutions but many unanswered questions remain.

Exercise 89.1: Repeat Exercise 87.3 for the boundary-layer approximation of the two-α solution. Compare $\overline{T}(z)$ for the single-α and two-α solution.

Exercise 89.2: Show that in the limit $\mu\to\infty$, $N(\mu)\to\infty$. Show that the largest boundary layer thickness δ_1, which always tends to zero when N is fixed and $\mu\to\infty$, tends to the value $\delta_1\to 1/e^2$ when $N(\mu)$ is also allowed to increase.

§ 90. An Improved Variational Theory Which Makes Use of the Fact That B is Small

In § 81 we noted that the Darcy-Prandtl number B^{-1} is very large in most porous materials; a value $B^{-1}=10^8$ would not be uncommon. It is therefore intelligent to set $B=0$ from the outset as we did in (70.10b). However, our analysis did not make use of the fact that $B=0$; indeed, if (78.3b) be granted, the analysis applies uniformly to all values of B.

When $B=0$ we may replace a variational problem with an integral side constraint, $\mathrm{R}\langle w\theta\rangle=\langle|\mathbf{u}|^2\rangle$, into a variational problem with a pointwise valid differential equation side constraint

$$\nabla^2 w-\mathrm{R}\nabla_2^2\theta=0\,. \tag{90.1}$$

This differential equation is the vertical component of the curl of the curl of (78.3b) when $B=0$. [4]

To form the variational problem we note that whether or not $B=0$, (83.2) holds and introducing the variables

$$\mathrm{R}\theta=\Theta \tag{90.2}$$

[4] S. Chan (1971) has exploited the same kind of simplification in his boundary-layer study of convection of an infinite Prandtl number fluid.

we find, from (83.2) that

$$\mathrm{R}=\frac{\langle|\nabla\Theta|^2\rangle}{\langle w\Theta\rangle}+\mu\frac{\langle(\overline{w\Theta}-\langle w\Theta\rangle)^2\rangle}{\langle w\Theta\rangle^2}\equiv\mathscr{I}[w,\Theta;\mu] \tag{90.3}$$

where, as before,

$$\mu=\langle w\Theta\rangle=\mathrm{R}(Nu-1)\,. \tag{90.4}$$

The variational problem which makes use of $B=0$ may be stated as follows

$$G(\mu)=\min_{\mathbf{H}^\dagger}\mathscr{I}[w,\Theta;\mu] \tag{90.5 a}$$

where

$$\mathbf{H}^\dagger=\{w,\Theta:\nabla^2 w-\nabla_2^2\Theta=0,\ w=\Theta=0|_{z=0,1},\ \overline{w}=\overline{\Theta}=0,\ w,\Theta\in AP(x,y)\}\,. \tag{90.5 b}$$

Among the elements $(w,\Theta)\in\mathbf{H}^\dagger$ are those which satisfy the normalizing condition

$$N\equiv\{w,\Theta:\mu=\langle w\Theta\rangle\}\,.$$

Every statistically stationary solution of the DOB equations with $B=0$ is simultaneously an element of $\mathbf{H}^\dagger\cap\mathbf{N}$. It follows that statistically stationary convection with heat transport μ cannot exist when

$$\mathrm{R}<G(\mu)\,. \tag{90.6}$$

Moreover, since the space $\mathbf{H}^\dagger$ has fewer elements than the space $\overline{\mathbf{H}}$ previously defined, we have

$$F(\mu)\leqslant G(\mu)\,. \tag{90.7}$$

The constrained variational problem, in $\mathbf{H}^\dagger$, should give a better estimate of the bounding heat transport curve (cf. Exercise 90.1).

Just as in our earlier analysis in § 84, we may note here that the nonlinear problem does not disallow superposition of separable solutions in polynomials of eigenfunctions of the membrane equation. Thus, introducing (84.3) with $\Theta=\theta^{(N)}$ into (90.3) and (90.5) we find that

$$\begin{aligned}\mathscr{I}_N(w_n;\alpha_n;\mu)=&\left(\textstyle\sum_n\langle(D\theta_n)^2+\alpha_n^2\theta_n^2\rangle\right)/\textstyle\sum_n\langle w_n\theta_n\rangle\\&+\mu\langle\textstyle\sum_n(w_n\theta_n-\langle w_n\theta_n\rangle)^2\rangle/\left(\textstyle\sum_n\langle w_n\theta_n\rangle\right)^2\end{aligned} \tag{90.8 a}$$

where

$$L_n w_n=(D^2-\alpha_n^2)\,w_n=-\alpha_n^2\theta_n,\qquad \alpha_n^2\neq 0 \tag{90.8 b}$$

and

$$w_n=L_n w_n=0\quad\text{at}\quad z=0,1 \tag{90.8 c}$$

The functions θ_n are defined by (90.8 b) and may be replaced with $-L_n w_n/\alpha_n^2$ in the functional $\mathscr{I}$.

We may now reformulate the variational problem (90.5 a) as follows

$$G(\mu)=\min_N G_N(\mu) \tag{90.9}$$

$$G_N(\mu)=\min_{\alpha_l} \hat{G}_N(\mu,\alpha_l) \tag{90.10}$$

where

$$\hat{G}_N(\mu,\alpha_l)=\min_{w_n} \mathscr{I}_N(w_n;\alpha_l,\mu) \tag{90.11}$$

and w_n is a differentiable function of $z\in[0,1]$ which satisfies (90.8 c).

The Euler equations for the minimum $\hat{G}_N$ of $\mathscr{I}_N$ are

$$\frac{\sigma}{\alpha_n^2} L_n^3 w_n-\sigma p L_n w_n+\mu F_n(\overline{w\theta})w_n=0 \tag{90.12 a}$$

where θ_n is defined by (90.8 b)

$$\overline{w\theta}=\sum_n w_n\theta_n=-\sum_n w_n L_n w_n/\alpha_n^2$$

$$\sigma=\langle\sum_n w_n\theta_n\rangle, \qquad \gamma=\langle\sum_n \nabla\theta_n^2\rangle$$

$$F_n(\overline{w\theta})\circ=L_n(\circ\,\overline{w\theta})+\overline{w\theta}\,L_n\circ$$

and

$$p=2(\hat{G}_N+\mu)-\gamma/\sigma\,.$$

The boundary conditions

$$w_n=D^2w_n=D^4w_n=0 \tag{90.12 b}$$

follow from (90.8 c) and $D^4w_n=0$ arises as a natural boundary condition (see Appendix B.3) in the calculation leading to (90.12 a).

Though the variational problem formulated in this section is more difficult than the one formulated in §§ 84 and 85, many of the mathematical properties of the two problems are shared (see Exercise 90.1).

Exercise 90.1 (Gupta and Joseph, 1973):

(a) Show that the minimizing function for (90.10) satisfy the relation

$$\langle D^2w_n(D^2\theta_n-\alpha_n\theta_n^2)\rangle-\langle\theta_n[-D^4w_n+\alpha_n^2D^2w_n+\alpha_n^4\theta_n]\rangle$$
$$+\frac{\langle|\nabla\theta|^2\rangle\langle w_nD^2w_n\rangle}{\langle w\theta\rangle}-2\mu\frac{\langle\overline{w\theta}w_nD^2w_n\rangle}{\langle w\theta\rangle}+\mu\frac{\langle w_nD^2w_n\rangle\langle\overline{w\theta^2}\rangle}{\langle w\theta\rangle^2}=0\,.$$

(b) Show that every solution of (90.12) satisfies the following orthogonality condition.

$$(\alpha_l^2-\alpha_n^2)\left[\frac{(\alpha_l^2-\alpha_n^2)^2}{\alpha_l^2\alpha_n^2}\langle Dw_lDw_n\rangle-(\alpha_l^2+\alpha_n^2)\langle\theta_l\theta_n\rangle-\langle D\theta_lD\theta_n\rangle+p\langle w_nw_l\rangle-\frac{2\mu}{\sigma}\langle\overline{w\theta}w_nw_l\rangle\right]=0\,.$$

(c) Prove that $G(0)=4\pi^2$, $dG(0)/d\mu=\frac{1}{2}$, $dNu/dG=\frac{1}{2}\pi^2$ and, hence, show that the bounding heat transport curve is given by $F(\mu)=G(\mu)$ up to terms of order $O(\mu^2)$.

(d) Prove that the bounding heat transport curve is continuously differentiable at each point of bifurcation.

(e) Formulate the variational problem in the following alternative way: Given R find an upper bound for the values of the heat transported $(Nu-1)$ by any statistically stationary solution of the DOB equations.

§ 91. Numerical Computation of the Single-α and Two-α Solution. Remarks about the Asymptotic Limit $\mu\to\infty$

The boundary value problem (90.1) was integrated numerically by a quasi-linearization technique coupled with the use of Conte's orthonormalization method. These methods and further details are described by Gupta and Joseph (1973).

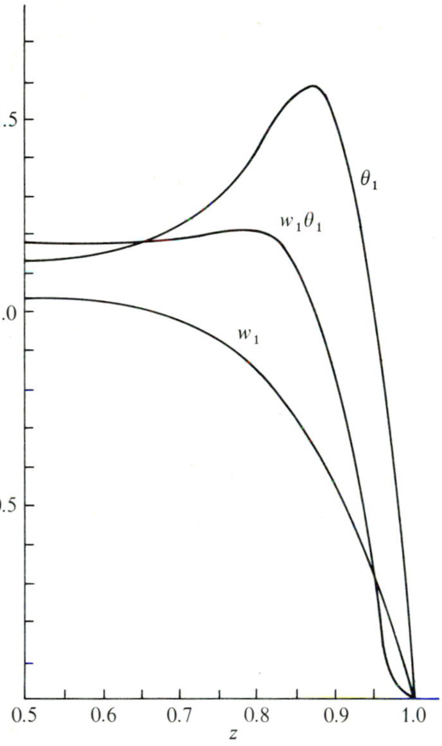

Fig. 91.1

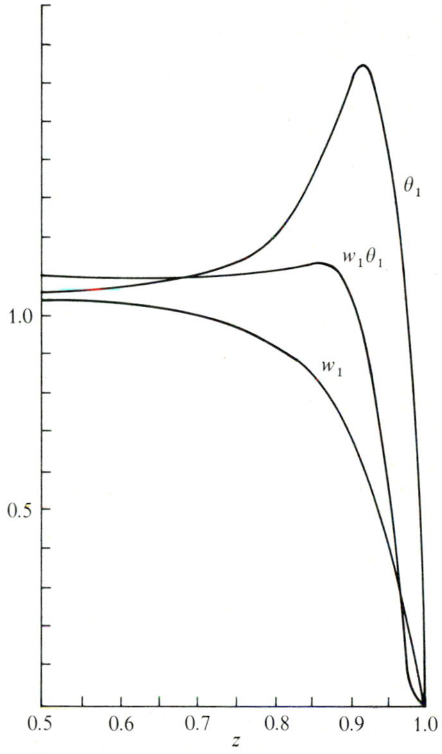

Fig. 91.2

Fig. 91.1: Single-α solutions ($\mu\simeq 843.4$, $G=201.4$, $Nu=5.07$). A pronounced boundary layer structure with $w_1\theta_1\approx 1$ in the interior has not yet developed (Gupta and Joseph, 1973)

Fig. 91.2: Single-α solutions ($\mu=2706$, $G=382.1$, $Nu=8.08$). A boundary layer structure with $w_1\theta_1\approx 1$ in the interior is better developed than in Fig. 91.1. The boundary layer of $w_1\theta_1$ is pushed closer to the wall by the overshoot of θ_1 (Gupta and Joseph, 1973)

(a) The single-α solution. The graphs of the single-α solution at two values of μ are exhibited in Figs. 91.1 and 91.2. These figures should be compared with Fig. 88.1 which gives the graph of θ_1 for the simpler variational problem in which it is not assumed that $B=0$. In the present case $w_1 \neq \theta_1$ and the overshoot of the function θ_1 allows $w_1\theta_1$ to assume a minimizing value (near one) over most of the channel without raising the values of the derivatives of w near the wall. The bounding heat transport curve $G_1(\mu)$ is exhibited in Figs. 89.1, 92.1 and 92.2.

(b) The two-α solution. The analysis of the two-α problem follows procedures already described in § 88. As in § 88 it was assumed that the functions $w_2(z)$ and $\theta_2(z)$ are antisymmetric and $w_1(z)$ and $\theta_1(z)$ symmetric functions of z. The graphs of a two-α solution are exhibited in Fig. 91.3 (cf. Figs. 88.2a, b). The bounding heat transport curve $G_2(\mu)$ is exhibited in Figs. 89.1 and 92.2.

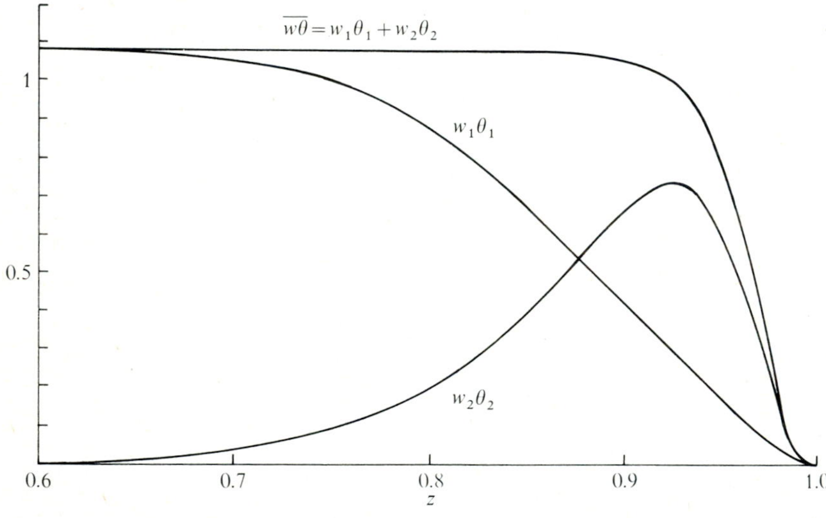

Fig. 91.3: Two-α solutions ($\mu=3680$, $G=428.7$, $Nu=9.58$). Here the boundary layer is such that $\overline{w\theta}=w_1\theta_1+w_2\theta_2\simeq$ at interior points. This boundary layer structure is already fairly well developed (Gupta and Joseph, 1973)

The same type of "stability" analysis which was described in § 88 was used to determine the point of bifurcation from a single-α to a two-α solution. The values of parameters at the point of bifurcation are

$$\mathrm{R}=G_{12}\simeq 221.5, \quad \mu_{12}\simeq 958, \quad Nu=5.33, \quad \alpha_2^2\simeq 111.7.$$

(c) The bifurcation property. In Fig. 91.4 we have plotted the variation of some measures of the amplitude of w_2 with small values of $\mu-\mu_{12}\geqslant 0$. This graph shows that w_2 (and θ_2) tends to zero so that

$$\lim_{\mu \downarrow \mu_{12}} \frac{w_2}{\sqrt{\mu - \mu_{12}}}$$

is bounded.

(d) Multi-α solutions. To study solutions with large values of N it is necessary to treat the variational problem when μ is large. It seems appropriate, in this limit, to treat the problem by boundary layer methods. Chan (1971) gave such a boundary layer analysis for the problem of regular convection in a fluid

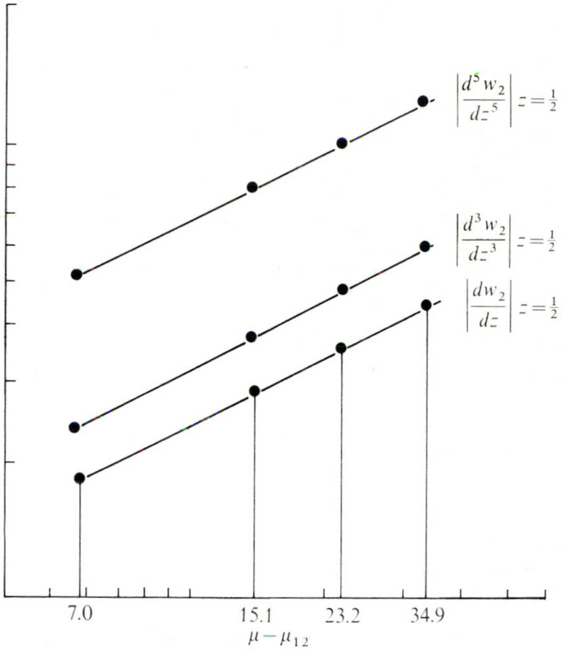

Fig. 91.4: The variation of two-α solution of Gupta and Joseph (1973) with μ near the point of bifurcation $(\mu, \hat{\mathscr{R}}) = (\mu_{12}, G_2) \simeq (957.5, 221.5)$. The points are taken from the first four entries of Table 3 of Gupta and Joseph. Suppose that $|dw_2/dz|_{z=1/2} = \sqrt{\mu - \mu_{12}}\, f(\mu)$ where $f(\mu)$ is slowly varying. Then $\log|dw_2/dz| = \frac{1}{2}\log(\mu - \mu_{12}) + \log f(\mu)$ should appear as a straight line of slope $\frac{1}{2}$ in a log-log plot, as shown (Joseph, 1974A)

with an infinite Prandtl number. Gupta (1972) applied Chan's analysis to the present problem and compared the results with what could be learned from numerical analysis. Agreement is obtained in a qualitative sense but a quantitative agreement seems to be lacking. The boundary-layer analysis is based on the hypothesis that, when μ is large,

$$\overline{w\theta} = \sum_{i=1} w_i \theta_i = 1$$

everywhere but in a small region of the boundary. This is a boundary layer in $\overline{w\theta}$ rather than in the separate factors w_i and θ_i. This kind of structure is already evident in Figs. 91.1, 91.2 and 91.3.

The boundary-layer analysis here relies more heavily on various *ad-hoc* assumptions than the boundary-layer analysis of § 89 and important questions are as yet unanswered. Though the results of Gupta's application of Chan's analysis are perhaps too uncertain to be reported here in full, the following two points deserve emphasis.

(a) A well known dimensional argument which leads to the law $Nu \propto R^{1/3}$ as $R \to \infty$ in the case of regular convection leads to $Nu \propto R$ as $R \to \infty$ in the case of porous convection. An infinite limiting Rayleigh number can obviously be achieved by making d, the plate separation, tend to infinity while $\Delta T = T_2 - T_1$ is kept fixed. But in this case the heat flux $(K\Delta T/d)Nu$ must be independent of d, being the heat flux into a semi-infinite region through a lower heated surface. Therefore, the Nusselt number must vary linearly with d. But the Nusselt number depends on d only through the Rayleigh number which also varies linearly with d. Therefore, Nu must vary linearly with R as $R \to \infty$.

The argument just given depends, of course, upon the assumption that Nu depends on d only through R and not on other parameters which also depend on d. This would be true if the DOB equations were valid in the limit of large R. However, we cannot expect these equations to hold when the scale of the motions is as small as a typical void in the porous bed. Despite this, some experiments do seem to be consistent with the requirements of the dimensional arguments in the limit of large R.

It is appropriate however to compare the prediction of the dimensional argument with the variational arguments since these both assume that the DOB equations govern when R is large. The prediction of the dimensional argument on the slope of the heat transport curve is supported by the result of asymptotic analysis of the variational problem.

(b) Asymptotic results for $N=1$ and $N=2$ do not compare well quantitatively with the corresponding numerical results even though they become parallel on a log-log scale.

§ 92. The Heat Transport Curve: Comparison of Theory and Experiment

A very well established feature of thermal convection in a pure fluid is that the heat flux $Nu(\mathrm{R})$ varies in linear segments with R. Such qualitative changes in the shape of the heat transport curve were first observed for the Bénard problem by Schmidt and Saunders (1938) and their significance was first recognized by Malkus (1954). It is natural to identify the points of transition between the "straight line" segments as points where the nature of convection changes. Krishnamurti (1970A, B) has carefully examined this transition phenomenon anew and essentially confirmed and extended the earlier observations. She finds that the flow changes from steady two-dimensional rolls to another steady cellular motion as the first point of bifurcation is passed.

A similar transition phenomenon has been observed for convection in a porous layer by Combarnous and LeFur (1969), by Buretta (1972) and by Buretta and Berman (1976). In the experiments of Buretta, heat-transfer measurements are obtained from different porous layers made up of packed glass beads (with 3, 6 and 14.3 mm diameters) in a cylindrical container of small aspect ratio (height/diameter ranges from $\frac{1}{10}$ to $\frac{1}{4}$).

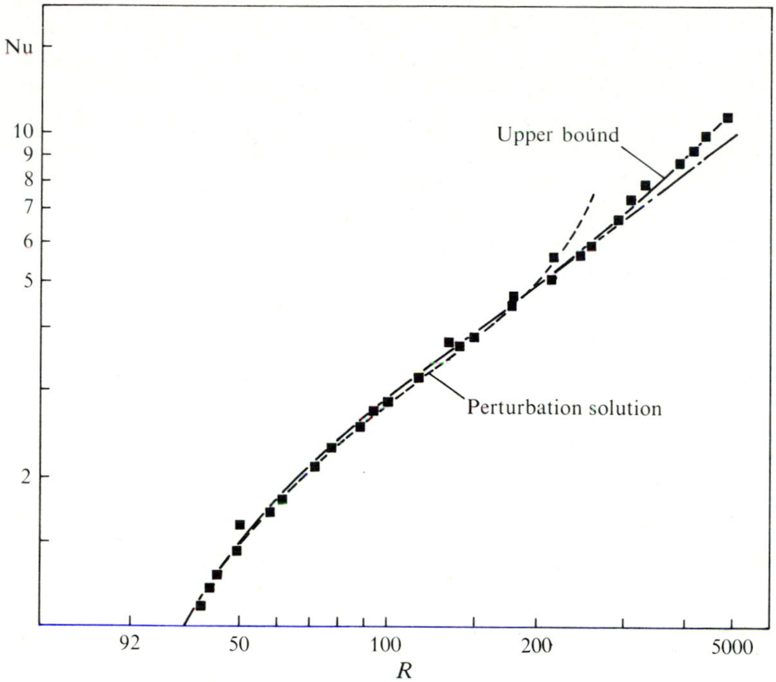

Fig. 92.1: Heat transport curves and experimental data of Buretta (1972). The perturbation solution is due to Palm, Weber and Kvernvold (1972)

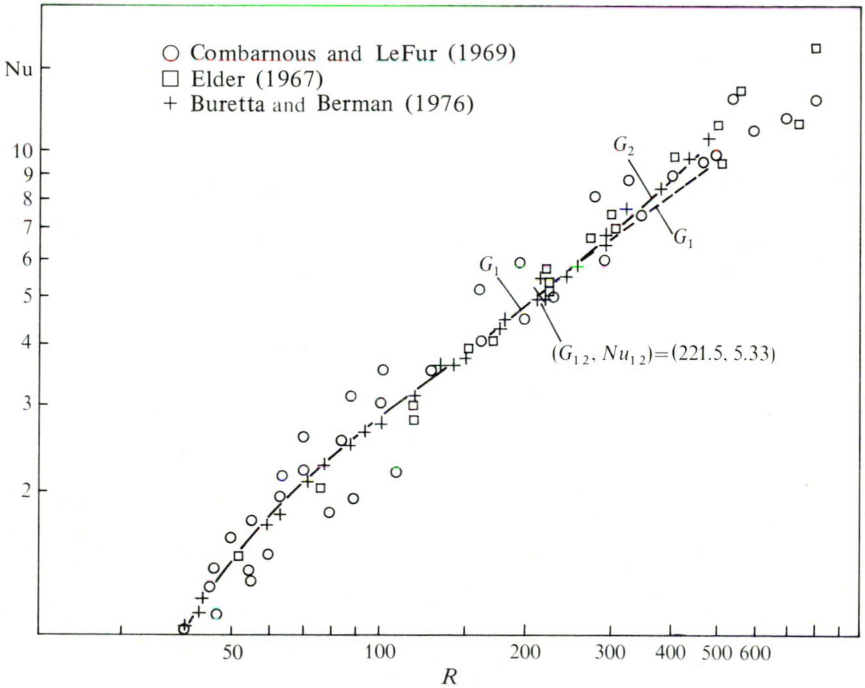

Fig. 92.2: Comparison of the bounding heat transport curve with experiments

In Fig. 92.1 we have compared all of the measurements of the heat transport in Buretta's experiments in four out of five runs[5] with the theoretical upper bound calculated numerically. The agreement between the experiments and theory is striking. The data of Combarnous and LeFur also seems to be in good agreement with the theory but the published experimental data are plotted on such a small scale as to make detailed comparison with theory very difficult (see Fig. 92.2[6]).

The nature of the agreement between theory and experiment is of particular interest. In making this comparison the continuity properties of the theoretical heat transport curve become important. We note that the slope of the bounding heat transport curve

$$\frac{dG(\mu)}{d\mu}=\frac{\langle(w_1\theta_1+w_2\theta_2-\langle w_1\theta_1+w_2\theta_2\rangle)^2\rangle}{\langle w_1\theta_1+w_2\theta_2\rangle^2}$$

must be continuous across the point of bifurcation ($\mu\leqslant\mu_{12}, w_2=\theta_2=0$). On the other hand, since the derivatives $dw_2(z;\mu)/d\mu$ and $d\theta_2(z;\mu)/d\mu$ are finite at $\mu=\mu_{12}$ (see Gupta and Joseph, 1973) a jump in the value of $d^2G/d\mu^2$ is expected as μ is increased past μ_{12}.

Returning now to the experiments, we note that the observed changes in the slope of the heat-transport curve are not precise enough to establish the nature of the discontinuity at the point of bifurcation. Indeed, transitions associated with secondary bifurcations may, in some cases, have the same continuity properties as the bounding heat transport curve. There are, of course, instabilities of a "snap through" type such as is observed in the breakdown of laminar pipe flow (cf. Joseph and Sattinger, 1972). Such instabilities are usually associated with hysteresis phenomena and they would ordinarily lead to actual discontinuities in the heat transport curve rather than to discontinuities in slope.

It follows, assuming the continuity of slope in the experiments, that the bifurcation would go undetected for some range of $R>G_{12}$ and the experimentally observed transition value would be too high. In view of this the values of 280—300 of Combarnous and LeFur and the 245 of Buretta and Berman are in good agreement with our numerically calculated value $G_{12}=221.5$.

Having made this case for agreement between theory and experiment we must caution that "upper bounds" would be expected to coincide with the heat transport curve only if the solutions of the DOB equations were to coincide with solutions of the Euler equations for the variational problem. With this caution in mind we consider the comparison of several heat transport curves with each other (Fig. 92.1) and with experimental data taken from three different papers (Fig. 92.2). The heat transport curve which is shown in Fig. 92.1 as a dotted line was computed by Palm, *et al.*, (1972) from a power series solution in ε, like (78.18) but carried to ε^6. This is an exact solution of the DOB equations for convection in rolls.

---

[5] The data from one run differed from the other four by about 10%.

[6] There is considerable uncertainty in the experimental data even when the experiments are carefully worked, and no experiment is guaranteed with an error of less than 10%.

Chapter XIII

Stability Problems for Viscoelastic Fluids

Molten plastics, petroleum oil additives and whipping creams are examples of incompressible viscoelastic fluids. The flow of such fluids requires effects without parallel in the flow of Newtonian fluids like air and water. Particularly striking effects occur when there are free surfaces which give a visual display of normal stresses which must exist in viscoelastic fluids when strong shearing motions are forced. For example, the rotation of a small rod in water induces a depression in the free surface around the rod; when the same rod rotates in certain viscoelastic fluids, the fluid will climb the rod (see Figs. 95.1, 2, 3). When a horizontal water jet is extruded from a capillary tube at a high Reynolds number, the jet will contract; when certain viscoelastic fluids are extruded at the same Reynolds number from the same tube, the jet may swell to several times its final diameter (see Fig. 95.6a, b). Tall Taylor cells in viscoelastic fluids may stand four times taller than cells which exist under equivalent conditions in Newtonian fluids (see Figs. 95.7, 8). There is a stability problem for each of the three examples shown in the figures (see § 95).

The mechanical response of viscoelastic fluids to forces is much more complicated than the response of Newtonian fluids. Not only is the response more complicated, but it also differs from fluid to fluid and even in the same fluid when undergoing different motions.

Though important stability problems for viscoelastic fluids abound, a general nonlinear theory of stability is not known. Even formal linearized theories of stability involve some controversy and the forms which conditional stability theorems must take have not been established. The difficulty in formulating a stability theory for viscoelastic fluids is the same one which brakes the development of a mechanical theory of flow—the complexity of the response. In this chapter we shall study stability and bifurcation of flows of viscoelastic fluids which Noll (1958) calls simple incompressible fluids.

Notation for Chapter XIII

Where possible, I have followed the notations used by Truesdell and Noll (1965). Some of the material in this chapter is taken from my (1974C) paper "Slow motion and viscometric motion: stability and bifurcation of the rest state of a simple fluid." About one-half of the material in this chapter is new and has not been published elsewhere. The first application of this new material can be found in my 1976 study (with experiments by G. S. Beavers) of the free surface on a simple fluid induced by the torsional oscillation of a rod of small diameter.

I am not going to list all of the symbols used in the chapter. Most of them are defined locally.

$\mathbf{X}$, X_i	position vector in the rest state, cartesian components of $\mathbf{X}$.
$\boldsymbol{\xi}(\mathbf{X},\tau,\varepsilon)$, ξ_i	position vector for particle $\mathbf{X}$ at time τ, cartesian components of $\boldsymbol{\xi}$.
$\boldsymbol{\chi}_t(\mathbf{x},\tau,\varepsilon)=\boldsymbol{\xi}$	relative position vector.
$\boldsymbol{\chi}_t(\mathbf{x},t,\varepsilon)=\mathbf{x}, x_i$	position for particle presently at $\mathbf{x}$, cartesian components of $\mathbf{x}$.
$\boldsymbol{\xi}=\mathbf{X}$, $\boldsymbol{\chi}_t=\mathbf{x}$	when $\varepsilon=0$.
$\left(\nabla_{\mathbf{X}}, \dfrac{\partial}{\partial X_i}\right)$	gradient, cartesian component.
$\left(\nabla_{\xi}, \dfrac{\partial}{\partial \xi_i}\right)$	gradient, cartesian component.
$\left(\nabla, \dfrac{\partial}{\partial x_i}\right)$	gradient, cartesian component.
$\mathbf{G}(s)$, $\mathbf{G}(s,\varepsilon)$, $\mathbf{G}_i(s)$	Histories; see (93.4b).
$\mathsf{G}(s)$, $\zeta(s)$, $\beta(s_1,s_2)$, $\gamma(s_1,s_2)$	Material functions.

We adopt the following convention for the components of a second order tensor. Let $\{\mathbf{e}_i\}$ be a cartesian basis. The dyadic product of $\mathbf{e}_i$ and $\mathbf{e}_j$ is conventionally designated as $\mathbf{e}_i\otimes\mathbf{e}_i$ or, as in this work, as $\mathbf{e}_i\mathbf{e}_j$. Second order tensors may be represented by a dyadic sum

$$\mathbf{B}=\mathbf{e}_l\mathbf{e}_m B_{lm} \quad \text{(summation convention holds)}$$

with components $B_{ij}=\mathbf{e}_i\cdot\mathbf{B}\cdot\mathbf{e}_j$. The gradient of a vector is defined

$$\nabla\mathbf{u}=\frac{\partial\mathbf{u}}{\partial x_l}\mathbf{e}_l, \qquad (\nabla\mathbf{u})_{ij}=\partial u_i/\partial x_j$$

so that $\mathbf{d}u=\nabla\mathbf{u}\cdot d\mathbf{x}$. The same type of representations may be defined for any orthonormal basis provided that $d\mathbf{x}$ is properly expressed in curvilinear coordinates.

To characterize a simple incompressible fluid it is necessary to give explicit form to the functional $\mathscr{F}\left[\mathop{\mathbf{G}(s)}_{s=0}^{\infty}\right]$ which represents the extra stress $\mathbf{S}$. Various types of approximations for $\mathscr{F}$ are given in the text. These approximations are related to functional expansions of $\mathscr{F}$ and may be called functional stresses. A list of the different functional stresses defined in the text and the number of the defining equations given below:

$\mathscr{F}$	$\equiv \mathscr{F}[\mathbf{G}(s)] \equiv \underset{s=0}{\overset{\infty}{\mathscr{F}}}[\mathbf{G}(s)]$
$\mathscr{F}_i$	(93.6)
$\mathscr{F}^{(N)}$	(94.14a, b, 94.15—94.17, 94.73)
$\mathscr{F}^{\langle n \rangle}$	(94.41, 94.42, 94.75)
$\mathbf{S}_n$	(94.5)
$\mathbf{S}_{(N)}$	(94.6, Notes to 94(b))
$\overline{\mathbf{S}}_{(N)}$	(94.9a)
$\mathbf{S}_n^{\langle l \rangle}$	(94.9b)
$\mathbf{S}_n^{(l)}$	bottom of pg. 225
$\mathbf{S}^{\langle n \rangle}$	(94.9c)

Most of the theory given in § 94 was developed after the book was sent to press. It would have been better to split the material given in § 94 into four sections (which now appear as subsections (a)—(d)) and to rearrange the sequence of some of the derivations. To avoid complicated rearrangements of the text, I have been forced to give a less than optimal, but I hope still clear, exposition of the new theory.

§ 93. Incompressible Simple Fluids. Functional Expansions and Stability

A simple fluid is a simple material. A simple material may be roughly characterized as follows: Let

$$\xi = \xi(\mathbf{X}, \tau)$$

be the position vector of the particle $\mathbf{X}$ at the time τ. The deformation of a body is completely specified when the position vectors ξ of its material points $\mathbf{X}$ are known. The cartesian components of the deformation gradient $\nabla_{\mathbf{X}}$ are designated as

$$\frac{\partial \xi_i}{\partial X_j} = [\nabla_{\mathbf{X}} \xi]_{ij} = F_{ij}.$$

The constitutive assumption for simple materials is that the stress $\mathbf{T}$ at a particle $\mathbf{X}$ which at the present time t is at the place

$$\mathbf{x} = \xi(\mathbf{X}, t)$$

is given by a tensor-valued functional

$$\mathbf{T} = \underset{\tau=-\infty}{\overset{t}{\bar{\bar{\mathscr{F}}}}}[\mathbf{F}(\tau)] \tag{93.1}$$

mapping the history $\mathbf{F}(\tau)$ into present time. The requirement that the form of the constitutive functional $\bar{\bar{\mathscr{F}}}$ should be unchanged under all proper orthogonal rotations (Green and Rivlin, 1957) or under all orthogonal rotations (Noll, 1958) leads to a more explicit representation for the stress **T**:

$$\mathbf{T}=\mathbf{F}^T(t)\cdot \underset{\tau=-\infty}{\overset{t}{\mathscr{F}}}[\mathbf{C}(\tau)]\cdot\mathbf{F}(t) \tag{93.2}$$

where

$$\mathbf{C}(\tau)=\mathbf{F}(\tau)^T\cdot\mathbf{F}(\tau)$$

is the Cauchy-Green strain tensor. In cartesian coordinates

$$T_{ij}=\frac{\partial x_i}{\partial X_r}\frac{\partial x_j}{\partial X_s}\underset{\tau=-\infty}{\overset{t}{\mathscr{F}_{rs}}}[\mathbf{C}(\tau)]$$

where

$$C_{ij}(\tau)=\frac{\partial\xi_m}{\partial X_i}\frac{\partial\xi_m}{\partial X_j}=[\mathbf{F}^T(\tau)\cdot\mathbf{F}(\tau)]_{ij}\,.$$

Noll (1958) distinguishes fluids from solids by noting that the stress in a fluid should be unchanged by a change of the reference configuration in which the material points **X** are defined[1]. In a solid, the state of the present stress depends on the choice of the reference configuration from which the deformations are measured. In Noll's theory the reference configuration is eliminated from the kinematic description of path lines and the present configuration is used as a reference:

$$\mathbf{X}=\mathbf{x}^{-1}(\mathbf{x},t)\,,$$

and

$$\xi=\xi(\mathbf{x}^{-1}(\mathbf{x},t),\tau)\equiv\chi_t(\mathbf{x},\tau) \tag{93.3}$$

is the position of the point at $\tau\leqslant t$ which at the present time is at

$$\mathbf{x}=\chi_t(\mathbf{x},t)\,.$$

We write $\xi=\xi(\tau)$ and $\mathbf{x}=\xi(t)$, for short, and introduce the relative deformation gradient

$$\mathbf{F}_t(\tau)=\nabla\chi_t\,,\qquad [\mathbf{F}_t(\tau)]_{ij}=\partial\xi_i/\partial x_j\,,$$
$$\mathbf{F}_t(t)=\mathbf{1}\,.$$

It follows from Noll's theory (see Truesdell and Noll, 1965) that the stress in an incompressible simple fluid can be represented as

$$\mathbf{T}+p\mathbf{1}=\mathbf{S}=\underset{s=0}{\overset{\infty}{\mathscr{F}}}[\mathbf{G}(s)]\,,\qquad \mathscr{F}[0]=0 \tag{93.4a}$$

[1] The ideas developed in this chapter are set in the context of Noll's (1958) theory of the simple fluid and depend directly on ideas expressed and implied in the papers of Rivlin and Ericksen (1955), Green and Rivlin (1957), Coleman and Noll (1961) and Pipkin (1964).

where $s=t-\tau$, p contains all terms which are scalar multiples of the unit tensor $\mathbf{1}$ and

$$\mathbf{G}(s)=\mathbf{F}_t^T(\tau)\cdot\mathbf{F}_t(\tau)-\mathbf{1}, \qquad G_{ij}(s)=\frac{\partial\xi_m}{\partial x_i}\frac{\partial\xi_m}{\partial x_j}-\delta_{ij}, \tag{93.4b}$$

$$\mathbf{G}(0)=0.$$

We call $\mathbf{S}$ the extra stress, $\mathscr{F}$ the response functional and $\mathbf{G}(s)$ the history.

Although a simple "... fluid may have definite memory of all its past experience, it reacts to those experiences only by comparing them with its present configuration. Like an administrator, the fluid has a filtering memory: Oblivious to the difference between the present and any fixed or established past state of affairs, it remembers its youthful distortions only by continually re-evaluating them in terms of present expediency." (Truesdell, 1974)

(a) Functional Expansions of $\mathscr{F}$, Stability and Bifurcation

Newcomers to the study of the fluid dynamics of simple fluids are surprised that it is necessary to maintain a distinction between constitutive relations which apply to one and the same fluid undergoing different types of motion. This is to say that the stresses in a simple fluid depend on the history of the deformation and may take on entirely different forms when the histories are different. The study of simple fluids is made difficult by the requirement that one must first characterize the material by specifying $\mathscr{F}$. Response functionals for particular physical fluids which are valid in all of the motions which the fluid could undergo are generally unknown, and with the notable exception of Newtonian fluids, may be unknowable. It is possible to circumvent this difficulty by postulating mathematical models or by restricting $\mathbf{G}(s)$ to a class on which $\mathscr{F}[\mathbf{G}(s)]$ reduces to something manageable. Manageable $\mathscr{F}$'s arise for example on histories of viscometric flows, slow steady motions and small amplitude time-dependent motions. The notion of mathematical models is logically no different than a constitutive assumption; in this sense the simple fluid is a model. In representing physical fluids it is, however, desirable to avoid the use of too special models leading to too special forms for the response $\mathscr{F}$. In the study of stability and bifurcation it is natural to eschew overly specialized fluid models and to look at motions which can be obtained as perturbations of histories $\mathbf{G}_0(s)$ on which manageable $\mathscr{F}[\mathbf{G}_0(s)]$ can be specified in sufficient generality. This procedure requires that we consider functional derivatives of the functional $\mathscr{F}$ at the point $\mathbf{G}_0(s)$.

Suppose the history is given in the following form

$$\mathbf{G}(s)=\mathbf{G}_0(s)+\varepsilon_1\mathbf{G}_1(s)+\varepsilon_2\mathbf{G}_2(s)+\cdots+\varepsilon_N\mathbf{G}_N(s). \tag{93.5}$$

The $\mathbf{G}_n(s)$ can be thought to arise from path lines of prescribed, kinematically admissible but otherwise arbitrary, velocity fields $\mathbf{U}_n(\mathbf{x},\tau)$, $\tau\leqslant t$. Consider the tensor-valued function

$$\boldsymbol{\Theta}(\varepsilon_1,\varepsilon_2,\ldots,\varepsilon_N)=\mathscr{F}[\mathbf{G}_0+\varepsilon_1\mathbf{G}_1+\cdots+\varepsilon_N\mathbf{G}_N].$$

We call the multilinear form

$$\mathscr{F}_{(l)}[\mathbf{G}_0+\varepsilon_1\mathbf{G}_1+\ldots+\varepsilon_N\mathbf{G}_N:\mathbf{G}_1,\mathbf{G}_2,.,.,.,\mathbf{G}_l]=\frac{\partial^l\boldsymbol{\Theta}}{\partial\varepsilon_1\partial\varepsilon_2\ldots\partial\varepsilon_l}$$

the lth order mixed functional derivative of $\mathscr{F}$ at the point $\mathbf{G}(s)$. The mixed functional derivative is multilinear in the tensors $\mathbf{G}_i$ following the colon. We call

$$\mathscr{F}_{(l)}[\mathbf{G}_0+\varepsilon\mathbf{G}_1:\underbrace{\mathbf{G}_1,\mathbf{G}_1,\ldots,\mathbf{G}_1}_{l\text{ times}}]=\frac{\partial^l\mathscr{F}[G_0+\varepsilon G_1]}{\partial\varepsilon^l}$$

the lth order functional derivative at the point $\mathbf{G}_0(s)+\varepsilon\mathbf{G}_1(s)$. We next consider the functional expansion which is induced by analytic histories[2]

$$\mathbf{G}(s,\varepsilon)=\mathbf{G}_0(s)+\varepsilon\mathbf{G}_{(1)}(s)+\frac{\varepsilon^2}{2}\mathbf{G}_{(2)}(s)+\frac{\varepsilon^3}{3!}\mathbf{G}_{(3)}(s)+\ldots.$$

Then

$$\begin{aligned}\mathscr{F}[\mathbf{G}(s,\varepsilon)]=\mathscr{F}[\mathbf{G}_0]+\varepsilon\mathscr{F}_{(1)}[\mathbf{G}_0:\mathbf{G}_1]+\frac{\varepsilon^2}{2}\{\mathscr{F}_{(2)}[\mathbf{G}_0:\mathbf{G}_{(1)},\mathbf{G}_{(1)}]\\+\mathscr{F}_{(1)}[\mathbf{G}_0:\mathbf{G}_{(2)}]\}+\frac{\varepsilon^3}{3!}\{\mathscr{F}_{(3)}[\mathbf{G}_0:\mathbf{G}_{(1)},\mathbf{G}_{(1)},\mathbf{G}_{(1)}]\\+3\mathscr{F}_{(2)}[\mathbf{G}_0:\mathbf{G}_{(1)},\mathbf{G}_{(2)}]+\mathscr{F}_{(1)}[\mathbf{G}_0:\mathbf{G}_{(3)}]\}+0(\varepsilon^4).\end{aligned}$$

2 The assumption of analyticity in ε is convenient but not necessary. We compute derivatives of the solution with respect to ε at $\varepsilon=0$. If the computation is carried say to second order then two derivatives with respect to ε are required. Differentiation of the stress requires the introduction of functional derivatives like Gateaux and Fréchet derivatives. These derivatives depend on the topology of the space of histories on the domain of the functional. Coleman and Noll [1960, 1961] use the Hilbert space $L_2[0,\infty]$ with weighted "fading memory" norm which decays rapidly as $s\to\infty$. The work of Coleman and Noll [1960, 1961] is important in that it provides the first example of how functional expansions can be used to approximate Noll's $\mathscr{F}$ in special asymptotic limits. Many different choices for the topology of the space of histories are of course possible. It is not possible however to determine which choices are mathematically correct without first dealing with the problem of finding solutions of the equations of motion. A space of initial histories may be assumed and shown to have desirable properties. It has then to be shown that the solution of equations of motion, with prescribed data on the boundary, exist and evolve in a space compatible with the one presumed for the initial history. This hard problem has not been extensively studied in continuum mechanics.

It is necessary to add that physical fluids know nothing about normed spaces. Such fluids may be usefully characterized by constants, functions and other quantities which may be measured in the laboratory. Spaces could be expected to arise in specifying the nature of the convergence which is associated with solutions of equations of motion but, as we have already noted, not much is known about such solutions even for fluids whose constitutive response is characterized by an equation much less general than $\mathbf{S}=\mathscr{F}[\mathbf{G}(s)]$.

In view of the foregoing I think that it is perhaps not wise, and certainly is not necessary, in trying to develop a algorithms for solving flow problems, to commit oneself at the start to some special history space. Of course, a complete mathematical theory should eventually include a catalogue of allowed history spaces and their properties.

The notation for functional derivatives used in this chapter suggests Gateaux derivatives. Questions of continuity which would require us to distinguish one derivative from another will not be discussed here.

Introducing the notation $(\cdot)_{(n)}=n!(\cdot)_n$ we have

$$\begin{aligned}\mathscr{F}[\mathbf{G}(s,\varepsilon)]=\mathscr{F}[\mathbf{G}_0]+\varepsilon\mathscr{F}_1[\mathbf{G}_0:\mathbf{G}_1]+\varepsilon^2\{\mathscr{F}_2[\mathbf{G}_0:\mathbf{G}_1,\mathbf{G}_1]\\ +\mathscr{F}_1[\mathbf{G}_0:\mathbf{G}_2]\}+\varepsilon^3\{\mathscr{F}_3[\mathbf{G}_0:\mathbf{G}_1,\mathbf{G}_1,\mathbf{G}_1]\\ +3\mathscr{F}_2[\mathbf{G}_0:\mathbf{G}_1,\mathbf{G}_2]+\mathscr{F}_1[\mathbf{G}_0:\mathbf{G}_3]\}+0(\varepsilon^4)\,.\end{aligned}\tag{93.6}$$

(b) Generation of the History of a Motion

Histories, $\mathbf{G}(s)$, are not really given; they are generated as solutions of the equations which govern the motion. Of course, a given history may always be regarded as a solution of the equations of motion driven by a suitably contrived body force field. It is then necessary to find the subsequent motion when the forcing is not contrived but, instead, is given *a priori*, as data. The subsequent motion satisfies the initial-history problem.

$$\begin{aligned}&\rho\left(\frac{\partial\mathbf{U}}{\partial t}+\mathbf{U}\cdot\nabla\mathbf{U}\right)=-\nabla p+\nabla\cdot\mathscr{F}[\mathbf{G}(s,\varepsilon)]+\mathbf{f}_1(\mathbf{x},t,\varepsilon)\,,\\ &\operatorname{div}\mathbf{U}=0,\qquad \mathbf{U}\big|_{\partial\mathscr{V}}=\mathbf{f}_2(\mathbf{x},t,\varepsilon)\,,\\ &\mathbf{U}(\mathbf{x},\tau,\varepsilon)\ \text{is prescribed in}\ \tau\leqslant t\,.\end{aligned}\tag{93.7}$$

There is at present no existence-uniqueness theory for this problem, even for steady solutions.

The parameter ε is defined by the details of the problem. In problems where motion is forced externally the magnitude of the forcing can be scaled with ε; $\varepsilon=0$ then corresponds to the rest state $\mathbf{G}_0(s)=0$. Small motions driven by steady or unsteady forcing may then be constructed in a series of powers of ε by methods described under § 94(b). In the problem of the free surface induced by torsional oscillation of a rod, $\varepsilon=\omega\Theta/2$ where ω is the frequency of oscillation and Θ is the *angle of twist*. In stability and bifurcation problems, ε is defined as a measure of the amplitude of the disturbance of some given motion with history $\mathbf{G}_0(s)$.

(c) Stability and Bifurcation of Steady Flow

It is instructive to consider the problem of stability and bifurcation of steady flow. The steady flow $\mathbf{U}(\mathbf{x})$ is solenoidal and satisfies

$$\rho\mathbf{U}\cdot\nabla\mathbf{U}=-\nabla p_0+\nabla\cdot\mathscr{F}[\mathbf{G}_0(s)]\quad\text{in }\mathscr{V}$$

and (93.8)

$$\mathbf{U}\big|_{\partial\mathscr{V}}=\mathbf{f}(\mathbf{x})\,.$$

Problem (93.8) cannot be solved without prior specification of a relation between $\mathscr{F}$ and $\mathbf{G}_0(s)$. Exact, explicit solutions of (93.8) are known for the special shearing motions which Coleman (1962) has called viscometric (Rivlin, 1956; Coleman and Noll, 1959). In steady flow the differential equation for path lines is separable

$$\frac{d\xi(\tau)}{d\tau}=\mathbf{U}(\xi), \qquad \xi(t)=\mathbf{x}$$

so that

$$\xi(\tau)-\mathbf{x}=\boldsymbol{\phi}(\mathbf{x},s), \qquad \boldsymbol{\phi}(\mathbf{x},0)=0 \tag{93.9}$$

and

$$\nabla\xi=\mathbf{1}+\nabla\boldsymbol{\phi}(\mathbf{x},s)=\mathbf{F}_t(\tau), \qquad \tau=t-s\,.$$

The form of $\boldsymbol{\phi}(\mathbf{x},s)$ is given by (93.12) and

$$\mathbf{G}_0(s)=\mathbf{F}_t^T(t-s)\cdot\mathbf{F}_t(t-s)-\mathbf{1}\,.$$

Supposing now that (93.8) has been solved for $\mathbf{U}(\mathbf{x})$ and that $\mathbf{G}_0(s)$ has been computed, we consider the evaluation of disturbance $\varepsilon\mathbf{u}$ of $\mathbf{U}$. The motion $\mathbf{U}(\mathbf{x})+\varepsilon\mathbf{u}(\mathbf{x},t,\varepsilon)$ generates the history $\mathbf{G}(s,\varepsilon)$. Looking forward to the analysis of given in § 94 we assume $\mathbf{G}(s,\varepsilon)$ in the series form (93.5). On taking account of the fact that $\mathbf{U}+\varepsilon\mathbf{u}$, $p_0+\varepsilon p$ and $G(s,\varepsilon)$ satisfy (93.7) and that $\mathbf{U}$, p_0 and $\mathbf{G}_0(s)$ satisfy (93.8), we find that the solenoidal fields $\mathbf{u}(\mathbf{x},t,\varepsilon)$ which evolve from prescribed initial histories $\mathbf{u}(\mathbf{x},\tau,\varepsilon)$, $\tau\leqslant t$, satisfy

$$\rho\left[\frac{\partial\mathbf{u}}{\partial t}+\mathbf{U}\cdot\nabla\mathbf{u}+\mathbf{u}\cdot\nabla\mathbf{U}+\varepsilon\mathbf{u}\cdot\nabla\mathbf{u}\right]=-\nabla p+\nabla\cdot\mathscr{F}_1[\mathbf{G}_0\colon\mathbf{G}_1]$$
$$+\varepsilon\{\nabla\cdot\mathscr{F}_2[\mathbf{G}_0\colon\mathbf{G}_1,\mathbf{G}_1]+\nabla\cdot\mathscr{F}_1[\mathbf{G}_0,\mathbf{G}_2]\}+0(\varepsilon^2) \tag{93.10}$$

and

$$\mathbf{u}(\mathbf{x},t,\varepsilon)\big|_{\partial\mathscr{V}}=0\,.$$

The advantage of introducing the functional expansion is that the functional derivatives are multinomial forms which have a much simpler structure than the original $\mathscr{F}$. Problem (93.10) is still formidable and its mathematical foundations have not been established, even in the linearized case considered below*.

In the linear theory of stability we set $\varepsilon=0$ and consider the reduced problem:

$$\rho\left[\frac{\partial\mathbf{u}}{\partial t}+\mathbf{U}\cdot\nabla\mathbf{u}+\mathbf{u}\cdot\nabla\mathbf{U}\right]=-\nabla p+\nabla\cdot\mathscr{F}_1[\mathbf{G}_0\colon\mathbf{G}_1]\,, \tag{93.11}$$

$$\mathbf{u}(\mathbf{x},t,0)\big|_{\partial\mathscr{V}}=0\,,$$

$$\mathbf{u}(\mathbf{x},\tau,0)\ \text{ is prescribed in } \mathscr{V} \text{ for } \ \tau\leqslant t\,.$$

* In a recent work, Slemrod (1976) has established conditions which insure the existence of a unique solution $\mathbf{U}=0$ of the linearized problem (93.11) when $\mathbf{G}_0=\mathbf{U}=0$ and $\mathscr{F}_1[0\colon\mathbf{G}(s)]$ is of integral type (see the first equation in Exercise 94.4). He also gives an existence theorem for forced linear problems, like (94.35). Slemrod's work provides a good mathematical underpinning for the initiating perturbation in the theory developed in § 94(b).

To study this problem we need to know the form of the first functional derivative of $\mathscr{F}$ at the point $\mathbf{G}_0$. General representations for $\mathscr{F}_1[\mathbf{G}_0:(\cdot)]$ are presently unknown when $\mathbf{G}_0(s)$ is the history of an arbitrary steady motion. General representations for $\mathscr{F}_1[\mathbf{G}_0:(\cdot)]$ when $\mathbf{G}_0(s)$ is the history of a viscometric flow (like those discussed in Exercise 93.4) have been given by Pipkin and Owen (1967) and by Pipkin (1968). These representations can be used in the study of the linearized stability of viscometric flows. According to Pipkin and Owen (p. 836) "The relation between the stress perturbation and the strain perturbation history is like a linear viscoelastic stress-strain relation for an anisotropic material whose symmetries are those of the unperturbed motion. It is found that in the most general case, twenty linear functionals are required in order to describe all interactions allowed by symmetry. In the case of incompressible fluids, the required number of functionals is reduced to thirteen."

Bifurcation is even more complicated; even at the lowest non-trivial order (terms of $0(\varepsilon)$ in 93.10) it is necessary to obtain general representations for the second functional derivative $\mathscr{F}_2[\mathbf{G}_0:(\cdot),(\cdot)]$ of $\mathscr{F}$ at the point $\mathbf{G}_0$. Such representations are unknown even when $\mathbf{G}_0$ is the history of a viscometric flow. The situation is much clearer when $\mathbf{G}_0(s)=0$ is the history of rest state. We consider this case next.

Exercise 93.1: Assume that the relative position vector $\chi_t(x,\tau)$ is an analytic function τ. Show that

$$\chi_t(\mathbf{x},\tau)=\mathbf{x}+\sum_{m=1}\frac{(-s)^m}{m!}\overset{(m-1)}{\mathbf{U}}(\mathbf{x},t) \tag{93.12}$$

where $\overset{(m)}{\mathbf{U}}$ is the mth material time derivative of $\mathbf{U}(\mathbf{x},t)$. Conclude that for steady flow $\chi_t(\mathbf{x},\tau)=\chi(\mathbf{x},s)$ is independent of t.

Exercise 93.2: Show that

$$\mathbf{F}(\tau)=\mathbf{F}_t(\tau)\cdot\mathbf{F}(t),$$

$$\mathbf{C}_t(\tau)=\mathbf{F}^{-1}(t)^T\cdot\mathbf{C}(\tau)\cdot\mathbf{F}^{-1}(t),$$

and

$$\frac{d\mathbf{F}}{dt}\cdot\mathbf{F}^{-1}(t)=\nabla\mathbf{U}\equiv\mathbf{L},\qquad \mathbf{U}(\mathbf{x},t)=\frac{d\mathbf{x}}{dt}.$$

These relations, and those given in Exercise 94.1, are, for example, derived in Truesdell and Noll (1965, pp. 45—56) and Astarita and Marrucci (1974, Chapter 3).

Exercise 93.3: Consider a spiral flow of a simple fluid between concentric cylinders. Let (r,θ,x) be polar coordinates in the cylinder. Show that there exists an axisymmetric spiral flow $\mathbf{U}=\mathbf{e}_x U(r)+\mathbf{e}_\theta V(r)$. Show that there exists a two-dimensional rotation around each vector $\mathbf{e}_r$ and in the tangent plane to the cylinder $r=$constant such that

$$2D'_{rx}=\kappa$$

is the only nonzero component of the stretching tensor $\mathbf{D}$. Let the angle measured from x in the tangent plane be α. Show that

$$2D_{r\theta}=\kappa\sin\alpha,\qquad 2D_{rx}=\kappa\cos\alpha$$

and

$$\kappa = \sqrt{(2D_{r\theta})^2 + (2D_{rx})^2} = 2D'_{rx}.$$

Exercise 93.4 (Rivlin, 1956; Markovitz, 1957; Coleman and Noll, 1961): What is the form which $\mathscr{F}$ must take in Newtonian fluids? Show that if $\mathbf{G}(s)$ is the history of a spiral flow, the stress $\mathscr{F}$ may be calculated when three functions of the local rate of shear κ are known[3].

§ 94. Stability and Bifurcation of the Rest State

To study the stability of the rest state, we consider the initial value problem for a disturbance of the rest state in a closed container $\mathscr{V}$; thus,

$$\rho\left[\frac{\partial \mathbf{U}}{\partial t} + \mathbf{U}\cdot\nabla\mathbf{U}\right] + \nabla p = \nabla\cdot\mathscr{F}, \quad \nabla\cdot\mathbf{U} = 0, \quad \mathbf{U}|_{\partial\mathscr{V}} = 0 \tag{94.1}$$

where $\mathbf{U}(\mathbf{x}, t)$ is the disturbance and $\mathscr{F}$ is evaluated on the history of $\mathbf{U}$ which is presumed given up to time t. The energy of this disturbance satisfies

$$\frac{\rho}{2}\frac{d}{dt}\langle|\mathbf{U}|^2\rangle = -\langle \mathbf{A}_1[\mathbf{U}]:\mathscr{F}\rangle, \quad A_1[\mathbf{U}] = \nabla\mathbf{U} + \nabla\mathbf{U}^T \tag{94.2}$$

where the integral $\langle \mathbf{A}_1 : \mathscr{F}\rangle$ is called the stress-power. For Newtonian fluids this integral may be written as (see Exercise 3.1)

$$\langle \mathbf{A}_1 : \mathscr{F}\rangle = \mu\langle \mathbf{A}_1 : \mathbf{A}_1\rangle = 2\mu\langle|\nabla\mathbf{U}|^2\rangle.$$

The rest state of a Newtonian fluid is globally and monotonically stable (see Exercise 4.4) because the stress-power is positive definite.

It has been shown by Truesdell (1952) and by Noll (unpublished, see Truesdell and Noll [p. 511]) that it is impossible for the stress power to be positive for all kinematically admissible motions and for all possible fluids. Their results, those of Coleman (1962) and of Dunn and Fosdick (1974) indicate that there are kinematically admissible motions which lead to negative values for the stress-power even when the response functional appears to be of a realistic type.

---

[3] Spiral flows are a special class of flows in the class of viscometric flows. Viscometric flows are locally equivalent to a state of pure shearing. These flows are so constrained that their history is of a special form; on these special histories the response functional $\mathscr{F}$ can be characterized through three scalar functions of κ. Viscometric flows are discussed in detail by Coleman, Markovitz and Noll (1966), Truesdell and Noll (1965), Rivlin and Sawyers (1971), Pipkin and Tanner (1973) and Truesdell (1974). Spiral (helical) flow of a simple fluid is discussed in detail in the first two works mentioned; the last three works are recent reviews of the theory of flow of non-Newtonian fluids; the last two works concentrate on viscometric flows and give complementary points of view of the mechanics of these flows.

It is not possible, of course, to decide about the asymptotic stability of the rest state without considering the destiny of disturbances which satisfy Eqs. (94.1). However, since kinematically admissible disturbances are admissible as initial conditions for (94.2), it does not seem possible to guarantee monotonic stability even in the class of dynamically admissible solutions of (94.1).

It appears unquestionably true that response functionals which lead to the instability of the rest state give an incorrect description of physical fluids. This point of view seems to have been clearly expressed first by A. Craik (1968). Craik considers the stability of the rest state of a viscoelastic fluid, confined between horizontally infinite parallel planes, to infinitesimal, two-dimensional disturbances. He remarks that "On physical grounds, we may assert that any physically realistic models should possess the property that a layer of fluid at rest between horizontal plane rigid boundaries is in stable equilibrium." This is an intuitively correct idea; the state of rest should be the terminal form of every solution of the initial value problem for a fluid which fills a container whose walls are at rest when there are no body forces present to drive a motion. I would expect that only response functionals which lead to the stability of the rest state describe physical fluids.

To study (94.1) we must know something about the stress $\mathscr{F}$. Manageable $\mathscr{F}$'s arise on slow motions and on unsteady motions of small amplitude. Small amplitude motions are unavoidable in the analysis of stability and bifurcation and they form the basis for the analysis given here.

(a) Slow Motion

Now we shall consider a frequently used but misguided procedure for studying stability. In this procedure a constitutive assumption which is correct for slow motions is used to test for stability to disturbances which may be assumed to be small but not slow. The theory of slow motions arises out of a perturbation of the rest state $\mathbf{G}_0(s)$ with a small velocity field proportional to $\varepsilon\mathbf{u}(\mathbf{x},t)$ and an even smaller higher acceleration field proportional to $\varepsilon^{m+1}\overset{(m)}{\mathbf{u}}(\mathbf{x},t)$ where $\overset{(m)}{\mathbf{u}}(\mathbf{x},t)$, $m\geqslant 0$, are bounded fields. One basis for the theory of slow motion is the retardation theorem of Coleman and Noll (1960). The expansion may also be obtained formally from the Rivlin-Ericksen fluids (see the bibliographical note at the end of the sub-section). In the theory of slow motions the functional expansions are given in terms of Rivlin-Ericksen tensors[4]. These tensors arise as derivatives of the relative Cauchy-Green strain tensor $\mathbf{C}_t(\tau)=\mathbf{F}_t^T(\tau)\cdot\mathbf{F}_t(\tau)$ evaluated at time $\tau=t$

$$\mathbf{A}_n=\left[\frac{d}{d\tau^n}\mathbf{C}_t(\tau)^T\right]_{\tau=t}.$$

[4] See Notes for 94(b).

The tensors A_n may be expressed in terms of the stretching tensor $\mathbf{A}_1 = \nabla\mathbf{U} + \nabla\mathbf{U}^T$ through the recursion formula (see Exercise 94.1)

$$\mathbf{A}_{n+1} = \frac{\partial \mathbf{A}_n}{\partial t} + \mathbf{U}\cdot\nabla\mathbf{A}_n + \mathbf{A}_n\cdot\nabla\mathbf{U} + (\mathbf{A}_n\cdot\nabla\mathbf{U})^T, \tag{94.3 a}$$

$$(\mathbf{A}_n\cdot\nabla\mathbf{U})_{ij} = (\mathbf{A}_n)_{il}\frac{\partial U_l}{\partial x_j}.$$

It follows that the tensor $\mathbf{A}_n$ may be regarded as mapping velocity vectors into second-order tensors. We indicate this functional dependence by writing $\mathbf{A}_n = \mathbf{A}_n[\mathbf{U}(\mathbf{x},t)]$. The history of motions analytic in $s = t - \tau$ may be referred to the $\mathbf{A}_n$ by expanding the history

$$\mathbf{G}(s) = \sum_{n=1} (-1)^n \frac{s^n}{n!} \mathbf{A}_n[\mathbf{U}(\mathbf{x},t)]. \tag{94.3 b}$$

Coleman and Noll (1960) have shown that if $\mathscr{F}$ has a fading memory in a sense defined by them, and the history $\mathbf{G}_\varepsilon(s) = \mathbf{G}(\varepsilon s)$ is the retardation of a given flow $\mathbf{G}(s)$, and $\mathbf{G}(s)$ is analytic so that

$$\mathbf{G}_\varepsilon(s) = \sum_{n=1} \frac{s^n}{n!} (\partial^n \mathbf{G}_\varepsilon(s)/\partial s^n)\big|_{s=0} = \sum_{n=1} \frac{(-s)^n}{n!} \hat{\mathbf{A}}_n$$

where $\hat{\mathbf{A}}_n = \varepsilon^n \mathbf{A}_n$ and the $\mathbf{A}_n$ are the Rivlin-Ericksen tensors evaluated on the velocity $\mathbf{U}(\mathbf{x},t)$, then $\mathbf{S}$ is given by

$$\mathbf{S} = \mathop{\mathscr{F}}_{s=0}^{\infty}[\mathbf{G}_\varepsilon(s)] = \sum_{n=1} \mathbf{S}_n[\hat{\mathbf{A}}_n, \hat{\mathbf{A}}_{n-1}, \ldots, \hat{\mathbf{A}}_1]. \tag{94.4}$$

. The tensor-valued function $\mathbf{S}_n$ of $\mathbf{A}_n$ can be written explicitly (for example, see Truesdell and Noll, 1965). The first four of the $\mathbf{S}_n$ are

$$\mathbf{S}_1[\mathbf{A}_1] = \mu\mathbf{A}_1, \tag{94.5 a}$$

$$\mathbf{S}_2[\mathbf{A}_1, \mathbf{A}_2] = \alpha_1\mathbf{A}_2 + \alpha_2\mathbf{A}_1^2, \tag{94.5 b}$$

$$\mathbf{S}_3[\mathbf{A}_1, \mathbf{A}_2, \mathbf{A}_3] = \beta_1\mathbf{A}_3 + \beta_2(\mathbf{A}_2\mathbf{A}_1 + \mathbf{A}_1\mathbf{A}_2) + \beta_3(tr\,\mathbf{A}_2)\mathbf{A}_1, \tag{94.5 c}$$

$$\begin{aligned}\mathbf{S}_4[\mathbf{A}_1, \mathbf{A}_2, \mathbf{A}_3, \mathbf{A}_4] = {}& \gamma_1\mathbf{A}_4 + \gamma_2(\mathbf{A}_3\mathbf{A}_1 + \mathbf{A}_1\mathbf{A}_3) + \gamma_3\mathbf{A}_2^2 \\ & + \gamma_4(\mathbf{A}_2\mathbf{A}_1^2 + \mathbf{A}_1^2\mathbf{A}_2) + \gamma_5(tr\,\mathbf{A}_2)\mathbf{A}_2 + \gamma_6(tr\,\mathbf{A}_2)\mathbf{A}_1^2 \\ & + [\gamma_7\, tr\,\mathbf{A}_3 + \gamma_8(tr\,\mathbf{A}_1\mathbf{A}_2)]\,\mathbf{A}_1.\end{aligned} \tag{94.5 d}$$

The coefficients $\mu, \alpha_1, \alpha_2, \beta_1, \beta_2, \beta_3, \gamma_1, \gamma_2, \ldots, \gamma_8$ are constants or, more generally, functions of the temperature. The $\mathbf{S}_n$ are linear combinations of functional derivatives evaluated on the rest history $\mathbf{G}_0(s)=0$ (see Exercise 94.4). The tensor $\mathbf{S}_n$ is linear in $\mathbf{A}_n$

$$\mathbf{S}_n=\hat{\alpha}_n\mathbf{A}_n+\mathscr{I}_n[\mathbf{A}_{n-1},\ldots,\mathbf{A}_1] \tag{94.5e}$$

where $\hat{\alpha}_n$ is a constant. The partial sums

$$\mathbf{S}_{(N)}=\sum_{n=1}^{N}\mathbf{S}_n[\mathbf{A}_n,\ldots,\mathbf{A}_1] \tag{94.6}$$

are stress tensors for the Coleman-Noll fluids of grade N. On retarded motions

$$\mathbf{S}_{(N)}=\sum_{n=1}^{N}\mathbf{S}_n[\hat{\mathbf{A}}_n,\ldots,\hat{\mathbf{A}}_1]=\sum_{n=1}^{N}\varepsilon^n\mathbf{S}_n[\mathbf{A}_n,\ldots,\mathbf{A}_1]$$

is an approximation to (94.4) with an error of order ε^{N+1}.

Retarded motions are slow motions for which partial time derivatives, holding ξ fixed, are also small. Retarded motions may be viewed as perturbations of rest state. They have slow speeds and even slower accelerations. Let $\xi'(\mathbf{X},\tau')=\chi_{t'}(\mathbf{x},\tau')$ be the position vector for the particle $\mathbf{X}$ in some slow motion and let ε be the parameter measuring how slow is slow (ε may be regarded as the scale of the forces driving the flow; when $\varepsilon=0$ the fluid is at rest). The velocity of the slow flow

$$\mathbf{U}(\xi',\tau')=\frac{d\xi'(\mathbf{X},\tau')}{d\tau'}=\varepsilon\frac{d\xi(\mathbf{X},\tau)}{d\tau},$$

where $\tau'=\tau/\varepsilon$ and $\xi(\mathbf{X},\tau)=\xi'(\mathbf{X},\tau')$, tends to zero with ε because the velocity

$$\mathbf{u}(\xi,\tau)=\frac{d\xi(\mathbf{X},\tau)}{d\tau}$$

is bounded. It follows that the mth acceleration gradient in the slow motion

$$\overset{(m)}{\mathbf{U}}(\xi',\tau')=\overset{(m+1)}{\xi'}(\mathbf{X},\tau')=\varepsilon^{m+1}\overset{(m+1)}{\xi}(\mathbf{X},\tau)=\varepsilon^{m+1}\overset{(m)}{\mathbf{u}}(\xi,\tau)$$

is ε^{m+1} times a bounded field. The higher accelerations in a slow flow must be tiny. We may regard $\tau'=\tau/\varepsilon$ as a time scale in which the motion is not slow. For example, fixing $\varepsilon=1/3600$, we have τ' in seconds and τ in hours.

The position of the particle presently at $\mathbf{x}$ in the slow flow is given by

$$\begin{aligned}\chi'_{t'}(\mathbf{x},t'-s)-\mathbf{x}&=\sum_{m=1}\frac{(-s)^m}{m!}\overset{(m-1)}{\mathbf{U}}(\mathbf{x},t')=\sum_{m=1}\frac{(-\varepsilon s)^m}{m!}\overset{(m-1)}{\mathbf{u}}(\mathbf{x},t)\\&=\chi_t(\mathbf{x},t-\varepsilon s)\end{aligned} \tag{94.7}$$

and its history is given by

$$\mathbf{G}'(s)=\sum_{m=1}\frac{(-s)^m}{m!}\mathbf{A}_m[\mathbf{U}(\mathbf{x},t')]=\sum_{m=1}\frac{(-\varepsilon s)^m}{m!}A_m[\mathbf{u}(x,t)=\mathbf{G}_\varepsilon(s)\,. \tag{94.8}$$

It follows from (94.7) and (94.8) that the motion which Coleman and Noll retard is the bounded field associated with $\mathbf{u}(\mathbf{x},t)$ which arises from the slow flow when the time scale is changed.

The most important applications of slow motions are to slow steady motions[5]. In steady motions it is not necessary to rescale the time; (94.7) and (94.8) follow directly from rescaling the velocity $\mathbf{U}(\mathbf{x})=\varepsilon\mathbf{u}(\mathbf{x})$. The concept of a slow motion is much better suited for the discussion of slow *steady* motions. An important simplifying feature for steady flow is that for these flows, $\partial\mathbf{A}_n/\partial t=0$ and

$\mathbf{A}_n, \mathbf{S}_n$ *are homogeneous tensor polynomials of degree n in* $\mathbf{U}(x)$.

For slow steady motions, $\mathbf{A}_n[\mathbf{U}]=\mathbf{A}_n[\varepsilon\mathbf{u}]=\varepsilon^n\mathbf{A}_n[\mathbf{u}]\equiv\varepsilon^n\mathbf{a}_n$ and

$$\mathbf{S}=\sum_{n=1}^{\infty}\varepsilon^n\bar{\mathbf{S}}_n[\mathbf{u},\ldots,\mathbf{u}]=\lim_{N\to\infty}\bar{\mathbf{S}}_{(N)},\qquad \bar{\mathbf{S}}_{(N)}=\sum_{n=1}^{N}\varepsilon^n\bar{\mathbf{S}}_n \tag{94.9a}$$

where

$$\bar{\mathbf{S}}_n[\mathbf{u},\ldots,\mathbf{u}]=\mathbf{S}_n[\mathbf{A}_n[\mathbf{u}],\ldots,\mathbf{A}_1[\mathbf{u}]]=\mathbf{S}_n[\mathbf{a}_n,\ldots,\mathbf{a}_1]\,.$$

Suppose now that the slow steady motion is analytic in ε and that

$$\mathbf{u}(\mathbf{x};\varepsilon)=\sum_{l=1}\varepsilon^{l-1}\mathbf{u}^{\langle l\rangle}\,.$$

Then,

$$\bar{\mathbf{S}}_n[\mathbf{u},\ldots,\mathbf{u}]=\mathbf{S}_n\Big[\sum_{r_1=1}^{\infty}\varepsilon^{r_1-1}\mathbf{u}^{\langle r_1\rangle},\ldots,\sum_{r_n=1}^{\infty}\varepsilon^{r_n-1}\mathbf{u}^{\langle r_n\rangle}\Big]\equiv\sum_{l=1}^{\infty}\varepsilon^{l-1}\mathbf{S}_n^{\langle l\rangle} \tag{94.9b}$$

where

$$\mathbf{S}_n^{\langle l\rangle}=\sum^{(l)}\bar{\mathbf{S}}_n[\mathbf{u}^{\langle r_1\rangle},\mathbf{u}^{\langle r_2\rangle},\ldots,\mathbf{u}^{\langle r_n\rangle}]$$

where $\sum^{(l)}$ is a summation for a fixed integer l over all sets of integers $r_i\geqslant 1$ such that

$$l=1-n+\sum_{i=1}^{n}r_i\,.$$

For example,

$$\mathbf{S}_2^{\langle l\rangle}=\sum^{(l)}\bar{\mathbf{S}}_2[\mathbf{u}^{\langle r_1\rangle},\mathbf{u}^{\langle r_2\rangle}]$$

where $\sum^{(l)}$ is a summation over all sets of integers $r_1\geqslant 1, r_2\geqslant 1$ such that $r_1+r_2=l+1$. It follows from (94.9) that

$$\mathbf{S}=\sum_{n=1}^{\infty}\varepsilon^n\mathbf{S}^{\langle n\rangle}$$

[5] For slow steady motions the stress tensor $\bar{\mathbf{S}}_{(N)}$ may also be obtained trivially by direct reduction of the Rivlin-Ericksen fluids of complexity N (see Notes for 94(b)). We say that hypothetical fluids whose stress is given by the tensors $\bar{\mathbf{S}}_{(N)}$ are CENR fluids of grade N.

where the $\mathbf{S}^{\langle n\rangle}$ are partial derivatives

$$\mathbf{S}^{\langle n\rangle}=\sum_{q+l=1+n}\mathbf{S}_q^{\langle l\rangle}=\frac{1}{n!}\left.\frac{\partial^n\mathbf{S}}{\partial\varepsilon^n}\right|_{\varepsilon=0} \tag{94.9c}$$

evaluated on the motion $\mathbf{U}=\varepsilon\mathbf{u}(\mathbf{x};\varepsilon)$.

In unsteady perturbations of the rest state we may imagine that the velocities are small,

$$\mathbf{U}(\mathbf{x},t)=\varepsilon\mathbf{u}(\mathbf{x},t)\,,$$

and the accelerations are small, $\overset{(m)}{\mathbf{U}}=\varepsilon\overset{(m)}{\mathbf{u}}$, but not nearly so small, $\overset{(m)}{\mathbf{U}}=\varepsilon^{m+1}\overset{(m)}{\mathbf{u}}$, as in retarded motions. Moreover, a glance at (94.3b) shows that the Rivlin-Ericksen tensors are not homogeneous polynomials in $\mathbf{u}$ when $\mathbf{u}$ is unsteady. In fact, using (94.3b), we find that

$$\mathbf{A}_n[\varepsilon\mathbf{u}(\mathbf{x},t)]=\varepsilon\frac{\partial^{n-1}}{\partial t^{n-1}}\mathbf{a}_1+\ldots+\varepsilon^n\mathbf{a}_n \tag{94.10}$$

where $\mathbf{a}_1=\mathbf{A}_1[\mathbf{u}]$.

To linearize the response, we let $\varepsilon\to 0$; then, using (94.10), we find the linearization

$$\mathbf{S}\to\mathscr{F}\left[\varepsilon\sum_{n=1}^{\infty}\frac{(-s)^n}{n!}\frac{\partial^{n-1}}{\partial t^{n-1}}\mathbf{a}_1\right] \tag{94.11}$$

of $\mathscr{F}[\mathbf{G}(s)]$ where $\mathbf{G}(s)$ is the history associated with $\mathbf{U}(\mathbf{x},t)$.

Again using (94.10) and (94.5, 6) we find the linearization

$$\mathbf{S}_{(N+1)}\to\mu\varepsilon\mathbf{a}_1+\varepsilon\sum_{n=1}^{N}\hat{\alpha}_n(\partial^n\mathbf{a}_1/\partial t^n) \tag{94.12}$$

$(N>0)$ of the stress tensor

$$\mathbf{S}_{(N+1)}=\sum_{n=1}^{N+1}\mathbf{S}_n[\mathbf{A}_n,\mathbf{A}_{n-1},\ldots,\mathbf{A}_1] \tag{94.13}$$

of the fluids of grade $N+1$.

The fluids of grade N are sometimes viewed as models of certain physical fluids. Defenders of this view will argue that the Newtonian fluid (Coleman-Noll fluid of grade 1) is an example of a case where the model is exact. Another view is that Newtonian fluids are an exception, that most non-Newtonian fluids can be approximated by a fluid of grade N only in the sense of retardation. I argue for the superiority of the second view in matters concerning stability. The point is that the retardation approximation (94.12) may be obtained from (94.11) only under the assumption that accelerations are as small as $\varepsilon^{m+1}\overset{(m)}{\mathbf{u}}(\mathbf{x},t)$. This assumption does not stem from the requirements of the stability problems and the time should not be retarded. In fact, in stability studies we must allow at least for the possibility of small amplitude oscillations of arbitrary frequency; these certainly do not imply retarded times.

(b) Time-Dependent Perturbations of the Rest State

It is possible to form a general and tractable theory of perturbation of the rest state from constitutive equations for fluids of integral type. These constitutive equations do not artificially restrict the allowed motions to too small a class, as do slow motions. The amplitude of the motions on which $\mathscr{F}$ is presumed to reduce to integral form are small in a sense to be specified later. But the accelerations are not restricted beyond consistency with small amplitudes; arbitrary frequencies are allowed if the amplitude is small. The common asymptotic approximations leading to fluids of complexity N (see Notes for § 94 (b)) and of grade N can be regarded as special cases of fluids of integral type. In particular, the fluids of grade N emerge automatically out of the integrals when the data is such as to generate the history of a slow motion.

Green and Rivlin (1957) have noted that the response functional for simple materials may be uniformly approximated by integral polynomials when the domain of the response functional is a suitable function space, and when $\mathscr{F}$ is continuous with respect to a measure of continuity appropriate to that space. Mild conditions of continuity have been stated by Chacon and Rivlin (1964). The mathematical basis for the representation of $\mathscr{F}$ in terms of multiple integrals over polynomials in the histories is the Stone-Weierstrass theorem. This theorem is generalization for functionals of the well-known Weierstrass approximation theorem for continuous functions. The kernel functions $K_{ij\ldots mn}$ of the approximating polynomials

$$\begin{aligned}\mathscr{F}_{ij}^{(N)} = {} & \textstyle\int_0^\infty K_{ijkl}(s)\,G_{kl}(s)\,ds + \int_0^\infty \int_0^\infty K_{ijklmn}(s_1,s_2)\,G_{kl}(s_1)\,G_{mn}(s_2)\,ds_1\,ds_2 \\ & + \textstyle\int_0^\infty \ldots \int_0^\infty K_{ijkl\ldots mn}(s_1,s_2,\ldots,s_N)\,G_{kl}(s_1)\ldots G_{mn}(s_n)\,ds_1\,ds_2\ldots ds_N\end{aligned} \tag{94.14a}$$

are restricted by the symmetries of $\mathscr{F}$ (see Spencer and Rivlin, 1960 and Truesdell and Noll, 1965, p. 98–100). The coefficients (here, kernels) of the approximating Weierstrass polynomials are not generally Taylor coefficients, so that the kernels K depend on N and are therefore not material parameters.

We shall now assume, following Coleman and Noll (1961) and Pipkin (1964), that

$$\mathscr{F}^{(N)} = \textstyle\sum_{n=1}^{N} \dfrac{1}{n!}\,\mathscr{F}_{(n)}[0\colon \underbrace{\mathbf{G},\ldots,\mathbf{G}}_{n \text{ times}}] \tag{94.14b}$$

is a functional derivative of $\mathscr{F}$ at the rest state and that

$$\mathscr{F} = \lim_{N\to\infty} \mathscr{F}^{(N)}\,. \tag{94.14c}$$

We follow the terminology of Truesdell and Noll (1965) and call $\mathscr{F}^{(N)}$ the Nth order fluid of integral type. The main consequence of (94.14b) is that the kernels in the Green-Rivlin tensors (94.14a) are to be regarded as material functions, independent of N. A good discussion of multiple integral representations for incompressible materials through order three has been given by Pipkin (1964). Eq. (94.14c) may be regarded as a basic constitutive hypotheses for the solution of problems which can be constructed as a perturbation of the rest state.

The isotropic forms of the tensor $\mathscr{F}^{(N)}$ follow easily from the following requirement: Each integrand of $\mathscr{F}^{(N)}$ is an isotropic polynomial of degree $m \leqslant N$ in the strain tensors $\mathbf{G}(s_l)$; we require these polynomials to be isotropic tensor-valued *m-linear* forms in the G_{kl} when the m "times" $s_1, s_2, \ldots, s_m$ are all different. When the "times" are all equal we may use invariance theory for polynomials for a single tensor to deduce that the kernels K are isotropic tensors of even order which are therefore expressible in products of Kronecker's delta (see Exercise 94.7). One finds that

$$\mathscr{F}^{(1)} = \int_0^\infty \zeta(s)\,\mathbf{G}(s)\,ds\,, \tag{94.15}$$

$$\begin{aligned}\mathscr{F}^{(2)} = \mathscr{F}^{(1)} + \int_0^\infty \int_0^\infty \{&\beta(s_1,s_2)\,\mathbf{G}(s_1)\cdot\mathbf{G}(s_2)\\ &+ \alpha(s_1,s_2)\,[\operatorname{tr}\mathbf{G}(s_1)]\,\mathbf{G}(s_2)\}\,ds_1\,ds_2\,,\end{aligned} \tag{94.16}$$

$$\mathscr{F}^{(N)} = \mathscr{F}^{(N-1)} + N\text{-fold integral over all } N\text{-linear products of } \mathbf{G}(s). \tag{94.17}$$

We may, without loss of generality, take the coefficients of symmetric products $\mathbf{G}(s_1)\cdot\mathbf{G}(s_2)\ldots\mathbf{G}(s_l)$ as symmetric. For example we may take $\beta(s_1,s_2)=\beta(s_2,s_1)$ because if $\beta(s_1,s_2)$ had an antisymmetric part it would integrate to zero in the expressions for $\mathscr{F}^{(m)}$, $m \geqslant 2$.

Various common constitutive approximations which are used in the study of the flow of simple fluids can be regarded as arising as special cases of the constitutive assumption (94.14c). Fluids of complexity N will be discussed in the notes at the end of § 94(b). Fluids of grade N arise from integral fluids of order N when the motion is slow and its history is expressed in powers of ε, the slowness parameter, as in (94.8). For fluids of grade two this was first noted by Coleman and Markovitz (1964); in this case

$$\mathscr{F}^{(2)}[\mathbf{G}'(s)] = \varepsilon\mu\mathbf{A}_1[\mathbf{u}(\mathbf{x},t)] + \varepsilon^2\{\alpha_1\mathbf{A}_2[\mathbf{u}(\mathbf{x},t)] + \alpha_2\mathbf{A}_1^2[\mathbf{u}(\mathbf{x},t)]\} + \mathbf{0}(\varepsilon^3)\,. \tag{94.18}$$

Terms which arise from the coefficient of $\alpha(s_1,s_2)$ in (94.16) are $\mathbf{0}(\varepsilon^3)$ because

$$\operatorname{tr}\mathbf{G}(s,\varepsilon) = \mathbf{0}(\varepsilon^2) \qquad \text{whenever} \quad \operatorname{tr}\mathbf{A}_1 = 0\,. \tag{94.19}$$

The coefficients μ, α_1 and α_2 are related to the material functions by the Coleman-Markovitz (1964) formulas:

$$\mu = -\int_0^\infty s\zeta(s)\,ds \tag{94.20a}$$

$$\alpha_1 = \int_0^\infty \frac{s^2}{2}\zeta(s)\,ds \tag{94.20b}$$

and

$$\alpha_2 = \int_0^\infty \int_0^\infty \beta(s_1,s_2)\,s_1 s_2\,ds_1\,ds_2\,. \tag{94.20c}$$

If $\zeta(s)<0$, then $\mu>0$ and $\alpha_1<0$. Analysis of the third order fluid (Exercise 94.8) proceeds along identical lines.

We call attention to the fact that (94.19) implies that the material function $\alpha(s_1,s_2)$ is redundant in the 2nd order retardation approximation. This is true generally, without retardation, whenever (94.16) is interpreted as arising from $\mathscr{F}[\mathbf{G}(s,\varepsilon)]$ in a perturbation for small strains, in the sense of Coleman and Noll (1961), or as a perturbation of the rest state in almost any sense (Pipkin, 1964) and especially in the sense of the perturbation theory to be developed below.

We consider that the history $\mathbf{G}(s,\varepsilon)$ is given by the series $\mathbf{G}(s,\varepsilon)=\varepsilon\mathbf{G}_1(s)+\varepsilon^2\mathbf{G}_2(s)+O(\varepsilon^3)$ where $\mathbf{G}_1(s)$ and $\mathbf{G}_2(s)$ are to be determined. Since $\beta(s_1,s_2)$ is symmetric in s_1 and s_2 it may be written as

$$\beta(s_1,s_2)=\frac{\partial^2\gamma(s_1,s_2)}{\partial s_1\,\partial s_2} \quad \text{where} \quad \gamma(s_1,s_2)=\gamma(s_2,s_1)\,.$$

We call $\gamma(s_1,s_2)$ the quadratic shear relaxation modulus. The linear shear relaxation modulus G is given by

$$\zeta(s)=\frac{d\mathsf{G}}{ds}\,.$$

Then, after integrating (94.16) by parts, using the second formula of Exercise 94.1, we find that

$$\begin{aligned}\mathscr{F}^{(2)}[\mathbf{G}(s,\varepsilon)] &= \textstyle\int_0^\infty \mathsf{G}(s)\mathbb{J}(t-s,\varepsilon)\,ds\\ &\quad+\textstyle\int_0^\infty\int_0^\infty\gamma(s_1,s_2)\mathbb{J}(t-s_1,\varepsilon)\cdot\mathbb{J}(t-s_2,\varepsilon)\,ds_1\,ds_2+O(\varepsilon^2)\end{aligned} \tag{94.21}$$

where

$$\mathbb{J}(t-s,\varepsilon)=\mathbf{F}_t^T(t-s,\varepsilon)\,\mathbf{A}_1[\mathbf{U}(\xi,t-s,\varepsilon)]\cdot\mathbf{F}_t(t-s,\varepsilon)\,.$$

Our program now is to purge redundant $O(\varepsilon^3)$ terms from (94.17) and (94.21). To do this, we find it convenient to develop a kinematics for perturbations from the rest state. Let $\mathbf{X}$ be the position of a particle when there is no motion, $\varepsilon=0$. The position of a particle which was at $\mathbf{X}$ when $\varepsilon=0$ is given by $\xi=\xi(\mathbf{X},\tau,\varepsilon)$ for $\tau\leqslant t$ where $\mathbf{x}=\xi(\mathbf{X},t,\varepsilon)$ is the present position of that same particle. Let ξ be analytic in ε. Then

$$\xi[\mathbf{X},\tau,\varepsilon]=\mathbf{X}+\textstyle\sum_{n=1}\varepsilon^n\xi^{[n]}(\mathbf{X},\tau), \qquad \xi^{[n]}(\mathbf{X},t)=\mathbf{x}^{[n]} \tag{94.22}$$

and

$$\mathbf{U}(\xi,\tau,\varepsilon)=\varepsilon\mathbf{u}(\xi,\tau,\varepsilon)=\textstyle\sum_{n=1}\varepsilon^n\dot{\xi}^{[n]}(\mathbf{X},\tau)=\sum_{n=1}\varepsilon^n\mathbf{U}^{[n]}(\mathbf{X},\tau) \tag{94.23}$$

is the velocity.

It is useful to define derivatives in which $\mathbf{X}$ is fixed and derivatives in which $\mathbf{x}$ is fixed. It is essential in the calculations which follow that functions and their arguments be carefully defined. Let $f(\mathbf{x}(\varepsilon),t,\varepsilon)$ and $g(\mathbf{X},t,\varepsilon)$ be arbitrary smooth functions. When $\varepsilon\to 0$, $\mathbf{x}\to\mathbf{X}$ (also $\xi(\tau,\varepsilon)\to\mathbf{X}$). We define:

$$g^{[n]}(\mathbf{X},t) = \frac{1}{n!}\frac{\partial^n g}{\partial \varepsilon^n}\bigg|_{\varepsilon=0}, \quad \mathbf{X} \text{ is held constant}$$

$$f^{\langle n\rangle}(\mathbf{X},t) = \frac{1}{n!}\frac{\partial^n f}{\partial \varepsilon^n}\bigg|_{\varepsilon=0}, \quad \begin{array}{l}\mathbf{x}(\varepsilon) \text{ is held fixed during differentiation.}\\ \text{After differentiation } \mathbf{x}(0)=\mathbf{X}\end{array} \tag{94.24}$$

$$f^{[n]}(\mathbf{X},t) = \frac{1}{n!}\left(\frac{\partial}{\partial\varepsilon}+\frac{d\mathbf{x}}{d\varepsilon}\cdot\nabla_{\mathbf{x}}\right)^n f\bigg|_{\varepsilon=0}.$$

The mappings $\mathbf{X}\leftrightarrow\mathbf{x}$, $\mathbf{X}\leftrightarrow\xi$ are one to one.

The deformation gradient

$$[\mathbf{F}(\tau,\varepsilon)]_{ij} = \frac{\partial\xi_i(\mathbf{X},\tau,\varepsilon)}{\partial X_j}$$

may also be developed in series

$$[\mathbf{F}(\tau,\varepsilon)]_{ij} = \delta_{ij} + \textstyle\sum_{n=1}\varepsilon^n \dfrac{\partial\xi_i^{[n]}}{\partial X_j}(\mathbf{X},\tau).$$

Using the fact that $\mathbf{F}(\tau,\varepsilon)\cdot\mathbf{F}^{-1}(\tau,\varepsilon)=1$, we find that

$$[\mathbf{F}^{-1}(\tau,\varepsilon)]_{ij} = \delta_{ij} - \varepsilon\frac{\partial\xi_i^{[1]}}{\partial X_j} + \varepsilon^2\left[\frac{\partial\xi_i^{[1]}}{\partial X_l}\frac{\partial\xi_i^{[1]}}{\partial X_i} - \frac{\partial\xi_i^{[2]}}{\partial X_j}\right] + O(\varepsilon^3). \tag{94.25}$$

We now compute

$$\begin{aligned}\mathbf{F}_t(\tau,\varepsilon) = \mathbf{F}(\tau,\varepsilon)\cdot\mathbf{F}^{-1}(t,\varepsilon) = \mathbf{1} + \varepsilon[\mathbf{F}^{[1]}(\tau) - \mathbf{F}^{[1]}(t)]\\ + \varepsilon^2[\mathbf{F}^{[2]}(\tau) - \mathbf{F}^{[2]}(t) - \mathbf{F}^{[1]}(t)(\mathbf{F}^{[1]}(\tau) - \mathbf{F}^{[1]}(t)] + O(\varepsilon^3)\end{aligned} \tag{94.26}$$

where, for example,

$$\mathbf{F}^{[2]}(\tau) = \nabla_{\mathbf{X}}\xi^{[2]}(\mathbf{X},\tau).$$

Finally, we form

$$\mathbf{G}(s,\varepsilon) = \mathbf{F}_t^T(\tau,\varepsilon)\cdot\mathbf{F}_t(\tau,\varepsilon) - \mathbf{1} = \varepsilon\mathbf{G}_1(s) + \varepsilon^2\mathbf{G}_2(s) + O(\varepsilon^3) \tag{94.27}$$

where

$$\mathbf{G}_1(s) = 2[\mathbf{E}^{[1]}(t-s) - \mathbf{E}^{[1]}(t)] \tag{94.28}$$

is the infinitesimal strain and

$$\begin{aligned}\mathbf{G}_2(s) = 2[\mathbf{E}^{[2]}(t-s) - \mathbf{E}^{[2]}(t)] + \mathbf{F}^{T[1]}(t-s)\cdot\mathbf{F}^{[1]}(t-s)\\ + \mathbf{F}^{T[1]}(t)\cdot\mathbf{F}^{[1]}(t) + \mathbf{F}^{[1]}(t)\cdot\mathbf{F}^{[1]}(t) + \mathbf{F}^{T[1]}(t)\cdot\mathbf{F}^{T[1]}(t)\\ - 2[\mathbf{F}^{T[1]}(t)\cdot\mathbf{E}^{[1]}(t-s) + \mathbf{E}^{[1]}(t-s)\cdot\mathbf{F}^{[1]}(t)]\end{aligned} \tag{94.29}$$

is the second order infinitesimal strain where

$$\mathbf{E}^{[n]}=\tfrac{1}{2}[\mathbf{F}^{T[n]}+\mathbf{F}^{[n]}].$$

Using (94.27), (94.28) and (94.29) we find that (94.16) reduces to

$$\begin{aligned}\mathscr{F}^{(2)}[\mathbf{G}(s,\varepsilon)]&=\varepsilon\int_0^\infty\zeta(s)\,\mathbf{G}_1(s)\,ds+\varepsilon^2\int_0^\infty\zeta(s)\,\mathbf{G}_2(s)\,ds\\&\quad+\varepsilon^2\int_0^\infty\int_0^\infty\beta(s_1,s_2)\,\mathbf{G}_1(s_1)\cdot\mathbf{G}_1(s_2)\,ds_1\,ds_2+O(\varepsilon^3).\end{aligned}\tag{94.30}$$

In (94.30) we have obtained an irreducible representation of the 2nd order approximation of the constitutive equation of an incompressible simple fluid of the integral type. But (94.30) is not yet in the form which is most convenient for computations. The canonical forms (94.41, 42, 50, 51) arise from the search for the most convenient computational algorithm for generating histories as solutions of the equations of motion. To motivate the analysis, let us construct a solution to the following forced problem as a perturbation of the state of rest

$$\rho\left(\frac{\partial\mathbf{U}}{\partial t}+\mathbf{U}\cdot\nabla\mathbf{U}\right)=-\nabla p+\nabla\cdot\mathscr{F}[\mathbf{G}(s)]+\varepsilon\mathbf{f}(\mathbf{x},t)\tag{94.31 a}$$

where

$$\mathbf{U}(\mathbf{x},t)\in\mathbf{H}=\{\mathbf{U}:\operatorname{div}\mathbf{U}=0,\ \mathbf{U}(\mathbf{x},t)|_{\partial\mathscr{V}}=0\}.\tag{94.31 b}$$

We suppose that $\mathbf{f}(\mathbf{x},t)=\mathbf{f}(\mathbf{x},t+T)$ so that the motion is subjected to periodic forcing and we look for periodic solutions. Eqs. (94.31a) and $\operatorname{div}\mathbf{U}=0$ in $\mathscr{V}$ are in the form

$$\mathscr{I}(\mathbf{x},t,\varepsilon)=0,\qquad \mathbf{x}\in\mathscr{V},\qquad \varepsilon\in I\tag{94.32}$$

where I is some open interval of the origin. Since $\mathscr{I}=0$ is an identity in ε,

$$\mathscr{I}^{[n]}(\mathbf{X},t)=0,\qquad \boldsymbol{\xi}(\mathbf{X},\tau,\varepsilon)=\mathbf{X}\quad\text{when}\quad\varepsilon=0,\quad\text{for all}\quad\tau\leqslant t\tag{94.33}$$

and since $\mathscr{I}=0$ is also an identity in $\mathbf{x}$,

$$\begin{aligned}\mathscr{I}^{[1]}(\mathbf{X},t)&=\mathscr{I}^{\langle 1\rangle}(\mathbf{X},t)+\mathbf{x}^{[1]}\cdot\nabla_{\mathbf{X}}\mathscr{I}(\mathbf{X},t,0)\\&=\mathscr{I}^{\langle 1\rangle}(\mathbf{X},t)=0.\end{aligned}$$

We find, by induction, that

$$\mathscr{I}^{\langle n\rangle}(\mathbf{X},t)=0.\tag{94.34}$$

The same simplification does not hold at the boundary $\partial\mathscr{V}$ of $\mathscr{V}$ since equation (94.31b) is not an identity in $\mathbf{x}$ when $\mathbf{x}\in\partial\mathscr{V}$. However, $\mathbf{U}^{[n]}(\mathbf{X},t)=0$ when $\mathbf{X}\in\partial\mathscr{V}$ and, since $\boldsymbol{\xi}^{[n]}(\mathbf{X},\tau)=0$ when $\mathbf{X}\in\partial\mathscr{V}$ we find using Eqs. (94.46) that

$\mathbf{U}^{\langle n\rangle}(\mathbf{X},t)=\mathbf{U}^{[n]}(\mathbf{X},t)=0$ when $\mathbf{X}\in\partial\mathscr{V}$. It follows that the trivial solution characterizes the rest state.

$$\mathbf{U}^{\langle 0\rangle}=\mathbf{G}(s,0)=\nabla p^{\langle 0\rangle}=0\,.$$

At first order

$$\rho\frac{\partial\mathbf{U}^{\langle 1\rangle}}{\partial t}=-\nabla_{\mathbf{X}}p^{\langle 1\rangle}+\nabla_{\mathbf{X}}\cdot[\mathscr{F}[\mathbf{G}(s,\varepsilon)]]^{\langle 1\rangle}+\mathbf{f}(\mathbf{X},t)\,. \tag{94.35}$$

At second order

$$\rho\left[\frac{\partial\mathbf{U}^{\langle 2\rangle}}{\partial t}+\mathbf{U}^{\langle 1\rangle}\cdot\nabla_{\mathbf{X}}\mathbf{U}^{\langle 1\rangle}\right]=-\nabla_{\mathbf{X}}p^{\langle 2\rangle}+\nabla_{\mathbf{X}}\cdot[\mathscr{F}[\mathbf{G}(s,\varepsilon)]]^{\langle 2\rangle} \tag{94.36}$$

where

$$\mathbf{U}^{\langle 1\rangle}\in\mathbf{H}\quad\text{and}\quad\mathbf{U}^{\langle 2\rangle}\in\mathbf{H}\,.$$

Problems (94.35) and (94.36) are framed in terms of partial derivatives with respect to ε evaluated at $\varepsilon=0$. The same property holds at higher orders.

We turn next to the computation of $\mathscr{F}^{\langle 1\rangle}$ and $\mathscr{F}^{\langle 2\rangle}$. It is easily established that if $\mathbf{G}(s,0)=0$ and $N\geqslant l$ then $(\mathscr{F}^{(N)})^{\langle l\rangle}=(\mathscr{F}^{(l)})^{\langle l\rangle}$. The constitutive assumption (94.14c) then implies that $\mathscr{F}^{\langle l\rangle}=(\mathscr{F}^{(l)})^{\langle l\rangle}$. Then, using (94.21), we may write

$$\mathscr{F}_{ij}[\mathbf{G}(s,\varepsilon)]=\int_0^\infty \mathsf{G}(s)Q_{ij}(\tau,\varepsilon)ds+\int_0^\infty\int_0^\infty\gamma(s_1,s_2)Q_{il}(\tau_1,\varepsilon)\,Q_{lj}(\tau_2,\varepsilon)\,ds_1\,ds_2 \tag{94.37}$$

where

$$Q_{ij}(\tau,\varepsilon)=\frac{\partial\boldsymbol{\chi}_t(\tau,\varepsilon)}{\partial x_i}\cdot\frac{\partial\mathbf{U}[\boldsymbol{\xi}(\mathbf{X},\tau,\varepsilon),\tau,\varepsilon]}{\partial x_j}+\text{transpose}.$$

We must compute derivatives with respect to ε, holding $\mathbf{x}=\boldsymbol{\xi}(\mathbf{X},t,\varepsilon)$ fixed. With $\mathbf{x}$ fixed we have $d\boldsymbol{\xi}(\mathbf{X},t,\varepsilon)=0$ and

$$d\mathbf{X}=-\frac{\partial\mathbf{X}}{\partial x_l}\left(\frac{\partial x_l}{\partial\varepsilon}\right)_{\mathbf{X}}d\varepsilon\,.$$

Then

$$\left(\frac{\partial\boldsymbol{\xi}}{\partial\varepsilon}\right)_{\mathbf{x}}=-\frac{\partial\boldsymbol{\chi}_t}{\partial x_l}\left(\frac{\partial x_l}{\partial\varepsilon}\right)_{\mathbf{X}}+\left(\frac{\partial\boldsymbol{\xi}}{\partial\varepsilon}\right)_{\mathbf{X}}.$$

For any function $g(\boldsymbol{\xi}(\mathbf{X},\tau,\varepsilon),\tau,\varepsilon)=g(\boldsymbol{\chi}_t(\mathbf{x},\tau,\varepsilon),\tau,\varepsilon)=\tilde{g}(\mathbf{x},\tau,\varepsilon)$ we have

$$\left(\frac{\partial\tilde{g}}{\partial\varepsilon}\right)_{\mathbf{x}}=\frac{\partial g}{\partial\xi_i}\left(\frac{\partial\xi_i}{\partial\varepsilon}\right)_{\mathbf{x}}+\left(\frac{\partial g}{\partial\varepsilon}\right)_{\boldsymbol{\xi}}.$$

Higher derivatives at constant $\mathbf{x}$ are computed by repeated application of this formula. In the limit $\varepsilon \to 0$ we find that

$$\left(\frac{\partial \xi}{\partial \varepsilon}\right)_{\mathbf{x}} = \chi_t^{\langle 1 \rangle}(\mathbf{X}, \tau) = ((\xi^{[1]})) \equiv \xi^{[1]}(\mathbf{X}, \tau) - \xi^{[1]}(\mathbf{X}, t), \tag{94.38}$$

$$\left(\frac{\partial \tilde{g}}{\partial \varepsilon}\right)_{\mathbf{x}} = \tilde{g}^{\langle 1 \rangle}(\mathbf{X}, \tau) = g^{\langle 1 \rangle}(\mathbf{X}, \tau) + \xi^{\langle 1 \rangle} \cdot \nabla_{\mathbf{X}} g^{\langle 0 \rangle}(\mathbf{X}, \tau). \tag{94.39}$$

We are assuming that $g^{\langle 0 \rangle} = 0$. Then, with $\varepsilon \to 0$ and $g^{\langle 0 \rangle} = 0$ we find that

$$\tfrac{1}{2}\left(\frac{\partial^2 \tilde{g}}{\partial \varepsilon^2}\right)_{\mathbf{x}} = \tilde{g}^{\langle 2 \rangle}(\mathbf{X}, \tau) = g^{\langle 2 \rangle}(\mathbf{X}, \tau) + ((\xi^{[1]})) \cdot \nabla_{\mathbf{X}} g^{\langle 1 \rangle}(\mathbf{X}, \tau). \tag{94.40}$$

Application of (94.38, 39, 40) to (94.37) yields

$$\mathscr{F}[\mathbf{G}(s, \varepsilon)]^{\langle 1 \rangle} = \int_0^\infty \mathsf{G}(s)\, \mathbf{A}(s)\, ds \tag{94.41}$$

where

$$\mathbf{A}(s) = \mathbf{A}_1[\mathbf{U}^{\langle 1 \rangle}(\mathbf{X}, t-s)]$$

and

$$\begin{aligned} \mathscr{F}[\mathbf{G}(s, \varepsilon)]^{\langle 2 \rangle} = \int_0^\infty \mathsf{G}(s) \{ &\mathbf{A}_1[\mathbf{U}^{\langle 2 \rangle}(\mathbf{X}, t-s)] + ((\xi^{[1]})) \cdot \nabla_{\mathbf{X}} \mathbf{A}(s) \\ &+ \mathbf{A}(s) \cdot \nabla_{\mathbf{X}}((\xi^{[1]})) + [\mathbf{A}(s) \cdot \nabla_{\mathbf{X}}((\xi^{[1]}))]^T \}\, ds \\ &+ \int_0^\infty \int_0^\infty \gamma(s_1, s_2)\, \mathbf{A}(s_1) \cdot \mathbf{A}(s_2)\, ds_1\, ds_2 \end{aligned} \tag{94.42}$$

where $((\xi^{[1]}))$ is given by (94.38). The expression (94.42) for the second partial functional derivatives of the stress has striking points of similarity with the of 2nd CENR fluid grade to which it reduces when the motion is steady (see 94.52).

Returning now to the perturbation problems (94.35) and (94.36) we find that

$$\nabla_{\mathbf{X}} \cdot [\mathscr{F}[\mathbf{G}(s, \varepsilon)]]^{\langle 1 \rangle} = \int_0^\infty \mathsf{G}(s)\, \nabla_{\mathbf{X}}^2 \mathbf{U}^{\langle 1 \rangle}(\mathbf{X}, t-s)\, ds \tag{94.43}$$

and

$$\begin{aligned} \nabla_{\mathbf{X}} \cdot [\mathscr{F}[\mathbf{G}(s, \varepsilon)]]^{\langle 2 \rangle} &= \int_0^\infty \mathsf{G}(s)\, \nabla_{\mathbf{X}}^2 \mathbf{U}^{\langle 2 \rangle}(\mathbf{X}, t-s)\, ds \\ &+ \nabla_{\mathbf{X}} \cdot \int_0^\infty \mathsf{G}(s) \{ ((\xi^{[1]})) \cdot \nabla_{\mathbf{X}} \mathbf{A}(s) + \mathbf{A}(s) \cdot \nabla_{\mathbf{X}}((\xi^{[1]})) + [\mathbf{A}(s) \cdot \nabla_{\mathbf{X}}((\xi^{[1]}))]^T \}\, ds \\ &+ \nabla_{\mathbf{X}} \cdot \int_0^\infty \int_0^\infty \gamma(s_1, s_2)\, \mathbf{A}(s_1) \cdot \mathbf{A}(s_2)\, ds_1\, ds_2. \end{aligned} \tag{94.44}$$

We next consider the computation of the quantities $((\xi^{[n]}))$. Which arise from the expansion of the history. It follows from (94.23) that

$$\frac{d\xi^{[n]}}{d\tau} = \mathbf{U}^{[n]}(\mathbf{X}, \tau).$$

Hence

$$((\xi^{[n]})) = \int_t^\tau \mathbf{U}^{[n]}(\mathbf{X}, \tau')\, d\tau' . \tag{94.45}$$

Moreover,

$$\mathbf{U}^{[1]}(\mathbf{X}, \tau) = \mathbf{U}^{\langle 1 \rangle}(\mathbf{X}, \tau) , \tag{94.46a}$$

$$\mathbf{U}^{[2]}(\mathbf{X}, \tau) = \mathbf{U}^{\langle 2 \rangle}(\mathbf{X}, \tau) + \xi^{[1]}(X, \tau) \cdot \nabla_{\mathbf{X}} \mathbf{U}^{\langle 1 \rangle} , \tag{94.46b}$$

$$\mathbf{U}^{[3]}(\mathbf{X}, \tau) = \mathbf{U}^{\langle 3 \rangle}(\mathbf{X}, \tau) + \xi^{[2]}(\mathbf{X}, \tau) \cdot \nabla_{\mathbf{X}} \mathbf{U}^{\langle 1 \rangle} + \xi^{[1]} \cdot \nabla_{\mathbf{X}} \mathbf{U}^{\langle 2 \rangle} + \tfrac{1}{2} \xi_i^{[1]} \xi_j^{[1]} \frac{\partial^2 \mathbf{U}^{\langle 1 \rangle}}{\partial X_i \partial X_j} . \tag{94.46c}$$

With these computational preliminaries aside, we turn next to the computation of the solutions of problem (94.35, 43) and (94.36, 44). We first solve (94.35, 43) for $\mathbf{U}^{\langle 1 \rangle}(\mathbf{X}, t)$. Then, using (94.45) and (94.46), we compute $((\xi^{[1]}))$. Problem (94.36, 44) may then be stated explicitly may be solved for $\mathbf{U}^{\langle 2 \rangle}(\mathbf{X}, t)$. We could compute, sequentially, the next approximations using the procedure just outlined and the constitutive approximation $\mathscr{F} = \mathscr{F}^{(3)}$ (see Exercise 94.8).

We have made heavy use of the rest state as a reference configuration. This procedure serves two objectives: in problems involving the computation of free surfaces it allows one to solve the perturbation problems on a domain of simple configuration. In addition, the material mapping serves the main purpose, the sequential computation of the history which is required to specify the present motion. The introduction of the rest state as a reference configuration is natural for simple solids. We know, however, that the stress in a simple fluid does not depend on the reference configuration. The canonical forms of the stress for perturbations of the rest state in the configuration of the fluid at time t may be developed from the equations given below. In these equations $\mathbf{x}$ is a fixed parameter independent of ε and the relative position vector

$$\boldsymbol{\chi}_t(\mathbf{x}, \tau, \varepsilon) = \mathbf{x} + \textstyle\sum_{n=1} \varepsilon^n \boldsymbol{\chi}^{\langle n \rangle}(\mathbf{x}, \tau) \tag{94.47}$$

where $\boldsymbol{\chi}^{\langle n \rangle}(\mathbf{x}, \tau) \equiv \chi_t^{\langle n \rangle} \equiv \boldsymbol{\chi}^{\langle n \rangle}$, is used to monitor the history of the motion. Moreover,

$$\begin{aligned}
&\mathbf{U}(\boldsymbol{\chi}_t(\mathbf{x}, \tau, \varepsilon), \tau, \varepsilon) \equiv \tilde{\mathbf{U}}(\mathbf{x}, \tau, \varepsilon) = \textstyle\sum_{n=1}^{\infty} \varepsilon^n \tilde{\mathbf{U}}^{\langle n \rangle}(\mathbf{x}, \tau) = \sum_{n=1} \varepsilon^n \dfrac{\partial \boldsymbol{\chi}^{\langle n \rangle}(\mathbf{x}, \tau)}{\partial \tau} , \\
&\boldsymbol{\chi}^{\langle n \rangle}(\mathbf{x}, \tau) = \textstyle\int_t^\tau \tilde{\mathbf{U}}^{\langle n \rangle}(\mathbf{x}, \tau)\, d\tau , \qquad (94.48) \\
&\tilde{\mathbf{U}}^{\langle 1 \rangle}(\mathbf{x}, \tau) = \mathbf{U}^{\langle 1 \rangle}(\mathbf{x}, \tau) , \\
&\tilde{\mathbf{U}}^{\langle 2 \rangle}(\mathbf{x}, \tau) = \mathbf{U}^{\langle 2 \rangle}(\mathbf{x}, \tau) + \boldsymbol{\chi}^{\langle 1 \rangle} \cdot \nabla \mathbf{U}^{\langle 1 \rangle} , \\
&\tilde{\mathbf{U}}^{\langle 3 \rangle}(\mathbf{x}, \tau) = \mathbf{U}^{\langle 3 \rangle}(\mathbf{x}, \tau) + \boldsymbol{\chi}^{\langle 2 \rangle} \cdot \nabla \mathbf{U}^{\langle 1 \rangle} + \boldsymbol{\chi}^{\langle 1 \rangle} \cdot \nabla \mathbf{U}^{\langle 2 \rangle} + \tfrac{1}{2} \chi_i^{\langle 1 \rangle} \chi_j^{\langle 1 \rangle} \frac{\partial^2 \mathbf{U}^{\langle 1 \rangle}}{\partial x_i \partial x_j} ,
\end{aligned}$$

$$\frac{\partial \boldsymbol{\chi}_t(\mathbf{x},\tau,\varepsilon)}{\partial x_i} = \mathbf{e}_i + \sum_{m=1}^{\infty} \varepsilon^m \frac{\partial \boldsymbol{\chi}^{\langle m\rangle}}{\partial x_i},$$

$$\frac{\partial \tilde{\mathbf{U}}(\boldsymbol{\chi}_t(\mathbf{x},\tau,\varepsilon),\tau,\varepsilon)}{\partial x_j} = \sum_{m=1}^{\infty} \varepsilon^m \frac{\partial \tilde{\mathbf{U}}^{\langle m\rangle}(\mathbf{x},\tau)}{\partial x_j},$$

and

$$\begin{aligned}
\{\mathbf{F}_t^T(\tau,\varepsilon)\cdot \mathbf{A}_1[\mathbf{U}(\boldsymbol{\chi}_t,\tau,\varepsilon)]\cdot \mathbf{F}_t(\tau,\varepsilon)\}_{ij} &= \frac{\partial \boldsymbol{\chi}_t}{\partial x_i}\cdot\frac{\partial \mathbf{U}}{\partial x_j} + \frac{\partial \boldsymbol{\chi}_t}{\partial x_j}\cdot\frac{\partial \mathbf{U}}{\partial x_i} \\
= \sum_{l=1}^{\infty} \varepsilon^l \left\{\frac{\partial \tilde{U}_i^{\langle l\rangle}}{\partial x_j} + \frac{\partial \tilde{U}_j^{\langle l\rangle}}{\partial x_i}\right\} &+ \sum_{m=1}^{\infty}\sum_{l=1}^{\infty} \varepsilon^{m+l}\left\{\frac{\partial \boldsymbol{\chi}^{\langle m\rangle}}{\partial x_i}\cdot\frac{\partial \tilde{\mathbf{U}}^{\langle l\rangle}}{\partial x_j} + \frac{\partial \boldsymbol{\chi}^{\langle m\rangle}}{\partial x_j}\cdot\frac{\partial \tilde{\mathbf{U}}^{\langle l\rangle}}{\partial x_i}\right\} \\
= \varepsilon\left\{\frac{\partial U_i^{\langle 1\rangle}}{\partial x_j} + \frac{\partial U_j^{\langle 1\rangle}}{\partial x_i}\right\} & \qquad (94.49) \\
+ \varepsilon^2 \left\{\left[\frac{\partial}{\partial x_j}(U_i^{\langle 2\rangle} + \boldsymbol{\chi}^{\langle 1\rangle}\cdot\nabla U_i^{\langle 1\rangle}) + \frac{\partial \boldsymbol{\chi}^{\langle 1\rangle}}{\partial x_i}\cdot\frac{\partial \mathbf{U}^{\langle 1\rangle}}{\partial x_j}\right] + \text{transpose}\right\} &+ 0(\varepsilon^3).
\end{aligned}$$

Eq. (94.49) leads directly to the canonical forms of the stress for perturbations of rest state expressed in the coordinate $(\mathbf{x},t)$ of the present configuration. For example,

$$\mathbf{U}^{\langle 1\rangle}(\mathbf{x},t)\in\mathbf{H}, \qquad \mathbf{U}^{\langle 2\rangle}(\mathbf{x},t)\in\mathbf{H};$$

$$\rho\frac{\partial \mathbf{U}^{\langle 1\rangle}(\mathbf{x},t)}{\partial t} = -\nabla p^{\langle 1\rangle} + \int_0^\infty \mathsf{G}(s)\nabla^2\mathbf{U}^{\langle 1\rangle}(\mathbf{x},t-s)\,ds + \mathbf{f}(\mathbf{x},t); \tag{94.50}$$

and

$$\begin{aligned}
\rho\frac{\partial \mathbf{U}^{\langle 2\rangle}(\mathbf{x},t)}{\partial t} + \rho\mathbf{U}^{\langle 1\rangle}\cdot\nabla\mathbf{U}^{\langle 1\rangle} &= -\nabla p^{\langle 2\rangle} + \int_0^\infty \mathsf{G}(s)\nabla^2\mathbf{U}^{\langle 2\rangle}(\mathbf{x},t-s)\,ds \\
&+ \nabla\cdot\int_0^\infty \mathsf{G}(s)\{\boldsymbol{\chi}_t^{\langle 1\rangle}\cdot\nabla\mathbf{A}(s) + \mathbf{A}(s)\cdot\nabla\boldsymbol{\chi}_t^{\langle 1\rangle} + [\mathbf{A}(s)\cdot\nabla\boldsymbol{\chi}_t^{\langle 1\rangle}]^T\}\,ds \\
&+ \nabla\cdot\int_0^\infty\int_0^\infty \gamma(s_1,s_2)\,\mathbf{A}(s_1)\cdot\mathbf{A}(s_2)\,ds_1\,ds_2
\end{aligned} \tag{94.51}$$

where

$$\boldsymbol{\chi}_t^{\langle 1\rangle} = \boldsymbol{\chi}_t^{\langle 1\rangle}(\mathbf{x},t-s)$$

and

$$\mathbf{A}(s) = \mathbf{A}[\mathbf{U}^{\langle 1\rangle}(\mathbf{x},t-s)].$$

Eq. (94.48) shows how to compute $\boldsymbol{\chi}^{\langle n\rangle}(\mathbf{x},\tau)$ when the $\mathbf{U}^{\langle l\rangle}(\mathbf{x},\tau)$, $l\leq n$ are given. The perturbation fields can all be computed sequentially. When $\mathbf{f}=0$ these equations are the ones that should be used to study stability and bifurcation of the rest state. In the special case in which $\mathbf{U}(\mathbf{x},t)=\mathbf{U}(\mathbf{x})$ is a steady flow we find, using (93.12) that $\boldsymbol{\chi}_t^{\langle 1\rangle}(\mathbf{x},t-s) = -s\mathbf{U}^{\langle 1\rangle}(\mathbf{x})$ and

$$O = -\nabla p^{\langle 1\rangle} + \mu\nabla^2\mathbf{U}^{\langle 1\rangle}$$

and

$$\rho \mathbf{U}^{\langle 1 \rangle} \cdot \nabla \mathbf{U}^{\langle 1 \rangle} = -\nabla p^{\langle 2 \rangle} + \mu \nabla^2 \mathbf{U}^{\langle 2 \rangle} + \nabla \cdot \{\alpha_1 \mathbf{A}_2[\mathbf{U}^{\langle 1 \rangle}] + \alpha_2 \mathbf{A}_1[\mathbf{U}^{\langle 1 \rangle}] \cdot \mathbf{A}_1[\mathbf{U}^{\langle 1 \rangle}]\} \tag{94.52}$$

where

$$\mu = \int_0^\infty \mathsf{G}(s)\,ds\,, \qquad \alpha_1 = -\int_0^\infty s\mathsf{G}(s)\,ds$$

and

$$\alpha_2 = \int_0^\infty \int_0^\infty \gamma(s_1, s_2)\,ds_1\,ds_2\,.$$

These relations are consistent with (94.20).

When $\mathbf{x}$ and ε are independent, points are given by the position vector $\chi_t(\mathbf{x}, \tau, \varepsilon)$ where $\chi_t(\mathbf{x}, \tau, 0) = \chi_t(\mathbf{x}, t, \varepsilon) = \mathbf{x}$. The relation

$$\mathbf{x} = \xi(\mathbf{X}, t, \varepsilon)$$

defines a relation between $\mathbf{X}$ and ε when $\mathbf{x}$ is fixed. When $\mathbf{X}$ is fixed, points are given by the position vector $\xi(\mathbf{X}, \tau, \varepsilon)$ where $\xi(X, \tau, 0) = \mathbf{X}$ and $\mathbf{x} = \xi(\mathbf{X}, \tau, \varepsilon)$ is a function of ε. This function defines a map between $\mathscr{V}_\varepsilon(\mathbf{x})$ and $\mathscr{V}_0(\mathbf{X})$ where $\mathscr{V}_\varepsilon(\mathbf{x})$ designates the region of space occupied by the fluid and $\mathscr{V}_0(\mathbf{X})$ the region of space occupied by the fluid at rest. The difference between $\mathbf{x}$ and $\mathbf{X}$ in a fluid is not important when $\mathscr{V}_\varepsilon$ is independent of ε. In this case the canonical forms for the problems governing the perturbation coefficients $(\cdot)^{\langle n \rangle}$ are unchanged when $\mathbf{x}$ is replaced by $\mathbf{X}$. For example, using the kinematic formulas listed in Exercise (94.9), we can show that (94.50) and (94.51) with $\mathbf{x}$ replaced $\mathbf{X}$ are the same as (94.35) and (94.36) with stresses given by (94.43) and (94.44).

The difference between $\mathbf{x}$ and $\mathbf{X}$ in fluids is important when $\mathscr{V}_\varepsilon(\mathbf{x})$ changes with ε, as in free surface problems. In this case, the boundary conditions are posed on a set of points $\mathbf{x} \in \partial\mathscr{V}_\varepsilon(\mathbf{x})$ which changes with ε and these conditions are not form invariant: if $f(\mathbf{x}, t, \varepsilon) = 0$ when $\mathbf{x} \in \partial\mathscr{V}_\varepsilon(\mathbf{x})$ then $f^{[n]}(\mathbf{X}, t) = 0$ when $\mathbf{X} \in \partial\mathscr{V}_0(\mathbf{X})$, but it does not follow that $f^{\langle n \rangle}(\mathbf{X}, t) = 0$. If $\mathscr{V}_0(\mathbf{X})$ is a domain of convenient configuration it is advantageous to introduce the rest coordinates $\mathbf{X}$ and to solve perturbation problems posed on this convenient domain.

The algorithms which lead to the canonical forms for perturbation problems on $\mathscr{V}_0(\mathbf{X})$ require the introduction of an invertible map $\mathbf{x} \leftrightarrow \mathbf{X}$ parametrized by ε. For unsteady motions, $\mathbf{x} = \xi(\mathbf{X}, t, \varepsilon)$ serves admirably both as a material mapping and a domain mapping. In steady problems, like those discussed under (d) below, domain mappings which are not material mappings may be more convenient.

Notes for § 94(b)

An incompressible Rivlin-Ericksen fluid of complexity N satisfies a constitutive equation of the form

$$\mathbf{T}_N + p\mathbf{1} = \mathbf{f}(\mathbf{A}_N, \mathbf{A}_{N-1}, \ldots, \mathbf{A}_1)\,.$$

The fluid of complexity one is called a Reiner-Rivlin fluid:

$$\mathbf{T}_1 + p\mathbf{1} = \mu\mathbf{A}_1 + \mu_1\mathbf{A}_1^2$$

where μ and μ_1 are polynomials in $\operatorname{tr}\mathbf{A}_1^2$ and $\operatorname{tr}\mathbf{A}_1^3=0$. The fluid of complexity two is given by

$$\mathbf{T}_2+p\mathbf{1}=\mu\mathbf{A}_1+\alpha_1\mathbf{A}_1^2+\alpha_2\mathbf{A}_2+\alpha_3\mathbf{A}_2^2+\alpha_4(\mathbf{A}_1\cdot\mathbf{A}_2+\mathbf{A}_2\cdot\mathbf{A}_1)$$
$$+\alpha_5(\mathbf{A}_1^2\cdot\mathbf{A}_2+\mathbf{A}_2\cdot\mathbf{A}_1^2)+\alpha_6(\mathbf{A}_1\cdot\mathbf{A}_2^2+\mathbf{A}_2^2\cdot\mathbf{A}_1)+\alpha_7(\mathbf{A}_1^2\cdot\mathbf{A}_2^2+\mathbf{A}_2^2\cdot\mathbf{A}_1^2)$$

where μ and α_j are polynomials in the scalar invariants of $\mathbf{A}_1$ and $\mathbf{A}_2$. Rivlin (1956) and Spencer and Rivlin (1959) derived canonical forms for the stress matrix in terms of $\mathbf{A}_n$. Green and Rivlin (1957) showed that theory of Rivlin and Ericksen (1955) could be derived from integral approximations of a general functional for simple materials.

Coleman and Noll (1960) derived retarded time approximations for the simple fluid using Fréchet expansions equivalent to (93.6) (see Exercise 94.4). These approximations are given by the partial sums $\mathbf{S}_{(N)}$ in the equations following (94.6). The fluid of grade 1 is the Newtonian fluid

$$\mathbf{S}_{(1)}=\mu\mathbf{A}_1$$

where μ is a constant. The Coleman-Noll fluid of grade two is given by

$$S_{(2)}=\mu\mathbf{A}_1+\alpha_1\mathbf{A}_2+\alpha_2\mathbf{A}_1^2=\mathbf{S}_1+\mathbf{S}_2$$

where μ, α_1 and α_2 are constants.

When the motion is steady $\mathbf{A}_n=\varepsilon\mathbf{a}_n$ and in the limit as $\varepsilon\to 0$ the Rivlin and Ericksen fluid of complexity N coincides with the Coleman-Noll fluid of grade N. We have called these fluids with stress tensor $\overline{\mathbf{S}}_{(N)}$ which are ordered on the basis of the homogeneity of the $\overline{\mathbf{S}}_n$ which is induced by the relation $\mathbf{A}_n[\varepsilon\mathbf{u}]=\varepsilon^n\mathbf{A}_n[\mathbf{u}]$, CENR fluids of order N. The formulas for the CENR fluids of order 3 and 4 were first given and applied to a nontrivial calculation of flow by Langlois and Rivlin (1963).

We have seen that the fluids of grade N drop out of the integral order fluids as a special case when the motion is slow and that the CENR fluids then follow from the further restriction that the slow motions are steady. Fluids of complexity N are not forms which arise naturally in perturbations of the rest state but they do arise when (94.14b, c) is evaluated for the histories which lead to the fluids of complexity N. Since a fluid of complexity N is independent of the tensors $\mathbf{A}_l$ for $l>N$, it follows from (94.3b) that

$$\mathbf{G}(s)=\sum_{n=1}^{N}\frac{(-s)^n}{n!}\mathbf{A}_n[\mathbf{U}(\mathbf{x},t)]$$

is a polynomial, linear in $\mathbf{A}_1,\mathbf{A}_2,\ldots,\mathbf{A}_N$. The stresses $\mathscr{F}^{(m)}$ are, therefore, isotropic polynomials composed of products of the same N tensors with coefficients which are defined in terms of integrals over the kernels (material functions). Canonical forms for fluids of complexity N arise from the application of isotropic invariance theory (see Spencer and Rivlin, 1960) to these isotropic polynomials.

(c) Stability of the Rest State

We have now completed the preliminaries to the stability analysis of the rest state of a simple fluid. The assumptions which we will make in the analysis are essentially statements about the spectral problem of linearized theory. The spectral problem may be obtained by copying the procedure which is used and is correct for nonlinear problems when there is no memory: the recipe is to linearize, then substitute exponential solutions proportional to $e^{-\sigma t}$.

Linearization of (94.1), using (93.11) with $\mathbf{G}_0(s)=0$ and the constitutive hypothesis (94.14c), gives

$$\rho\frac{\partial\mathbf{u}}{\partial t}+\nabla p=\int_0^\infty \mathbf{G}(s)\nabla^2\mathbf{u}(t-s)\,ds,\qquad \nabla\cdot\mathbf{u}=0,\qquad \mathbf{u}|_{\partial\mathscr{V}}=0. \tag{94.53}$$

Substituting

$$\mathbf{u}(\mathbf{x},t)=e^{-\sigma t}\hat{\mathbf{u}}(\mathbf{x}), \qquad p(\mathbf{x},t)=e^{-\sigma t}\hat{p}(\mathbf{x}) \tag{94.54}$$

into (93.15) we have

$$-\rho\sigma\hat{\mathbf{u}}+\nabla\hat{p}=k(\sigma)\nabla^2\hat{\mathbf{u}}, \qquad \nabla\cdot\hat{\mathbf{u}}=0, \qquad \hat{\mathbf{u}}\big|_{\partial\mathscr{V}}=0 \tag{94.55}$$

where

$$k(\sigma)=\int_0^\infty \mathsf{G}(s)\,e^{\sigma s}\,ds\,. \tag{94.56}$$

Eqs. (94.55) and (94.56) define the *spectral problem* of the linearized theory of the rest state of a simple fluid.

We say that the rest state is stable if

$$\operatorname{re}(\sigma)>0 \tag{94.57a}$$

and is unstable if

$$\operatorname{re}(\sigma)<0\,. \tag{94.57b}$$

We are assuming a principle of linearized stability (see § 7). When (94.57a) holds, the rest state is conditionally stable. The principle of linearized stability holds for systems of nonlinear ordinary differential equations as well as for nonlinear partial differential equations of the Navier-Stokes type.

This principle has been partially established for nonlinear ordinary differential equations of the functional-differential equation type analogous to (94.53)[8]. Pending a deeper justification we shall assume this principle.

It is convenient to regard (94.55) as an eigenvalue problem with eigenvalues

$$\hat{\Lambda}=\rho\sigma/k(\sigma)\,. \tag{94.58}$$

The eigenvalues $\hat{\Lambda}$ *of* (94.55) *are real-valued: they form a discrete, denumerable set* $\hat{\Lambda}=\hat{\Lambda}_n$ $(n=1,2,\dots\infty)$ *which may be arranged as an increasing sequence.*

$$\hat{\Lambda}_1\leqslant\hat{\Lambda}_2\leqslant\dots\hat{\Lambda}_n, \qquad \lim_{n\to\infty}\hat{\Lambda}_n\to\infty\,.$$

Proof: The eigenvalues $\hat{\Lambda}$ may be characterized as the critical points of the Rayleigh quotient

$$\hat{\Lambda}_n=\min_{H_n}\langle|\nabla\mathbf{u}|^2\rangle/\langle|\mathbf{u}|^2\rangle$$

where H_n is the complement of the Hilbert space of solenoidal vectors which vanishes on $\partial\mathscr{V}$; the complement is orthogonal to the eigensubspaces of the first

---

[8] Private communication by R. K. Miller extending the results of Miller (1971). A good mathematical theory for the linearized problem (94.53) has recently been given by Slemrod. Slemrod proves that the rest state is asymptotically stable in a certain L_2 norm of gradients when the shear relaxation satisfies certain mild conditions (see Exercise 95.1).

n eigenvectors. The properties asserted by the theorem are guaranted by standard theorems about the variational characterization of eigenvalues of self-adjoint operators in Hilbert spaces (see B.3 of Appendix B).

Assuming the principle of linearized stability we may now assert that the rest state of a simple fluid is stable if we have $\mathrm{re}(\sigma_n)>0$ for all eigenvalues σ_n, where σ_n are the possibly complex-valued roots of the equation

$$k(\sigma_n)\hat{\Lambda}_n=\rho\sigma_n\,. \tag{94.59}$$

The real and imaginary parts of (94.59) may be written as

$$\hat{\Lambda}_n\textstyle\int_0^\infty \mathsf{G}(s)\,e^{\xi_n s}\cos\eta_n s\,ds=\rho\,\xi_n\,, \tag{94.60a}$$

$$\hat{\Lambda}_n\textstyle\int_0^\infty \mathsf{G}(s)\,e^{\xi_n s}\sin\eta_n\,ds=\rho\eta_n \tag{94.60b}$$

where

$$\sigma_n=\xi_n+i\eta_n\,.$$

If we consider solutions of (94.60) over all possible containers $\mathscr{V}$, we must allow $\hat{\Lambda}_n(\mathscr{V})$ to take on all positives values. Simple fluids whose shear relaxation modulus $\mathsf{G}(s)$ is such that for some n and some container $\xi_n<0$ are not physical fluids. Such unphysical simple fluids have an unstable rest state.

It is generally believed that the shear relaxation modulus $\mathsf{G}(s)$ is a positive and monotonically decreasing function. When $\mathsf{G}(s)>0$ there is no non-oscillatory solution ($\eta_n=0$) of (94.53) with $\xi_n<0$. Moreover, there can be no solution of (94.60b) with $\xi_n<0$ and $\eta_n\neq 0$ if $\mathsf{G}(s)$ decays rapidly enough. Assuming a principle of linearized stability, it follows that simple fluids with a positive and rapidly decaying shear modulus are conditionally stable.

We have argued that the histories which are appropriate in the study of the stability of the rest state of a simple fluid do not lead to a retardation approximation and to the fluids of grade N. Of course it is possible to study stability of the rest state of fluids of arbitrary grade by the spectral method; for example, the fluid of grade $N+1$ is also governed by (94.55) with

$$k(\sigma)=\mu+\textstyle\sum_{n=1}^{N+1}(-\sigma)^n\hat{\alpha}_n\,.$$

Hence, the eigenvalues σ_n are related to the eigenvalues $\hat{\Lambda}_n$:

$$\hat{\Lambda}_n=\rho\sigma_n/(\mu+\textstyle\sum_{l=1}^{N+1}(-\sigma_n)^l\hat{\alpha}_l)\,. \tag{94.61}$$

The simplest case is a second-order fluid; for this,

$$\sigma_n=\mu\hat{\Lambda}_n/(\rho+\hat{\alpha}_1\hat{\Lambda}_n)\,. \tag{94.62}$$

Since $\hat{\alpha}_1=\alpha_1$ is negative in polyisobutylene solutions and $\hat{\Lambda}_n\to\infty$, we find that there are eigenvalues $\sigma_n<0$ and it follows that the rest state of the second-order retardation approximation is unstable. Similar instability results will hold for the nth order approximation if the coefficients $\hat{\alpha}_n$ lie in a certain set[9].

If the fluid of grade N is accepted as a real constitutive equation (a "model" equation) for some fluid in all motions, then one is obliged to consider stability analyses of the type just given. Certainly the always unstable rest state of the fluid of grade 2 with $\alpha_1 < 0$ would seem to eliminate these fluids as models for real fluids in all motions. Their use in the sense of retardation is, however, in no way damaged by the instability result. It is always possible that a fluid of grade $N > 1$ closely describes a real fluid, if not in all motions, in "most" motions of interest. Merit in the use of the fluid of grade N as a model is as much a matter for experience as for analysis. Certainly the fluid of grade N whose coefficients μ and $\hat{\alpha}_n$ lie in the stable set is a better candidate for a fluid model than the fluid of grade N whose coefficients are in the unstable set. The stable and unstable sets of coefficients can be given by analysis but the values of the coefficients in real fluids must be determined experimentally.

On the other hand, it should be clear that the concept of a fluid model is not natural to stability studies. It is better to make the theory of the simple fluid practical by restricting considerations to motions which are both appropriate for stability analysis and lead to reductions in the complexity of the response. In the linearized case this leads to the theory of infinitesimal viscoelasticity rather than to fluids of grade N.

The interesting way in which a simple fluid may appear to obey different stress laws when undergoing different motions can be illustrated by considering the bifurcation problem for a simple fluid heated from below. Sokolov and Tanner (1972) studied the stability part of the problem by the advocated method. They find that the spectrum σ_n is real-valued for many models of the simple fluid. Given this "exchange of stability", bifurcation theory shows that steady convection bifurcates from steady conduction. To construct the steady solutions we use the retardation approximations and the CENR fluids. This leads, at 2nd order, to problems in the form of (94.52). On the other hand, if at criticality σ_1 is pure imaginary, and other technical conditions hold, time-periodic bifurcation is expected and the constitutive approximations leading to the CENR fluids are no longer satisfied. Instead, we must turn to a functional expansion for time periodic motions. This leads, at second order, to problems in the form of (94.52).

---

[9] Coleman and Mizel (1966) following earlier work of Coleman, Duffin and Mizel (1965) have considered the stability of shearing flows of a second-order fluid with $\alpha_1 < 0$. The coefficient α_1 is generally believed to be negative in polyisobutylene solutions which are used in experiments. The aforementioned stability problem with $\alpha_1 > 0$ has been studied by Ting (1963). Coleman, Duffin, and Mizel were interested in determining the stability properties of the fluid of second grade (which arises rigorously from the retardation theorem of Coleman and Noll (1960)) when the fluid of grade two is regarded as exact. They are careful to emphasize that the use of a more general constitutive relation might yield different results. They show that for certain critical channel widths, a flow which is initially a laminar shearing flow cannot remain so. Coleman and Mizel show that for these critical values of h, if there is any flow at all then the departure from shearing flow must appear instantly in the first time derivative of the velocity. The rest state is a special case of shearing flow. Craik (1968), using a linearized theory of stability, has shown that the second-order fluid is unstable but the linear viscoelastic fluid is stable to infinitesimal two-dimensional disturbances.

(d) Bifurcation of the Rest State of a Simple Fluid Heated from below

We consider a pool of liquid (a simple fluid) resting on a hot flat plate and confined by vertical insulating side walls. If the bottom plate is not too hot, the fluid will be motionless and the transport of heat from the bottom plate across the fluid will take place by heat conduction. At a critical temperature difference, the conduction solution will lose stability and some form of motion will begin in the fluid. The existence of motion will alter the shape of the free surface at the top of the fluid and will change the amount of heat transported from the amount which would be transported by conduction alone. We shall use bifurcation theory to compute how the shape of the free surface and the heat transport curve depend on the coefficients of the CENR fluids.

The liquid pool is confined to a cylinder whose arbitrary cross section is designated by the set of points $(x,y)\in\mathscr{A}$, independent of the vertical coordinate z. The bottom plate at $z=0$ is horizontal and is held at a fixed temperature $T_0+\Delta T$. The top of the fluid is given by $z=\mathscr{h}(x,y;\varepsilon)$ where $\varepsilon^2=Nu-1$ is the Nusselt number discrepancy. The simple fluid is assumed to be governed by the Oberbeck-Boussinesq equations as in the classical problem of Bénard; however, we are considering simple fluids whose stress response is expandable as in (94.9a). The region occupied by the fluid changes with the Nusselt number and is given by

$$\mathscr{V}_\varepsilon=[(x,y,z):(x,y)\in\mathscr{A},\ 0\leqslant z\leqslant \mathscr{h}(x,y;\varepsilon)]\,.$$

In $\mathscr{V}_\varepsilon$ we require that the OB equations (54.2) hold. Here, however, $\mathbf{S}$ is not Newtonian and the free surface $\mathscr{f}(x,y,z,\varepsilon)=z-\mathscr{h}(x,y;\varepsilon)=0$ is to be determined as a part of the solution. Since the fluid is incompressible, the conservation of volume implies that the mean height

$$\bar{\mathscr{h}}=\iint_{\mathscr{A}}\mathscr{h}(x,y;\varepsilon)\,dxdy\Big/\iint_{\mathscr{A}}dxdy \tag{94.63}$$

is independent of ε.

The construction of a bifurcating motion can be carried out when the temperature which is prescribed on the boundary is compatible with a static solution of the OB equations. A static solution is possible only in the case $T=T(z)$. In this case

$$T-T_0=\Delta T(1-z/\bar{\mathscr{h}}) \tag{94.64}$$

where, for convenience, we have set the reference temperature T_0 of T at a height $\bar{\mathscr{h}}$ above the hot plate. We shall require that (94.64) holds at all points of the free surface $z=\mathscr{h}(x,y;\varepsilon)$ of the liquid pool[10]. We might suppose that the cylinder

[10] Another possibility is that the temperature is also prescribed on the free surface at $z=\mathscr{h}(x,y;\varepsilon)$ (see Exercise 94.12). A third possibility is that the air insulates the water. This assumption rests on the fact that the thermal conductivity of liquids is often much greater than thermal conductivity of air (see Exercises 55.5 and 94.13).

has a top rigid surface above the free surface and that the temperature of the rigid top is prescribed and compatible with (94.64). If the air layer between the rigid surface and the free surface of the liquid is small, convection will be initiated in the liquid. We assume that the static distribution (94.64) holds in the air right up to the boundary $z=\mathscr{h}(x,y;\varepsilon)$.

The boundary conditions are that $\mathbf{u}=0$ on $z=0$ and on the boundary $\partial\mathscr{A}$ of the cylinder. The temperature is $T_0+\Delta T$ on $z=0$ and is $T_0+\Delta T(1-\mathscr{h}/\bar{\mathscr{h}})$ on $z=\mathscr{h}(x,y;\varepsilon)$. The side walls are insulated; $\mathbf{n}$ is the outward normal to the side wall $\partial\mathscr{A}$ and $\mathbf{n}_{\mathscr{A}}\cdot\nabla T=0$ on $\partial\mathscr{A}$. On the free surface the shear stresses must vanish, the jump in the normal components of the stress must balance surface tension forces and the kinematic condition for material surfaces $\mathscr{f}=d\mathscr{f}/dt=0$ must hold. As a compatibility equation with $\mathbf{u}|_{\partial\mathscr{A}}=0$, we require that $\mathscr{h}|_{\partial\mathscr{A}}=\bar{\mathscr{h}}$.

Sokolov and Tanner (1972) have studied the stability of the conduction solution (94.64) in the more restricted situation in which the free surface does not deform. They subject the rest state to initial histories of disturbances of infinitesimally small amplitude but arbitrary frequency. Their analysis is similar to the study of the rest state which was given in the previous section. They find that for many models of a simple fluid considered by them, a "principle of exchange of stability" holds and $\sigma(\Delta T_c)=0$ at criticality.

We are going to assume in our analysis that $\sigma(\Delta T_c)=0$ is a simple eigenvalue of the spectral problem. We then assume, following the lessons learned from bifurcation theory in the Newtonian case, that the bifurcating solution is steady.

To construct the bifurcating solution we use the CENR approximations for perturbations of the rest state. Here we have an example of the interesting fact that different forms of $\mathscr{F}$ are used to study stability, on the one hand, and bifurcation, on the other hand, in one and the same simple fluid.

To formulate the bifurcation theory, we first define a triad of orthonormal vectors at each point of the surface $z=\mathscr{h}$: two tangential vectors $\mathbf{m}/|\mathbf{m}|$ and $\mathbf{p}/|\mathbf{p}|$ lying in the surface $\mathscr{f}\equiv z-\mathscr{h}=0$ and the normal $\mathbf{n}=\nabla(z-\mathscr{h})/|\nabla(z-\mathscr{h})|$ where

$$\begin{bmatrix}\nabla(z-\mathscr{h})\\ \mathbf{m}\\ \mathbf{p}\end{bmatrix}=\begin{bmatrix}\mathbf{e}_z-\nabla_2\mathscr{h}\\ \mathbf{e}_y+\mathbf{e}_z\partial\mathscr{h}/\partial y\\ \mathbf{e}_x(1+|\partial\mathscr{h}/\partial y|^2)-\mathbf{e}_y\dfrac{\partial\mathscr{h}}{\partial x}\dfrac{\partial\mathscr{h}}{\partial y}+\mathbf{e}_z\dfrac{\partial\mathscr{h}}{\partial x}\end{bmatrix},$$

and $\nabla_2=\mathbf{e}_x\partial/\partial x+\mathbf{e}_y\partial/\partial y$.

We next introduce functions θ, ϕ, H and h:

$$\theta=T-T_0-\Delta T(1-z/\bar{\mathscr{h}}),$$

$$\phi=\theta\sqrt{\rho_0\alpha g\bar{\mathscr{h}}/\Delta T},$$

$$H=p-p_a+\rho_0 g\left[z-\bar{\mathscr{h}}-\frac{\alpha\Delta T}{2}(z-\bar{\mathscr{h}})^2\right],$$

$$h=\mathscr{h}(x,y;\varepsilon)-\bar{\mathscr{h}},$$

and the parameter

$$\Lambda=\sqrt{\rho_0\alpha g\Delta T/\bar{h}}. \tag{94.65a}$$

Finally, we define ε^2 as the Nusselt number discrepancy

$$Nu-1=\varepsilon^2. \tag{94.65b}$$

The Nusselt number relative to the wall at $z=0$ is defined as

$$Nu=\frac{k\dfrac{d\bar{T}(0)}{dz}}{-k\dfrac{\Delta T}{\bar{h}}}=1-\frac{1}{\Lambda}\frac{d\bar{\phi}(0)}{dz}$$

where the overbar designates horizontal averaging as in (94.63). To shorten the writing of problems, it is convenient to define a set

$$F=\{(\mathbf{u},\phi,h):\operatorname{div}\mathbf{u}=0,\ |\mathbf{u}|=\phi=0\ \text{ on }\ z=0,\ |\mathbf{u}|=\mathbf{n}_{\mathscr{A}}\cdot\nabla\phi=h=0\ \text{ on }\ \partial\mathscr{A},\ \bar{h}=0\}.$$

The boundary value problem for steady bifurcating solutions may be written as

$$\left.\begin{aligned}&-\mathbf{u}\cdot\nabla\mathbf{u}+\Lambda\phi\mathbf{e}_z-\nabla H+\nabla\cdot\mathbf{S}=0, &(94.65\text{a})\\ &-\mathbf{u}\cdot\nabla\phi+\Lambda w+\kappa\nabla^2\phi=0, &(94.65\text{b})\end{aligned}\right\}\ \text{in}\ \mathscr{V}_\varepsilon$$

$$\left.\begin{aligned}&\phi=\mathbf{u}\cdot\nabla f=\nabla f\cdot\mathbf{S}\cdot\mathbf{p}=\nabla f\cdot\mathbf{S}\cdot\mathbf{m}=0, &(94.65\text{c})\\ &\sigma\nabla_2\cdot\left\{\frac{\nabla_2 h}{(1+|\nabla_2 h|^2)^{1/2}}\right\}-\rho_0 g\left[h-\frac{\Delta T\alpha}{2}h^2\right]+H=\mathbf{n}\cdot\mathbf{S}\cdot\mathbf{n}, &(94.65\text{d})\end{aligned}\right\}\ \text{on}\ f=0$$

$$(\mathbf{u},\phi,h)\in F. \tag{94.65e}$$

The rest state is an exact solution of (94.64) which is defined by $Nu=1$ $(\varepsilon=0)$ and

$$\begin{aligned}(\mathbf{u},\mathbf{S},\phi,h,T,p)&=(\mathbf{u}^{\langle 0\rangle},\mathbf{S}^{\langle 0\rangle},\phi^{\langle 0\rangle},h^{\langle 0\rangle},T^{\langle 0\rangle},P^{\langle 0\rangle})\\ &=\left(0,0,0,\bar{h},\ \Delta T(1-z/\bar{h}),\ p_a-\rho_0 g\left[(z-\bar{h})-\frac{\alpha\Delta T}{2}(z-\bar{h})^2\right]\right).\end{aligned}$$

The rest solution exists for all values of Λ. When $\Lambda=\Lambda^{\langle 0\rangle}$ (corresponding to $\Delta T=\Delta T_c$ where ΔT_c is the critical temperature difference) then the rest solution loses its stability to motion.

To study the motion, we first define an invertible mapping which is analytic in a small parameter ε to be defined; for example,

$$\mathscr{V}_\varepsilon(x,y,z) \Longleftrightarrow \mathscr{V}_0(X,Y,Z)$$

under the transformation

$$x=X, \quad y=Y, \quad z=\mathscr{z}(X,Y,Z;\varepsilon)$$

where the stretching function $\mathscr{z}$ is invertible

$$Z=\hat{Z}(x,y,z;\varepsilon)=\mathscr{z}(X,Y,Z;0)$$

and maps boundary points into boundary points

$$0=\mathscr{z}(X,Y,0;\varepsilon)$$

and

$$\mathscr{h}=\mathscr{z}(X,Y,\bar{\mathscr{h}};\varepsilon)\,.$$

No further specification of the mapping than the one just given is required; the boundary values of the mapping are determined uniquely from the bifurcation problem pivoted around the rest state as a power series in the parameter $\varepsilon=\sqrt{Nu-1}$

$$\begin{bmatrix} \mathbf{u}(x,y,z;\varepsilon) \\ \phi(x,y,z;\varepsilon) \\ H(x,y,z;\varepsilon) \\ h(x,y,z;\varepsilon) \\ \Lambda(\varepsilon)-\Lambda^{\langle 0\rangle} \end{bmatrix} = \sum_{n=1}\varepsilon^n \begin{bmatrix} \mathbf{u}^{[n]}(X,Y,Z) \\ \phi^{[n]}(X,Y,Z) \\ H^{[n]}(X,Y,Z) \\ h^{\langle n\rangle}(X,Y) \\ \Lambda^{\langle n\rangle} \end{bmatrix}. \tag{94.66a}$$

In the notation of § 93, the extra stress is given by

$$\mathbf{S}=\sum_{n=1}\varepsilon^n \sum_{l+q=1+n} \mathbf{S}_q^{\langle l\rangle}[\mathbf{u}^{\langle r_1\rangle},\ldots,\mathbf{u}^{\langle r_q\rangle}]=\sum_{n=1}\varepsilon^n \mathbf{S}^{\langle n\rangle}(X,Y,Z)\,.$$

Here,

$$(\circ)^{[n]}=\frac{1}{n!}\left(\frac{\partial}{\partial\varepsilon}+\frac{d}{d\varepsilon}\frac{\partial}{\partial z}\right)^n(\circ)=\frac{1}{n!}\frac{d^n(\circ)}{d\varepsilon^n}$$

in the nth substantial derivative following the mapping and

$$(\circ)^{\langle n\rangle}=\frac{1}{n!}\frac{\partial^n(\circ)}{\partial\varepsilon^n}$$

is the nth partial derivative.

The functions on the right of (94.66a) are all defined in the reference domain $\mathscr{V}_0(X, Y, Z)$. To express the solution in the deformed domain $\mathscr{V}_\varepsilon(x, y, z)$ it is convenient to invert the mapping. This requires an explicit form for the mapping function $\mathscr{z}$. Since the construction giving $\mathscr{z}$ is independent of the interior values of $\mathscr{z}(X, Y, Z; \varepsilon)$ we may construct a scale mapping

$$z = \mathscr{z} = Z\mathscr{h}(X, Y; \varepsilon)/\bar{\mathscr{h}}$$

having the required properties. Then we may continue (94.66a) as

$$= \sum_{n=1} \varepsilon^n \begin{bmatrix} \mathbf{u}^{[n]}(x, y, z\bar{\mathscr{h}}/\mathscr{h}) \\ \phi^{[n]}(x, y, z\bar{\mathscr{h}}/\mathscr{h}) \\ H^{[n]}(x, y, z\bar{\mathscr{h}}/\mathscr{h}) \\ h^{\langle n\rangle}(x, y) \\ \Lambda^{\langle n\rangle} \end{bmatrix} = \sum_{n=1} \varepsilon^n \begin{bmatrix} \mathbf{u}^{\langle n\rangle}(x, y, z) \\ \phi^{\langle n\rangle}(x, y, z) \\ H^{\langle n\rangle}(x, y, z) \\ h^{\langle n\rangle}(x, y) \\ \Lambda^{\langle n\rangle} \end{bmatrix} \tag{94.66b}$$

(94.66b) gives two ways to express the solution in $\mathscr{V}_\varepsilon$. The series on the left of (94.66b) follows from inverting the mapping; the domain of the functions in unchanged; the series on the right expresses the solution in terms of partial derivatives which are first defined on $\mathscr{V}_0$ and then extended, by declaration, onto $\mathscr{V}_\varepsilon$.[11]

To generate the perturbation problems for the partial derivatives, we note that if $A(x, y, z; \varepsilon) = 0$ is an identity in $\mathscr{V}_\varepsilon$, then $A^{[n]}(x, y, z; \varepsilon) = 0$ and a simple induction argument (see Exercise B4.7) gives $A^{\langle n\rangle}(x, y, z; \varepsilon) = 0$. If $A(x, y, \mathscr{h}(x, y; \varepsilon)) = 0$, then $A^{[n]} = \dfrac{1}{n!}\left(\dfrac{\partial}{\partial\varepsilon} + h^{\langle 1\rangle}(x, y; \varepsilon)\dfrac{\partial}{\partial z}\right)^n A$. Using these properties we find that when $\nu \geqslant 1$:

$$\left.\begin{aligned} &\sum_{l+n=\nu} \mathbf{u}^{\langle n\rangle}\cdot\nabla\mathbf{u}^{\langle l\rangle} - \sum_{l+n=\nu} \Lambda^{\langle n\rangle}\phi^{\langle l\rangle}\mathbf{e}_z + \nabla H^{\langle \nu\rangle} - \nabla\cdot\mathbf{S}^{\langle\nu\rangle} = 0, &(94.67\text{a})\\ &\sum_{l+n=\nu} \mathbf{u}^{\langle n\rangle}\cdot\nabla\phi^{\langle l\rangle} - \sum_{l+n=\nu} \Lambda^{\langle n\rangle} w^{\langle l\rangle} - \kappa\nabla^2\phi^{\langle\nu\rangle} = 0. &(94.67\text{b}) \end{aligned}\right\} \text{ in } \mathscr{V}_0$$

$$(\mathbf{u}^{\langle\nu\rangle}, \phi^{\langle\nu\rangle}, h^{\langle\nu\rangle}) \in F, \tag{94.67c}$$

$$\phi^{[\nu]} = (\nabla\!\!\!/\cdot\mathbf{S}\cdot\mathbf{p})^{[\nu]} = (\nabla\!\!\!/\cdot\mathbf{S}\cdot\mathbf{m})^{[\nu]} = (w - \mathbf{u}\cdot\nabla_2 h)^{[\nu]} = 0, \tag{94.67d}$$

$$[\mathbf{n}\cdot\mathbf{S}\cdot\mathbf{n}]^{[\nu]} = H^{[\nu]} - \rho_0 g\left[h - \frac{\alpha\Delta T}{2}h^2\right]^{[\nu]} + \sigma\nabla_2\cdot\left[\frac{\nabla_2 h}{(1+|\nabla_2 h|^2)^{1/2}}\right]^{[\nu]} \tag{94.67e}$$

where (94.67d, e) are evaluated on $h = 0$. From the normalizing condition (94.65b) we find that

$$-\Lambda^{\langle\nu-2\rangle} = \frac{d\bar{\phi}^{\langle\nu\rangle}(0)}{dz} = \bar{\phi}^{\langle\nu\rangle}(\bar{\mathscr{h}}) - \frac{1}{\kappa}\sum_{l+n=\nu+1}\langle w^{\langle n\rangle}\phi^{\langle l\rangle}\rangle \tag{94.67f}$$

where $\Lambda^{\langle -1\rangle} \equiv 0$ and

$$\langle\circ\rangle = \frac{1}{\bar{\mathscr{h}}}\int_0^{\bar{\mathscr{h}}} \circ\, dZ$$

[11] If there is a bifurcating solution analytic in ε, then this extension will be possible (see Joseph, 1973; Joseph and Sturges, 1975).

To derive (94.67f) take the horizontal average of (94.67) noting that $\overline{w}^{\langle l\rangle}=0$. We find that

$$\frac{d}{dZ}\left(\sum_{l+n=\nu}\overline{w^{\langle n\rangle}\phi^{\langle l\rangle}}-\kappa\frac{d\bar{\phi}^{\langle\nu\rangle}(Z)}{dZ}\right)=0 \tag{94.68}$$

and (94.67f) follows after integrating (94.68). The boundary values for $\phi^{\langle\nu\rangle}(\bar{h})$ may be obtained in terms of lower-order partial derivatives by unfolding the first of Eqs. (94.67f).

When $\nu=1$ we find that $A^{[1]}=A^{\langle 1\rangle}+h^{\langle 1\rangle}\dfrac{\partial A^{\langle 0\rangle}}{\partial Z}$ for all $A(x,y,h(x,y;\varepsilon);\varepsilon)$. We note that $\mathbf{S}^{\langle 1\rangle}=\mathbf{S}_1^{\langle 1\rangle}=\mu\mathbf{A}_1^{\langle 1\rangle}=\mu\mathbf{A}_1[u^{\langle 1\rangle}]$ and use (94.64) to find that

$$\left.\begin{aligned}&\Lambda^{\langle 0\rangle}\phi^{\langle 1\rangle}\mathbf{e}_z-\nabla H^{\langle 1\rangle}+\nabla\cdot\mathbf{S}^{\langle 1\rangle}=0, && (94.69\text{a})\\ &\Lambda^{\langle 0\rangle}w^{\langle 1\rangle}+\kappa\nabla^2\phi^{\langle 1\rangle}=0, && (94.69\text{b})\end{aligned}\right\}\text{ in }\mathscr{V}_0$$

$$(\mathbf{u}^{\langle 1\rangle},\phi^{\langle 1\rangle},h^{\langle 1\rangle})\in F, \tag{94.69c}$$

$$\left.\begin{aligned}&\phi^{\langle 1\rangle}=S_{zy}^{\langle 1\rangle}=S_{zx}^{\langle 1\rangle}=w^{\langle 1\rangle}=0, && (94.69\text{d})\\ &S_{zz}^{\langle 1\rangle}-H^{\langle 1\rangle}=-\rho_0gh^{\langle 1\rangle}+\sigma\nabla_2^2h^{\langle 1\rangle} && (94.69\text{e})\end{aligned}\right\}\text{ when } h=0$$

and

$$-\Lambda^{\langle 0\rangle}=\bar{\phi}^{\langle 2\rangle}-\frac{1}{\kappa}\langle w^{\langle 1\rangle}\phi^{\langle 1\rangle}\rangle. \tag{94.69f}$$

The problem (94.69) is a self-contained eigenvalue problem and determines the eigenvalue $\Lambda^{\langle 0\rangle}$ which, by assumption, has multiplicity one. The eigenfunction belonging to $\Lambda^{\langle 0\rangle}$ is uniquely determined to within an arbitrary multiplicative constant which is determined uniquely by (94.69f). To see how (94.69f) determines the constant, we first note that $h^{\langle 1\rangle}$ is determined to within the same multiplicative constant by (94.69e). We then note that

$$A^{[2]}=A^{\langle 2\rangle}+h^{\langle 1\rangle}\frac{\partial A^{\langle 1\rangle}}{\partial Z}+\left(h^{\langle 2\rangle}+\frac{h^{\langle 1\rangle 2}}{2}\frac{\partial}{\partial Z}\right)A^{\langle 0\rangle}.$$

Therefore,

$$\phi^{[2]}=\phi^{\langle 2\rangle}+h^{\langle 1\rangle}\frac{\partial\phi^{\langle 1\rangle}}{\partial Z}=0$$

and elimination of $\bar{\phi}^{\langle 2\rangle}$ from (94.69f) gives

$$-\Lambda^{\langle 0\rangle}=-h^{\langle 1\rangle}\phi^{\langle 1\rangle}-\frac{1}{\kappa}\langle w^{\langle 1\rangle}\phi^{\langle 1\rangle}\rangle.$$

Since $\Lambda^{\langle 0\rangle}$ is known and $h^{\langle 1\rangle},\phi^{\langle 1\rangle}$ and $w^{\langle 1\rangle}$ are known to within an arbitrary multiplicative constant, (94.69f) determines the multiplicative constant.

The first nonlinear effects enter at second order:

$$\mathbf{S}^{\langle 2\rangle}=\mathbf{S}_1^{\langle 2\rangle}+\mathbf{S}_2^{\langle 1\rangle}=\mu A_1[\mathbf{u}^{\langle 2\rangle}]+\alpha_1\mathbf{A}_2[\mathbf{u}^{\langle 1\rangle}]+\alpha_2\mathbf{A}_1^2[\mathbf{u}^{\langle 1\rangle}]\equiv\mu\mathbf{A}_1^{(2)}+\alpha_1\mathbf{A}_2^{(1)}+\alpha_2\mathbf{A}_1^{(1)}\mathbf{A}_1^{(1)} \tag{94.70}$$

where the symbols with superscripts in parentheses

$$\mathbf{A}_l^{(m)}\equiv\mathbf{A}_l(\mathbf{u}^{\langle m\rangle}),\qquad \mathbf{S}_l^{(m)}=\mathbf{S}_l(\mathbf{u}^{\langle m\rangle})$$

mean that the tensors $\mathbf{A}_l$ or $\mathbf{S}_l$ should be evaluated on the field $\mathbf{u}^{\langle m\rangle}$; in particular, $\mathbf{S}_1^{\langle m\rangle}=\mathbf{S}_1^{(m)}=\mu\mathbf{A}_1^{(m)}$ and $\mathbf{S}_n^{\langle 1\rangle}=\mathbf{S}_n^{(1)}$.

$$\Lambda^{\langle 0\rangle}\phi^{\langle 2\rangle}\mathbf{e}_z+\mu\nabla\cdot\mathbf{A}_1^{(2)}-\nabla H^{\langle 2\rangle}=\mathbf{u}^{\langle 1\rangle}\cdot\nabla\mathbf{u}^{\langle 1\rangle}-\Lambda^{\langle 1\rangle}\phi^{\langle 1\rangle}\mathbf{e}_z-\nabla\cdot\mathbf{S}_2^{\langle 1\rangle}, \tag{94.71a}$$

$$\Lambda^{\langle 0\rangle}w^{\langle 2\rangle}+\kappa\nabla^2\phi^{\langle 2\rangle}=\mathbf{u}^{\langle 1\rangle}\cdot\nabla\phi^{\langle 1\rangle}-\Lambda^{\langle 1\rangle}w^{\langle 1\rangle}, \tag{94.71b}$$

$$(\mathbf{u}^{\langle 2\rangle},\phi^{\langle 2\rangle},h^{\langle 2\rangle})\in F, \tag{94.71c}$$

$$\phi^{\langle 2\rangle}+h^{\langle 1\rangle}\phi^{\langle 1\rangle}_{,Z}=w^{\langle 2\rangle}+h^{\langle 1\rangle}w^{\langle 1\rangle}_{,Z}-\mathbf{u}^{\langle 1\rangle}\cdot\nabla_2 h^{\langle 1\rangle}=0, \tag{94.71d}$$

$$A^{(2)}_{1zx}+h^{\langle 1\rangle}A^{(1)}_{1zx,z}-\nabla_2 h^{\langle 1\rangle}\cdot\mathbf{A}_1^{(1)}\cdot\mathbf{e}_x+A^{(1)}_{1zz}h^{\langle 1\rangle}_{,x}+\mu^{-1}S^{(1)}_{2zx}=0, \tag{94.71e}$$

$$A^{(2)}_{1zy}+h^{\langle 1\rangle}A^{(1)}_{1zy,z}-\nabla_2 h^{\langle 1\rangle}\cdot\mathbf{A}_1^{(1)}\cdot\mathbf{e}_y+A^{(1)}_{1zz}h^{\langle 1\rangle}_{,y}+\mu^{-1}S^{(1)}_{2xy}=0, \tag{94.71f}$$

$$\mu A^{(2)}_{1zz}+h^{\langle 1\rangle}(\mu A^{(1)}_{1zz}-H^{\langle 1\rangle})_{,z}+S^{(1)}_{2zz}-H^{\langle 2\rangle}=\sigma\nabla_2^2 h^{\langle 2\rangle}-\rho_0 g[h^{\langle 2\rangle}-\alpha\Delta T h^{\langle 1\rangle^2}]. \tag{94.71g}$$

The boundary value problem (94.71) can be solved only if the inhomogeneous terms satisfy the orthogonality relation

$$\begin{aligned}
&\mu\langle\mathbf{u}^{\langle 1\rangle}\cdot(\nabla\cdot\mathbf{A}_1^{(2)})-\mathbf{u}^{\langle 2\rangle}\cdot A_1^{(1)}\rangle+\kappa\langle\phi^{\langle 1\rangle}\nabla^2\phi^{\langle 2\rangle}-\phi^{\langle 2\rangle}\nabla^2\phi^{\langle 1\rangle}\rangle\\
&\qquad+\langle\mathbf{u}^{\langle 1\rangle}\cdot\nabla\cdot\mathbf{S}_2^{\langle 1\rangle}\rangle-\langle\mathbf{u}^{\langle 1\rangle}\cdot\nabla H^{\langle 2\rangle}-\mathbf{u}^{\langle 2\rangle}\cdot\nabla H^{\langle 1\rangle}\rangle+2\Lambda^{\langle 1\rangle}\langle w^{\langle 1\rangle}\phi^{\langle 1\rangle}\rangle\\
&\qquad=\langle\mathbf{u}^{\langle 1\rangle}\cdot(\mathbf{u}^{\langle 1\rangle}\cdot\nabla)\mathbf{u}^{\langle 1\rangle}\rangle+\langle\phi^{\langle 1\rangle}(\mathbf{u}^{\langle 1\rangle}\cdot\nabla)\phi^{\langle 1\rangle}\rangle=0.
\end{aligned}$$

This relation may be reduced further by integrating by parts and using the condition $A^{(1)}_{1zx}=A^{(1)}_{1zy}=0$:

$$[\mu\mathbf{e}_z\cdot\mathbf{A}_1^{(2)}\cdot\mathbf{u}^{\langle 1\rangle}-\kappa\phi^{\langle 2\rangle}\phi^{\langle 1\rangle}_{,Z}]_{Z=\bar{h}}-w^{\langle 2\rangle}[H^{\langle 1\rangle}-\mu A^{(1)}_{1zz}]_{Z=\bar{h}}-\langle\mathbf{S}_2^{\langle 1\rangle}\cdot\nabla\mathbf{u}^{\langle 1\rangle}\rangle+2\Lambda^{\langle 1\rangle}\langle w^{\langle 1\rangle}\phi^{\langle 1\rangle}\rangle=0. \tag{94.72a}$$

Second-order quantities $\mathbf{A}_1^{(2)},\phi^{\langle 2\rangle}$ and $w^{\langle 2\rangle}$ may be eliminated from (94.72) using (94.71d—g),

$$\begin{aligned}
2\Lambda^{\langle 1\rangle}\langle w^{\langle 1\rangle}\phi^{\langle 1\rangle}\rangle=\langle\mathbf{S}_2^{\langle 1\rangle}:\nabla\mathbf{u}^{\langle 1\rangle}\rangle+[\mu\{&h^{\langle 1\rangle}\mathbf{e}_z\cdot\mathbf{A}^{(1)}_{1,Z}-\nabla_2 h^{\langle 1\rangle}\cdot\mathbf{A}_1^{(1)}\\
&+A^{(1)}_{1zz}\nabla_2 h^{\langle 1\rangle}\}\cdot\mathbf{u}^{\langle 1\rangle}+\mathbf{e}_z\cdot\mathbf{S}_2^{\langle 1\rangle}\cdot\mathbf{u}^{\langle 1\rangle}-\kappa h^{\langle 1\rangle}(\phi^{\langle 1\rangle}_{,Z})^2\\
&+(h^{\langle 1\rangle}w^{\langle 1\rangle}_{,Z}-\mathbf{u}^{\langle 1\rangle}\cdot\nabla_2 h^{\langle 1\rangle})(H^{\langle 1\rangle}-\mu A^{(1)}_{1zz})]_{Z=\bar{h}}.
\end{aligned} \tag{94.72b}$$

Eq. (94.73) gives the slope of the heat-transport curve evaluated on the bifurcating solution at the point of bifurcation; this slope depends on the constants α_1 and α_2 through the tensor $\mathbf{S}_2^{\langle 1\rangle}=\mathbf{S}_2^{(1)}$; for example,

$$\begin{aligned}
\langle\mathbf{S}_2^{\langle 1\rangle}:\nabla\mathbf{u}^{\langle 1\rangle}\rangle&=\alpha_1\langle\mathbf{A}_1^{(1)}:\mathbf{A}_2^{(1)}\rangle+\alpha_2\langle\mathbf{A}_1^{(1)}\cdot A_1^{(1)}\rangle\\
&=(\alpha_1+\alpha_2)\langle\mathbf{A}_1^{(1)}:\mathbf{A}_1^{(1)}\cdot\mathbf{A}_1^{(1)}\rangle
\end{aligned}$$

where the last equality follows from symmetry and integration by parts using $w^{\langle 1\rangle}=0$. It follows that

$$\left.\frac{d\Lambda}{d\sqrt{Nu-1}}\right|_{Nu=1}=\Lambda^{\langle 1\rangle}(\mu,\alpha_1,\alpha_2)$$

where $\Lambda^{\langle 1\rangle}$ is given by (94.72b). In passing, we note that if the deflection of the free surface is neglected, then $S^{\langle 1\rangle}_{2zx}=S^{\langle 1\rangle}_{2zy}=0$ and $\Lambda^{\langle 1\rangle}=(\alpha_1+\alpha_2)\langle\mathbf{A}_1^{(1)}:\mathbf{A}_1^{(1)}\cdot\mathbf{A}_1^{(1)}\rangle$.

The values of $\Lambda^{\langle n \rangle}$ for $n \geqslant 2$ are obtained from the solvability requirement at higher orders. These values depend on the CENR coefficients which appear at higher orders.

In the same way, the shape of the free surface, which is given by the series (94.66), depends on the CENR constants; $h^{\langle 1 \rangle}$ is a function of μ alone, $h^{\langle 2 \rangle}$ depends on the second-order constants α_1 and α_2, $h^{\langle 3 \rangle}$ depends on the third-order coefficients, and so on.

The analysis just given assumes that conduction loses stability as a real simple eigenvalue σ_1 crosses the origin strictly. If, on the other hand, a pair of simple conjugate imaginary eigenvalues crosses the line $\operatorname{re}\sigma_1 = 0$ the bifurcating solution is expected to be time-periodic. Time-periodic bifurcations can be studied by the methods outlined in Chapter II using the computational algorithms developed under § 94(b).

Exercise 94.1 (see Exercise 93.2): Show that

$$\frac{d\mathbf{F}^{-1}(\tau)}{dt} = -\mathbf{F}^{-1}(\tau)\cdot\nabla_{\xi}\mathbf{U}(\xi,\tau), \qquad \frac{d}{d\tau}\left[\frac{\partial X_i}{\partial \xi_k}\right] = -\frac{\partial X_i}{\partial \xi_l}\frac{\partial U_l(\xi,\tau)}{\partial \xi_k};$$

$$\mathbf{A}_m[\mathbf{U}(\boldsymbol{\xi},\tau)] = \mathbf{F}^{-1^T}(\tau)\cdot\frac{d^m\mathbf{C}(\tau)}{d\tau^m}\cdot\mathbf{F}^{-1}(\tau), \qquad m \geqslant 1;$$

$$[\mathbf{A}_m[\mathbf{U}(\boldsymbol{\xi},\tau)]]_{ij} = \frac{\partial X_l}{\partial \xi_i}\frac{d^m C(\tau)_{lk}}{d\tau^m}\frac{\partial X_k}{\partial \xi_j};$$

$$\frac{d\mathbf{C}_t(\tau)}{d\tau} = \mathbf{F}_t(\tau)^T\cdot\mathbf{A}_1[\mathbf{U}(\boldsymbol{\xi},\tau)]\cdot\mathbf{F}_t(\tau) = \mathbf{F}_t(\tau)^T\cdot[\nabla_{\xi}\mathbf{U} + \nabla_{\xi}\mathbf{U}^T]\cdot\mathbf{F}_t(\tau);$$

$$\frac{d^m\mathbf{C}_t(\tau)}{d\tau^m} = \mathbf{F}_t(\tau)^T\cdot\mathbf{A}_m[\mathbf{U}(\xi,\tau)]\cdot\mathbf{F}_t(\tau); \qquad m \geqslant 1,$$

$$\left.\frac{d^m\mathbf{C}_t(\tau)}{d\tau^m}\right|_{\tau=t} = \mathbf{A}_m[\mathbf{U}(\mathbf{x},t)];$$

and prove (94.3a).

Exercise 94.2 (Power series expansion of the response functional in a special case): Consider the functional

$$\mathscr{F}[\mathbf{G}(s)] = \int_0^\infty f(s)\mathbf{H}[\mathbf{G}(s)]\,ds$$

where

$$\mathbf{H}(\mathbf{0}) = 0.$$

Suppose that $\mathbf{H}$ is an analytic function of the components of $\mathbf{G}(s)$, that $\mathbf{G}(s)$ is an analytic tensor-valued function of s and that

$$\int_0^\infty s^n f(s)\,ds$$

is bounded for all integers n.

(a) Show that

$$\mathbf{S} = \sum_{n=1}^\infty \mathbf{S}_n = \sum_{n=1}^\infty \frac{1}{n!}\left[\frac{d^n}{ds^n}\mathbf{H}[\mathbf{G}(s)]\right]_{s=0} \int_0^\infty s^n f(s)\,ds$$

where

$$\left.\frac{dH_{ij}}{ds}\right|_{s=0} = -\frac{\partial H_{ij}}{\partial G_{lk}}(\mathbf{A}_1)_{lK}$$

and

$$\left.\frac{d^2H_{ij}}{ds^2}\right|_{s=0} = \frac{\partial^2 H_{ij}}{\partial G_{lK}\partial G_{mn}}(\mathbf{A}_1)_{lK}(\mathbf{A}_1)_{mn} + \frac{\partial H_{ij}}{\partial G_{lK}}(\mathbf{A}_2)_{lK}\,.$$

(b) Use the fact that any isotropic tensor of even order can be written as a linear combination of all possible permutations of the Kronecker delta to show that

$$\mathbf{S}_2 = \alpha_1 \mathbf{A}_2 + \alpha_2 \mathbf{A}_1^2$$

and

$$\mathbf{S}_3 = \beta_1 \mathbf{A}_3 + \beta_2(\mathbf{A}_2\mathbf{A}_1 + \mathbf{A}_1\mathbf{A}_2) + \beta_3[\operatorname{tr}\mathbf{A}_2]\mathbf{A}_1$$

where $\alpha_1, \alpha_2, \beta_1, \beta_2, \beta_3$ are constants or, more generally, functions of the temperature.

Exercise 94.3: Show that the constitutive relation for a Newtonian fluid can be obtained from (94.16) when the fading memory of the fluid is infinitely short or when the motion is steady.

Exercise 94.4 (Functional expansions at the rest state): (a) Compare (93.6) and (94.30) to deduce that

$$\mathscr{F}_1[0\colon \mathbf{G}_1(s)] = \textstyle\int_0^\infty \zeta(s)\,\mathbf{G}_1(s)\,ds\,,$$

$$\mathscr{F}_1[0\colon \mathbf{G}_2(s)] = \textstyle\int_0^\infty \zeta(s)\,\mathbf{G}_2(s)\,ds\,,$$

and

$$\mathscr{F}_2[0\colon \mathbf{G}_1(s), \mathbf{G}_1(s)] = \textstyle\int_0^\infty \int_0^\infty \int s_1 s_2\, \beta(s_1,s_2)\,\mathbf{G}_1(s_1)\cdot \mathbf{G}_1(s_2)\,ds_1\,ds_2\,,$$

(b) Compare (93.6) and (94.41, 42) to deduce that

$$\mathscr{F}_1[0\colon \mathbf{G}^{\langle 1\rangle}(s)] = \textstyle\int_0^\infty \mathsf{G}(s)\,\mathbf{A}_1[\mathbf{U}^{\langle 1\rangle}(s)]\,ds\,,$$

$$\mathscr{F}_1[0\colon \mathbf{G}^{\langle 2\rangle}(s)] = \textstyle\int_0^\infty \mathsf{G}(s)\,\{\mathbf{A}_1[\mathbf{U}^{\langle 2\rangle}(s)]$$

$$+((\boldsymbol{\xi}^{[1]}))\cdot\nabla_{\mathbf{X}}\mathbf{A}_1[\mathbf{U}^{\langle 1\rangle}(s)] + [\mathbf{A}_1[\mathbf{U}^{\langle 1\rangle}(s)]\cdot\nabla_{\mathbf{X}}((\boldsymbol{\xi}^{[1]})) + \text{transpose}]\}\,ds$$

$$\mathscr{F}_2[0\colon \mathbf{G}^{\langle 1\rangle}(s), \mathbf{G}^{\langle 1\rangle}(s)] = \textstyle\int_0^\infty\int_0^\infty \gamma(s_1,s_2)\,\mathbf{A}_1[\mathbf{U}^{\langle 1\rangle}(s_1)]\cdot\mathbf{A}_1[\mathbf{U}^{\langle 1\rangle}(s_2)]\,ds_1\,ds_2\,.$$

What form do these functional derivatives take when evaluated in the present rather than the reference configuration?

(c) Compare (93.6) and (94.52) to deduce that for steady flow

$$\mathscr{F}_1[0\colon \mathbf{G}^{\langle 1\rangle}(s)] = \mu\mathbf{A}_1[\mathbf{U}^{\langle 1\rangle}(\mathbf{x})]\,,$$

$$\mathscr{F}_1[0\colon \mathbf{G}^{\langle 2\rangle}(s)] = \mu\mathbf{A}_1[\mathbf{U}^{\langle 2\rangle}(\mathbf{x})] + \alpha_1\mathbf{A}_2[\mathbf{U}^{\langle 1\rangle}(\mathbf{x})]$$

and

$$\mathscr{F}_2[0\colon \mathbf{G}^{\langle 1\rangle}(s), \mathbf{G}^{\langle 1\rangle}(s)] = \alpha_2\mathbf{A}_1[\mathbf{U}^{\langle 1\rangle}(\mathbf{x})]\cdot\mathbf{A}_1[\mathbf{U}^{\langle 1\rangle}(\mathbf{x})]\,.$$

Exercise 94.5: Using (94.12) find the set of parameters for which the fluid of grade four is unstable.

Exercise 94.6 (Craik 1968): If $e^{\sigma s}\mathsf{G}(s)\to 0$ as $s\to 0$, then

$$\int_0^\infty \mathsf{G}(s)\,e^{\sigma s}\,ds=\frac{1}{\sigma}\int_0^\infty \frac{d\mathsf{G}}{ds}(1-e^{\sigma s})\,ds\,.$$

Show that

$$\hat{\Lambda}\eta\int_0^\infty \frac{d\mathsf{G}}{ds}\,e^{\xi s}\sin\eta s\,ds$$

is positive if

$$f(s)=\frac{d\mathsf{G}}{ds}\,e^{\xi s}$$

is a positive monotonically decreasing function of s. Establish conditions on $\mathsf{G}(s)$ which will guarantee that there are no solutions of (94.59) with $\xi<0$.

Exercise 94.7: Isotropic forms of the integral fluids (94.14a) of order N. Each integrand is (94.14a) is a tensor polynomial multilinear in the Cauchy strains $G_{kl}(s_1)\,G_{mn}(s_2)\ldots$. This multilinear tensor-valued form is to be isotropic for all positive values of $s_1, s_2, \ldots, s_N$. In particular when $s_1=s_2=\cdots=s_N\equiv s$ the integrand is a tensor polynomial of a single tensor $\mathbf{G}(\hat{s})$. The coefficients of this tensor-valued polynomial

$$K_{ij\ldots klmn}(\hat{s},\hat{s},\ldots,\hat{s})$$

must then be an isotropic tensor of even order. Such tensors must necessarily be expressible in terms of products of Kronecker's delta with scalar coefficients $c(\hat{s},\hat{s},\ldots,\hat{s})$ which depend on the N equal times $\hat{s}$. Recalling then that the required forms are multilinear in the G_{nl} when the times are all different, show that apart from terms proportional to **1**

$$\begin{aligned}\mathscr{F}^{(3)}[\mathbf{G}(s)]=&\int_0^\infty \zeta(s)\,\mathbf{G}(s)\,ds+\int_0^\infty\int_0^\infty\{\beta(s_1,s_2)\,\mathbf{G}(s_1)\cdot\mathbf{G}(s_2)+\alpha(s_1,s_2)[\operatorname{tr}\mathbf{G}(s_1)]\,\mathbf{G}(s_2)\}\,ds_1\,ds_2\\&+\int_0^\infty\int_0^\infty\int_0^\infty\{\psi_1(s_1,s_2,s_3)\,\mathbf{G}(s_1)\cdot\mathbf{G}(s_2)\cdot\mathbf{G}(s_3)+\psi_2(s_1,s_2,s_3)[\operatorname{tr}\mathbf{G}(s_1)]\,\mathbf{G}(s_2)\cdot\mathbf{G}(s_3)\\&+\psi_3(s_1,s_2,s_3)[\operatorname{tr}\mathbf{G}(s_1)][\operatorname{tr}\mathbf{G}(s_2)]\,\mathbf{G}(s_3)+\psi_4(s_1,s_2,s_3)\operatorname{tr}[\mathbf{G}(s_1)\cdot\mathbf{G}(s_2)]\,\mathbf{G}(s_3)\}\,ds_1\,ds_2\,ds_3\,.\end{aligned}\tag{94.73}$$

where the ψ_i are material functions.

Exercise 94.8 (The reduction of the constitutive equation of third order on the history of a steady motion; Schowalter, 1976 Chapter 10). Assume that the history of some steady motion $\mathbf{U}(\mathbf{x})$ may be expanded into a series of Rivlin-Ericksen tensors. Show that

$$\begin{aligned}\mathscr{F}^{(3)}=&\mu\mathbf{A}_1+\alpha_1\mathbf{A}_2+\alpha_2\mathbf{A}_1^2+\beta_1\mathbf{A}_3+\beta_2(\mathbf{A}_2\mathbf{A}_1+\mathbf{A}_1\mathbf{A}_2)\\&+(\beta_4\operatorname{tr}\mathbf{A}_2+\beta_5\operatorname{tr}\mathbf{A}_1^2)\,\mathbf{A}_1+\beta_6\mathbf{A}_1^3+O(\mathbf{U}^4)\end{aligned}\tag{94.74}$$

where μ, α_1, α_2 are given by (94.20) and

$$\beta_1=\frac{-1}{3!}\int_0^\infty s^3\zeta(s)\,ds\,,$$

$$\beta_2=-\tfrac{1}{2}\int_0^\infty\int_0^\infty s_1s_2(s_1+s_2)\,\beta(s_1,s_2)\,ds_1\,ds_2\,,$$

$$\beta_4=-\tfrac{1}{2}\int_0^\infty\int_0^\infty s_1^2s_2\,\alpha(s_1,s_2)\,ds_1\,ds_2\,,$$

$$\beta_5=-\int_0^\infty\int_0^\infty\int_0^\infty s_1s_2s_3\,\psi_4(s_1,s_2,s_3)\,ds_1\,ds_2\,ds_3\,,$$

$$\beta_6=-\int_0^\infty\int_0^\infty\int_0^\infty s_1s_2s_3\,\psi_1(s_1,s_2,s_3)\,ds_1\,ds_2\,ds_3\,.$$

Use the Hamilton-Cayley theorem to prove that, if $\operatorname{tr}\mathbf{A}_1=0$, then

$$\mathbf{A}_1^3=\tfrac{1}{2}[\operatorname{tr}\mathbf{A}_1^2]\,\mathbf{A}_1+\mathbf{1}\det\mathbf{A}_1\,.$$

Note that $\operatorname{tr}\mathbf{A}_1^2=\operatorname{tr}\mathbf{A}_2$ and prove that, apart from terms proportional to **1**, the last three terms of (94.74) may be replaced by

$$\beta_3[\operatorname{tr}\mathbf{A}_2]\,\mathbf{A}_1\,,\qquad \beta_3=\beta_4+\beta_5+\tfrac{1}{2}\beta_6\,.$$

Exercise 94.9 (More kinematics for perturbations of the rest state): Show that

$$\boldsymbol{\xi}(\mathbf{X},\tau,\varepsilon)=\mathbf{x}+\textstyle\sum_{l=1}\varepsilon^l((\xi^{[l]}))=\boldsymbol{\chi}_t(\mathbf{x},\tau,\varepsilon)=\mathbf{x}+\textstyle\sum_{l=1}\varepsilon^l\chi^{\langle l\rangle}(\mathbf{x},\tau)$$

where

$$((\xi^{[l]}))=\xi^{[l]}(\mathbf{X},\tau)-\xi^{[l]}(\mathbf{X},t)\,.$$

Suppose that $\mathbf{X}$ and ε are independent. Expand

$$\mathbf{x}=\boldsymbol{\xi}(\mathbf{X},\tau,\varepsilon)=\mathbf{X}+\textstyle\sum_{l=1}\varepsilon^l\xi^{[l]}(X,t)$$

and show that

$$\boldsymbol{\chi}^{\langle 1\rangle}(\mathbf{X},\tau)=((\xi^{[1]}))\,,$$

$$\boldsymbol{\chi}^{\langle 2\rangle}(\mathbf{X},\tau)=((\xi^{[2]}))-\xi^{[1]}(\mathbf{X},t)\cdot\nabla_{\mathbf{X}}\boldsymbol{\chi}^{\langle 1\rangle}\,.$$

and $(n\geqslant 1)$

$$\boldsymbol{\chi}^{\langle n\rangle}(\mathbf{X},\tau)=((\xi^{[n]}))+\textstyle\int_t^\tau\{\tilde{\mathbf{U}}^{\langle n\rangle}(\mathbf{X},\tau')-\mathbf{U}^{[n]}(\mathbf{X},\tau')\}\,d\tau'\,.$$

Moreover, when $\varepsilon\to 0$

$$\begin{aligned}\tfrac{1}{3}\left(\frac{\partial^3\mathbf{U}(\boldsymbol{\xi},\tau,\varepsilon)}{\partial\varepsilon^3}\right)_{\mathbf{x}}&=\mathbf{U}^{\langle 3\rangle}(\mathbf{X},\tau)+\boldsymbol{\chi}^{\langle 2\rangle}(\mathbf{X},\tau)\cdot\nabla_{\mathbf{X}}\mathbf{U}^{\langle 1\rangle}(\mathbf{X},\tau)\\&\quad+\boldsymbol{\chi}^{\langle 1\rangle}\cdot\nabla_{\mathbf{X}}\mathbf{U}^{\langle 2\rangle}(\mathbf{X},\tau)+\tfrac{1}{2}\chi_i^{\langle 1\rangle}\chi_j^{\langle 1\rangle}\frac{\partial^2\mathbf{U}^{\langle 1\rangle}}{\partial X_i\partial X_j}\end{aligned}$$

and

$$\begin{aligned}\frac{\partial\mathbf{U}(\boldsymbol{\xi},\tau,\varepsilon)}{\partial x_i}\cdot\frac{\partial\boldsymbol{\chi}_t(\mathbf{x},\tau,\varepsilon)}{\partial x_j}&=\frac{\partial U_j^{\langle 1\rangle}(\mathbf{X},\tau)}{\partial X_i}\\&\quad+\varepsilon^2\left\{\frac{\partial U_j^{\langle 2\rangle}(\mathbf{X},\tau)}{\partial X_i}+\boldsymbol{\chi}^{\langle 1\rangle}\cdot\nabla_{\mathbf{X}}\left(\frac{\partial U_j^{\langle 1\rangle}}{\partial X_i}\right)+\left[\frac{\partial\boldsymbol{\chi}^{\langle 1\rangle}}{\partial X_i}\cdot\nabla_{\mathbf{X}}U_j^{\langle 1\rangle}+\text{transpose}\right]\right\}\\&\quad+\varepsilon^3\left\{\frac{\partial U_j^{\langle 3\rangle}(\mathbf{X},\tau)}{\partial X_i}+\frac{\partial\boldsymbol{\chi}^{\langle 1\rangle}}{\partial X_j}\cdot\frac{\partial}{\partial X_i}(\boldsymbol{\chi}^{\langle 1\rangle}\cdot\nabla_{\mathbf{X}})\mathbf{U}^{\langle 1\rangle}(\mathbf{X},\tau)+\frac{\partial\mathbf{U}^{\langle 2\rangle}}{\partial X_i}\cdot\frac{\partial\boldsymbol{\chi}^{\langle 1\rangle}}{\partial X_j}\right.\\&\quad+\frac{\partial\mathbf{U}^{\langle 1\rangle}}{\partial X_i}\cdot\frac{\partial\mathbf{U}^{\langle 2\rangle}}{\partial X_j}+\frac{\partial}{\partial X_i}(\boldsymbol{\chi}^{\langle 1\rangle}\cdot\nabla_{\mathbf{X}})\,U_j^{\langle 2\rangle}+\frac{\partial}{\partial X_i}(\boldsymbol{\chi}^{\langle 2\rangle}\cdot\nabla_{\mathbf{X}})\,U_j^{\langle 1\rangle}\\&\quad\left.+\tfrac{1}{2}\frac{\partial}{\partial X_i}\left(\chi_m^{\langle 1\rangle}\chi_l^{\langle 1\rangle}\frac{\partial^2U_j^{\langle 1\rangle}}{\partial X_m\partial X_l}\right)\right\}+O(\varepsilon^4)\,.\end{aligned}$$

Exercise 94.10 (Canonical form of the third order fluid for perturbations of the rest state): To simplify the notation let $\mathbf{A}=\mathbf{A}_1$, $\chi_t=\chi$ and define a second order tensor

$$\mathbf{B}\{\chi,\mathbf{A}\}=\chi\cdot\nabla\mathbf{A}+\mathbf{A}\cdot\nabla\chi+(\mathbf{A}\cdot\nabla\chi)^T,$$

$$B_{ij}\{\chi,\mathbf{A}\}=\chi\cdot\nabla A_{ij}+\frac{\partial\chi_l}{\partial X_i}A_{lj}+\frac{\partial\chi_l}{\partial X_j}A_{li}.$$

When the field $\mathbf{U}(\mathbf{x})$ is independent of the time

$$\mathbf{B}\{\mathbf{U},\mathbf{A}[\mathbf{U}]\}=\mathbf{A}_2[\mathbf{U}],\quad \mathbf{B}\{\mathbf{U},\mathbf{B}\{\mathbf{U},\mathbf{A}[\mathbf{U}]\}\}=\mathbf{A}_3[\mathbf{U}].$$

etc. Show that (94.73) may be written as

$$\begin{aligned}
\mathscr{F}^{\langle 3\rangle}[\mathbf{G}(s,\varepsilon)] &= \varepsilon\int_0^\infty \mathsf{G}(s)\,\mathbf{A}[\mathbf{U}^{\langle 1\rangle}(s)]\,ds\\
&+\varepsilon^2\Big\{\int_0^\infty \mathsf{G}(s)(\mathbf{A}[\mathbf{U}^{\langle 2\rangle}(s)]+\mathbf{B}\{\chi^{\langle 1\rangle},\mathbf{A}[\mathbf{U}^{\langle 1\rangle}(s)]\})\,ds\\
&+\int_0^\infty\int_0^\infty \gamma(s_1,s_2)\,\mathbf{A}[\mathbf{U}^{\langle 1\rangle}(s_1)]\cdot\mathbf{A}[\mathbf{U}^{\langle 1\rangle}(s_2)]\,ds_1\,ds_2\Big\}\\
&+\varepsilon^3\Big\{\int_0^\infty \mathsf{G}(s)(\mathbf{A}[\mathbf{U}^{\langle 3\rangle}(s)]+\mathbf{B}\{\chi^{\langle 1\rangle},\mathbf{A}[\mathbf{U}^{\langle 2\rangle}]\}+\tfrac{1}{2}\mathbf{B}\{\chi^{\langle 1\rangle},\mathbf{B}\{\chi^{\langle 1\rangle},\mathbf{A}[\mathbf{U}^{\langle 1\rangle}]\}\}\\
&+\mathbf{B}\{(\chi^{\langle 2\rangle}-\tfrac{1}{2}\chi^{\langle 1\rangle})\cdot\nabla_{\mathbf{X}}^{\langle 1\rangle}\chi^{\langle 1\rangle},\mathbf{A}[\mathbf{U}^{\langle 1\rangle}]\})\,ds\\
&+\int_0^\infty\int_0^\infty \gamma(s_1,s_2)\{\mathbf{A}[\mathbf{U}^{\langle 1\rangle}(s_1)]\cdot(\mathbf{A}[\mathbf{U}^{\langle 2\rangle}(s_2)+\mathbf{B}\{\chi^{\langle 1\rangle},\mathbf{A}[\mathbf{U}^{\langle 1\rangle}(s_2)]\})\\
&+(\mathbf{A}[\mathbf{U}^{\langle 2\rangle}(s_1)+\mathbf{B}\{\chi^{\langle 1\rangle},\mathbf{A}[\mathbf{U}^{\langle 1\rangle}(s_1)]\})\cdot\mathbf{A}[\mathbf{U}^{\langle 1\rangle}(s_2)]\}\,ds_1\,ds_2\\
&+\int_0^\infty\int_0^\infty 2\sigma(s_1,s_2)\frac{\partial\mathbf{U}^{\langle 1\rangle}(s_1)}{\partial X_j}\cdot\frac{\partial\chi^{\langle 1\rangle}(s_1)}{\partial X_j}\mathbf{A}[\mathbf{U}^{\langle 1\rangle}(s_2)]\,ds_1\,ds_2\\
&+\int_0^\infty\int_0^\infty\int_0^\infty(\psi_1(s_1,s_2,s_3)\,\mathbf{A}(s_1)\cdot\mathbf{A}(s_2)\cdot\mathbf{A}(s_3)\\
&+\psi_4(s_1,s_2,s_3)\,\mathrm{tr}[\mathbf{A}(s_1)\cdot\mathbf{A}(s_2)]\,\mathbf{A}(s_3))\,ds_1\,ds_2\,ds_3\Big\}+O(\varepsilon^4)
\end{aligned}\tag{94.75}$$

where $\mathbf{A}(s)=\mathbf{A}[\mathbf{U}^{\langle 1\rangle}(s)]$ and $\mathbf{U}^{\langle 1\rangle}(s)=\mathbf{U}^{\langle 1\rangle}(\mathbf{X},t-s)$ and all spatial derivatives are taken with respect to $\mathbf{X}$.

Exercise 94.11. Show that

$$\chi^{\langle 1\rangle}=-s\mathbf{U}^{\langle 1\rangle}(\mathbf{X})$$

and

$$\chi^{\langle 2\rangle}=-s\mathbf{U}^{\langle 2\rangle}(\mathbf{X})+\frac{s^2}{2}\mathbf{U}^{\langle 1\rangle}\cdot\nabla_{\mathbf{X}}\mathbf{U}^{\langle 1\rangle}$$

when the motion is steady. Show that (94.75) leads to the CENR fluids of order three when the motion is steady.

Exercise 94.12: Formulate the bifurcation problem for a viscoelastic fluid heated from below when the temperature is prescribed on the bottom and side walls by (94.64) and the temperature at the free surface is $T=T_0$. Show how this problem is different from the one considered in this section and how to modify the perturbation analysis to obtain the solution.

Exercise 94.13: Repeat the instructions of Exercise 94.12 when the free surface is insulated. (See Exercise 55.5.)

§ 95. Stability of Motions of a Viscoelastic Fluid

The most studied problems in the theory of stability of flow of viscoelastic fluids are about the instability of flow between rotating cylinders and the instability of shear flows of the Poiseuille and Couette types. Stability theory for the motion of simple fluids is still in such an early stage of development that a list of references may be more useful than a report of the content of partial and tentative results. Reviews of what is presently known about the theory of stability of motion and a fairly complete list of references can be found in the recent papers of Pearson (1976) and of Petrie and Denn (1976).

There are many interesting and as yet unsettled problems of instability of spiral flow. Perhaps the most important is the *melt-fracture instability*. When molten polymer is extruded through a die there is a limiting flow rate, dependent upon the polymer and the geometry of the die, beyond which a smooth extrudate cannot be obtained. This instability phenomenon is known as melt-fracture; the instability is important because it limits the rates at which plastic products can be processed. The origins of the melt-fracture instability are not yet well understood though many interesting mechanisms have been proposed.

Stability problems for the flow of viscoelastic fluids in which there are free surface are easily visualized. Three examples are (a) the climbing bubble instability (b) the symmetry breaking instability of oscillating bubble and (c) the striping instability.

(a) The Climbing Fluid Instability

It is well known that when a vertical rod of small diameter rotates in a non-Newtonian fluid, the fluid may climb the rod. This climbing, sometimes called the "Weissenberg effect", is an effect of normal stresses; the climbing (see Fig. 95.1) develops in response to the "hoop stress" or extra tension which develops as a result of shearing and is greatest near the rod where the shearing is greatest. Deep in the fluid the fundamental equilibrium balance is between the hoop stress and the radial pressure gradient; the radial pressure gradient cannot be maintained near the constant pressure free surface and the fluid rises near the rod to replace the missing pressure forces in balancing the hoop stresses induced by shearing.

The climbing fluid shown in Figs. 95.1 and 95.2 is steady and stable. At higher rates of rotation this steady configuration loses stability to a time-periodic, axisymmetric motion (Fig. 95.3). At still higher rates of rotation the axisymmetric time-periodic solution loses its stability to a nonaxisymmetric motion which periodically ruptures the free surface.

The climbing instability may be an interesting manifestation of Rayleigh's instability mechanism for rotating flows (see § 38). At low speeds, the distribution of angular momentum is the same as a potential vortex flow. In such a flow the angular momentum is constant and is stable by Rayleigh's criterion. The steady secondary motion is driven by normal stresses. It appears to take form as a single cell which carries the lower angular momentum from the outside of

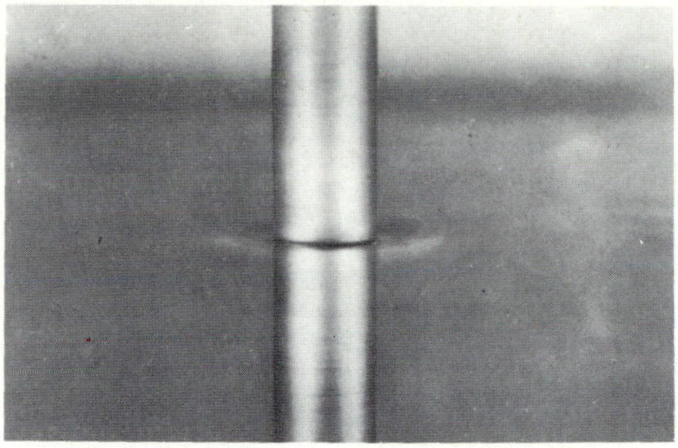

a

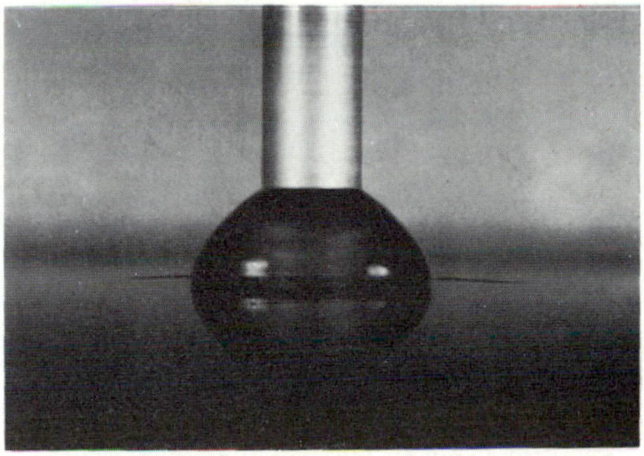

b

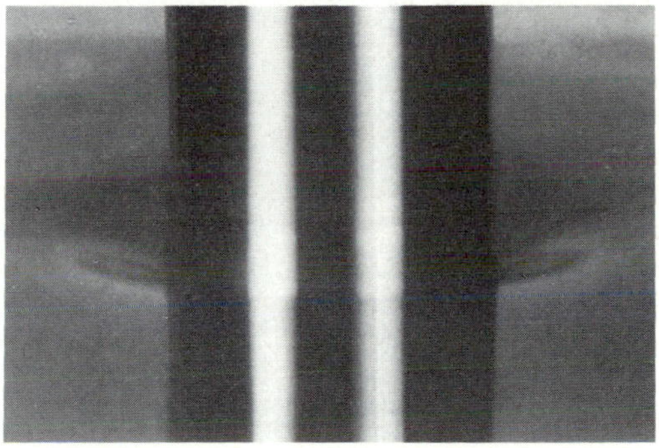

c

Fig. 95.1: The three rods are rotating at ten revolutions per second in a large vat of fluid at room temperature. The radius of the rods is 0.476 cm in (a) and (b) and 1.905 cm in (c). The fluid in (a) is a light Newtonian oil which cannot support normal stresses. Deep in the fluid the pressure balances inertia (centrifugal) forces. The free surface is the barometer of the interior pressure distribution and it sinks near the rod where the pressure is smallest. In (b) and (c) the vat is filled with STP (polyisobutylene in petroleum oil). The free surface is dominated by normal stresses when the rod radius is smaller than some critical radius $r<r_c\simeq 3\alpha_1+2\alpha_2$ and by inertia when $r>r_c$. The fluid will not climb when $r>r_c\simeq 4.6$ cm as is evident from the photographs in frames (b) and (c). G.S. Beavers and D.D. Joseph (1975)

the cell near the free surface toward the rod and the higher angular momentum away from the rod near the bottom of cell. At higher speeds the secondary motion is more intense and it forces an accumulation of low angular momentum fluid in the climbing bubble (see panels a, b, c of Fig. 95.3). The adverse distribution of angular momentum in the bubble is unstable by Rayleigh's criterion; the higher momentum fluid near the rod is pushed outwards by centripetal acceleration. The result is a "bulge" of "thrown outward" fluid (panel c of Fig. 95.3)

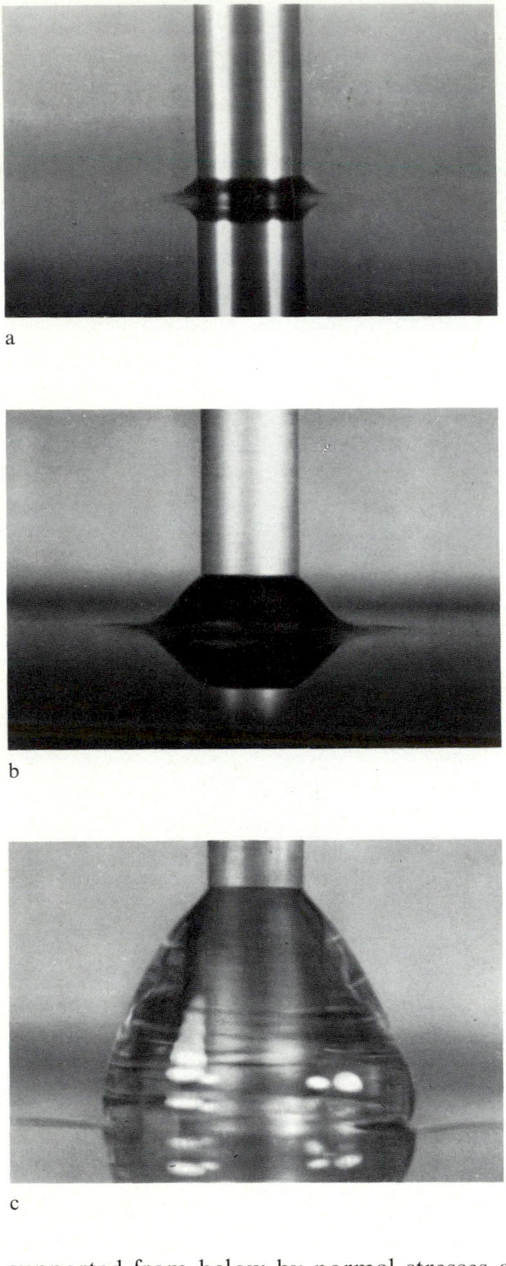

a

b

c

Fig. 95.2: The three rods have the same radius $a=0.476$ cm and rotate with the same speed $\omega=5$ revolutions per second in STP. The temperatures are (a) 10 °C, (b) 25 °C and (c) 50 °C. Large steady climbing configurations eventually lose stability to time-periodic motions (see Fig. 95.3). G. S. Beavers and D. D. Joseph (1975)

supported from below by normal stresses and from the side by the increasingly strong surface tension forces generated by the high curvatures on the "bulge". The bulge buildup proceeds until the accumulated fluid is sufficiently heavy to be dragged down by gravity. The bulge then falls into the body of the fluid dragging most of the bubble with it. The depleted bubble (panel f of Fig. 95.3) is now back at "go" and the same sequence of bubble growth, accumulation of low momentum fluid, and overturning by centripetal acceleration can be ini-

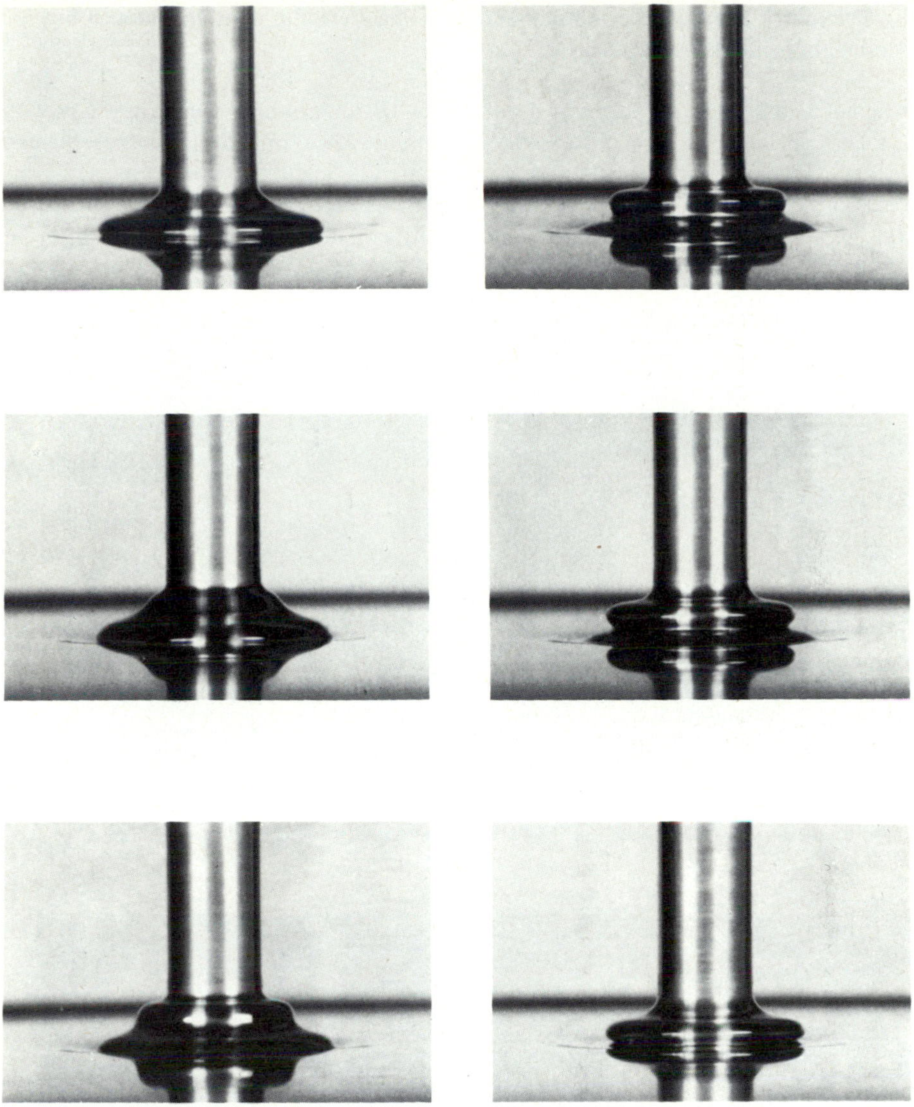

Fig. 95.3: Time-periodic motions of the climbing fluid. Rod radius = 0.635 cm; rotational speed 13.3 rev./sec; frequency of periodic motion = 0.4 cycles/sec. The time-periodic solution arises from instability and bifurcation of the steady configurations (Fig. 95.1). The time-periodic motion first appears as an infinitesimal perturbation of the steady climb. As the angular velocity is increased the amplitude of the oscillation increases and the frequency of the oscillation changes. Beavers and Joseph (1975)

tiated once again. The foregoing explanation is, of course, still tentative and no part of it can be stated categorically until analysis of the secondary motion is complete.

(b) Symmetry Breaking Instabilities of the Time-Periodic Motion Induced by Torsional Oscillations of a Rod

When a small rod of radius a is rotated with an angular frequency equal to $\Omega(a,t)=\varepsilon\sin\omega t$ in a viscoelastic fluid, the fluid will climb the rod (Joseph and Beavers, 1976). The climb is divided into a steady mean part and an oscillating part. The oscillating part of the climb is not visible to the naked eye, however, and the shape of the free surface is essentially constant over an oscillation cycle.

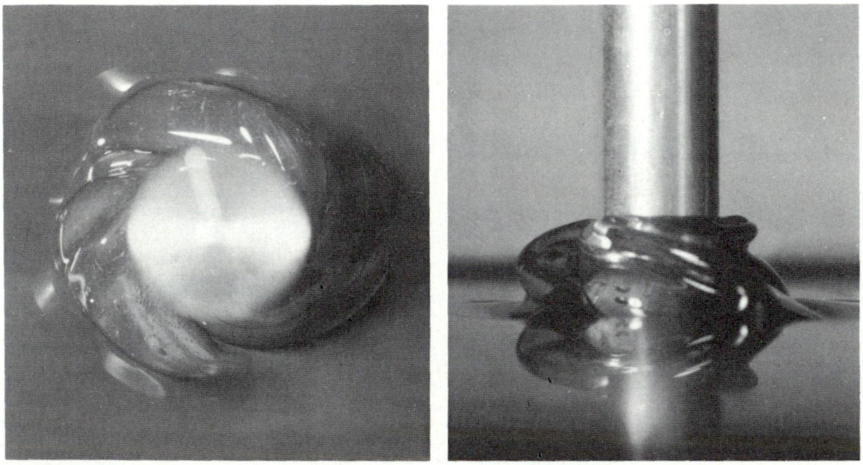

Fig. 95.4: Top view and side view of the three-lobe configuration bifurcating from an axisymmetric time-periodic flow: $\omega=9.2$ cycles/sec., $\Theta=235°$. Beavers and Joseph, 1976

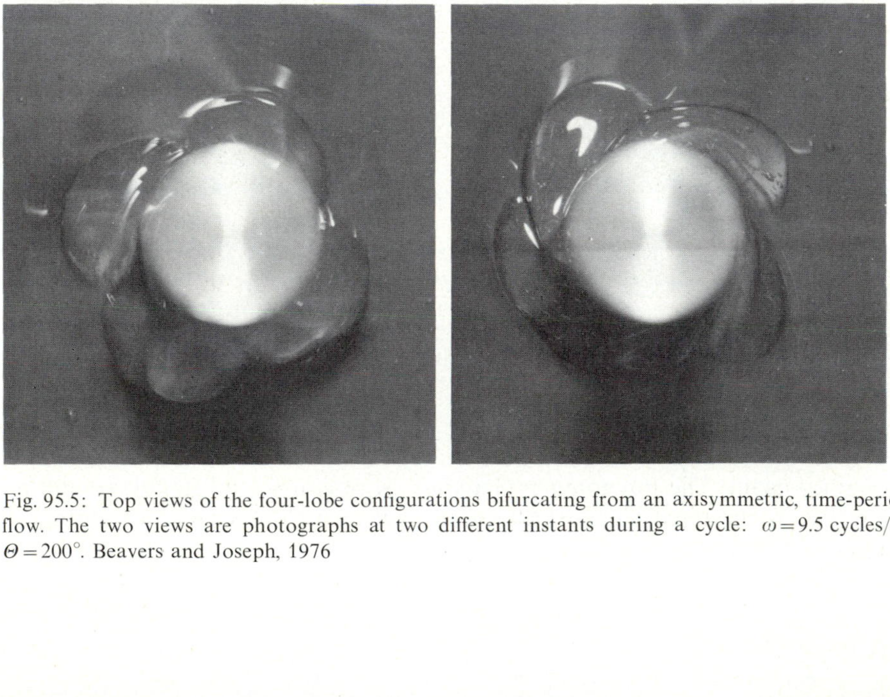

Fig. 95.5: Top views of the four-lobe configurations bifurcating from an axisymmetric, time-periodic flow. The two views are photographs at two different instants during a cycle: $\omega=9.5$ cycles/sec., $\Theta=200°$. Beavers and Joseph, 1976

It is easy to verify that the amplitude ε is given by $\varepsilon = \omega\Theta/2$, where Θ is the *angle of twist*, the maximum angular displacement of the rod. When ε is large enough, the axisymmetric climbing bubble loses its stability to another time-periodic motion with a different symmetry pattern. The new symmetry pattern has a certain integral number of lobes which are determined by operating conditions. In Figs. 95.4 and 95.5 we have displayed photographs of three and four-lobe configurations which bifurcate from axisymmetric time-periodic flow.

(c) The Striping Instability

This is an instability of the surface on liquid jets which are extruded from plane and circular ducts. The striping of the surface of the circular jet was observed by Rodenacker and was reported by Giesekus (1972). The photograph of Fig. 95.4a is taken from the Giesekus paper. We note that the jet shown in Fig. 95.6a swells to several times its former diameter. This magnitude of swelling is never

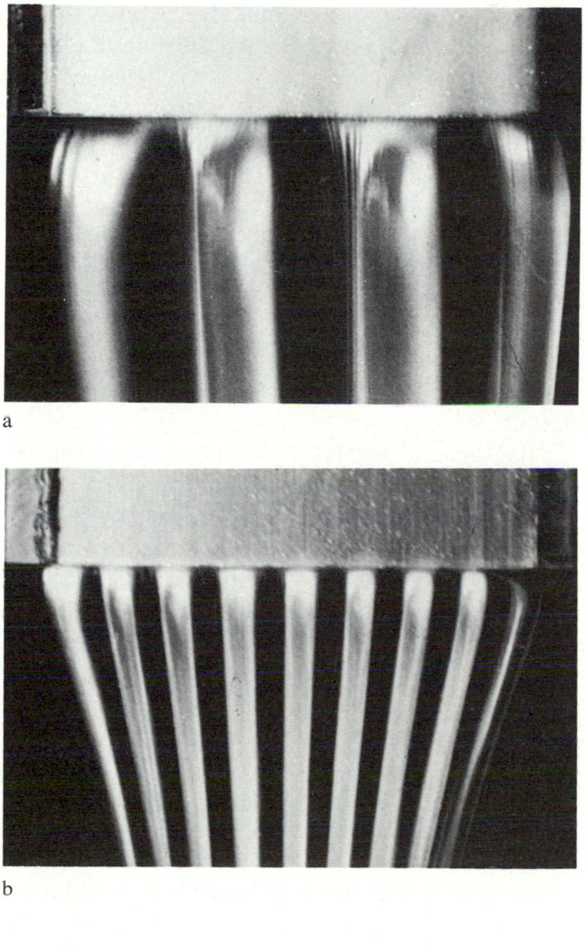

a

b

Fig. 95.6: (a) Jet discharging from a plane channel. Material: 4% solution of polyisobutylene in decalin; inside dimensions: $11 \times 4.2 \times 0.2$ cm; driving pressure difference: 0.75×10^5 Nm^{-2}. The large grooves seem to be generated as a time periodic instability leading to creation of grooves at the center and annihilation at the edges. (b) Jet discharging from a plane channel with wide grooves. Material: 4% solution of polyisobutylene in decalin; inside dimensions: $16.4 \times 3.9 \times 0.1$ cm; driving pressure difference 1.5×10^5 Nm^{-2}. H. Giesekus (1972)

found in Newtonian jets. According to Giesekus the striping "... phenomenon can be still better observed in the discharge from a plane channel ...".

If the rate of discharge (in a plane channel) is increased, at (a wall shear rate) $D_w \approx 2000\ \mathrm{sec}^{-1}$, (wall shear stress) $\tau_w \approx 6000$ dynes cm^{-2}, instantaneously a second kind of groove is generated leading to a pattern of stripes more widely separated (our Fig. 95.6 b), though the narrower grooves do not vanish. These wider stripes are not stationary but travel from the middle to both sides, i.e., they are generated steadily in the middle and annihilated at the edges."

(d) Tall Taylor Cells in Polyacrylamide

Steady axisymmetric vortices in Newtonian liquids are stable for an interval of angular velocities above the first critical value. The "squareness" of the cross-section of the vortices barely changes as the Taylor number is varied across the stable interval; moreover, the hysteresis effects, if present are not strong when the Taylor number does not exceed the second critical value where axisymmetric flow becomes unstable. Above this second critical value, hysteresis effects and non-uniqueness are the rule (Coles, 1965); now the number of cells and the number of waves around a cell depend on the history of the speed changes of the inner cylinder. In Coles' experiments in the Newtonian liquids there are nearly 28 square cells when the angular velocity is slightly above the first critical value; the maximum number of cells achieved is 32 and the minimum number (a rare event), is 18. Also, the cells in Coles' experiments never have height-width ratios greater than 16/9, even when the motion is doubly periodic.

Taylor vortex flow also arises through instability and bifurcation of Couette flow of viscoelastic liquids. Although it appears that the property of near "squareness" of the cross sections of cells holds in the experiments when the angular velocity is near its first critical value, the stable axisymmetric flows which do appear differ greatly from the axisymmetric motions in Newtonian liquids.

Beavers and Joseph (1974) report experiments with a polyacrylamide solution consisting of 54.80 percent by weight glycerine, 43.72 percent water, and 1.48 percent polyacrylamide crystals. This polyacrylamide solution is more pseudoplastic than most other viscoelastic fluids that have been used in the previously reported experiments. With this fluid, some new properties of Taylor vortices have been observed. The most important of these is that at a given rate of shearing the number of steady, axisymmetric Taylor vortices is not unique, and may vary between 14 and 4.

At low values of the rotational speed the instability in the polyacrylamide solution appears in the form of toroidal vortices of almost square cross-section, comparable in size to the cells for Newtonian fluids near the critical Taylor number. In Fig. 95.7 we have compared the initial instability configurations for a Newtonian fluid (the 16 cells in Fig. 95.7(a)) and the polyacrylamide solution (the 14 cells in Fig. 95.7(b). It is evident from these photographs that the structure of the Taylor cells in the polyacrylamide solution is different from the structure

Fig. 95.7: Taylor cells of approximately square cross section near the initial instability point: (a) Newtonian fluid (light oil), $\Omega = 1.2$ rev./sec, $\nu = 0.36\ \text{cm}^2/\text{sec}$, $\Omega/\nu = 3.34\ \text{cm}^{-2}$, 16 cells; (b) polyacrylamide solution, $\Omega = 6$ rev./sec, $\nu =$ approx. $10.8\ \text{cm}^2/\text{sec}$, $\Omega/\nu = 5.5\ \text{cm}^{-2}$, 14 cells (Beavers and Joseph, 1974)

Fig. 95.8: Tall Taylor cells in polyacrylamide solution at a rotational speed $\Omega = 22$ rev./sec. Temp. in the fluid $= 118\,^\circ\text{F}$, $\nu =$ approx. $8\ \text{cm}^2/\text{sec}$. (a) 8 cells; (b) speed increased from (a) to give 4 cells and then returned to 22 rev./sec; 4 cell configuration remains: $\Omega/\nu = 2.75\ \text{cm}^{-2}$ (Beavers and Joseph, 1974)

in the Newtonian fluid. Flow visualization, using tracer particles, is outstandingly better for the flow field which develops in the polyacrylamide solution.

When the rotational speed of the inner cylinder is increased beyond that which gives the configuration of Fig. 95.7(b), very dramatic changes in the cell pattern are observed. The number of cells decreases as the speed increases, going first to 12, then 10, 8, 6, and finally 4. When a stable 4 cell configuration has been reached, it will retain its stability as the speed is then decreased. This hysteresis is strikingly demonstrated in Figs. 95.8(a) and 95.8(b). Fig. 2(a) shows a stable 8 cell configuration at a rotational speed of 22 rev/sec, whereas Fig. 95.8(b) shows a stable 4 cell configuration at the same speed. The Taylor cell structure of Fig. 95.8(b) is obtained from that of Fig. 95.8(a) by first increasing the speed to transfer the stability of the 8 cell configuration to the 4 cell configuration, followed by a decrease in the speed.

The distinct change from a stable 14 cell configuration, with a cell aspect ratio of about 1, to a stable 4 cell configuration, with a cell aspect ratio of 4, seems not to have been observed before. Beavers and Joseph believe that the tall Taylor cells are another manifestation of normal stress effects in viscoelastic fluids.

The difference in behavior and cell character between non-Newtonian and Newtonian fluids can be further emphasized by comparing the tall cells in the polyacrylamide solution, Fig. 95.8(b) with the 16 cells in the Newtonian oil, Fig. 95.7(a). Since the Taylor numbers for the two flows are approximately the same, the cause of the difference observed must be sought elsewhere.

Exercise (added in proof) 95.1 (Slemrod, 1976):

Show that all solutions of the linearized problem (94.53) satisfy the following "energy" identity

$$\frac{dI}{dt} = -\langle \textstyle\int_0^\infty G''(s)[\nabla\delta\mathbf{u} : \nabla\delta\mathbf{u}]\,ds\rangle$$

where

$$\delta\mathbf{u} = \mathbf{u}(\mathbf{x}, t-s) - u(\mathbf{x}, t),$$

$$I = \left\langle \left| \frac{\partial \mathbf{u}(\mathbf{x}, t)}{\partial t} \right|^2 \right\rangle - \langle \textstyle\int_0^\infty G'(s)[\nabla\delta\mathbf{u} : \nabla\delta\mathbf{u}]\,ds\rangle,$$

$$G(s) \in C^2[0, \infty],$$

$$G(s) \to 0 \quad \text{as} \quad S \to \infty$$

and the angle brackets denote volume averaged integrals. Slemrod shows that $\mathbf{u}(\mathbf{x}, t) \equiv 0$ is asymptotically stable if $G(s)$ and $-G'(s)$ are strictly positive

$$G''(s) \geq 0 \quad \text{and} \quad -\textstyle\int_0^\infty s^2 G'(s)\,ds < \infty\,.$$

Chapter XIV

Interfacial Stability

We turn now to problems involving the stability of interfaces separating fluid bodies. The mechanics of such problems, the laws which govern the motion of the fluid where the interface meets a solid boundary, are only partially understood. Some of the difficulties at the foundation are more easily understood from the study of the equation which governs the energy of the two fluids (Dussan V., 1975). We shall use the energy equation to obtain a dynamic analysis of the stability of equilibrium configurations of surfaces separating two fluids at rest. We compare this dynamic analysis with a static analysis involving the second variation of the free energy. Following Dussan V. (1975) we shall show that the configuration which the static method judges stable when disturbances are small are unstable when the disturbances are large.

§ 96. The Mechanical Energy Equation for the Two Fluid System

A closed container is filled with two immiscible incompressible fluids each of which fills a connected region of space separated by a common interface. The geometry of this configuration is sketched in Fig. 96.1. The volumes, surfaces and lines which will be used in the analysis are all defined in Fig. 96.1.

The equations governing the motion in the two fluids are

$$\rho_i \frac{d}{dt}\mathbf{u}_i = \nabla\cdot\mathbf{T}_i - \rho_i\mathbf{g}, \qquad \mathbf{x}\in\mathscr{V}_i, \qquad i=1,2 \quad \text{(no summation)}. \tag{96.1}$$

The subscripts are associated with properties of the fluids in $\mathscr{V}_1$ and $\mathscr{V}_2$, ρ_i is the density, $\mathbf{T}_i = -p_i\mathbf{1} + \mathbf{S}_i$ is the stress tensor and $\mathbf{S}_i$ is the extra stress or stress deviator. Take the scalar product of (96.1) with $\mathbf{u}_i$, integrate over $\mathscr{V}_i$ and use Reynolds' transport theorem and the divergence theorem to obtain the energy identities $(i=1,2)$

$$\frac{d}{dt}\int_{\mathscr{V}_i}\frac{\rho u^2}{2} = \int_{\partial\mathscr{V}_i}\mathbf{u}\cdot\mathbf{T}\cdot\mathbf{n}_i - \int_{\mathscr{V}_i}\operatorname{tr}\mathbf{T}\mathbf{D}[\mathbf{u}] - \int_{\mathscr{V}_i}\rho\mathbf{g}\cdot\mathbf{u} \tag{96.2}$$

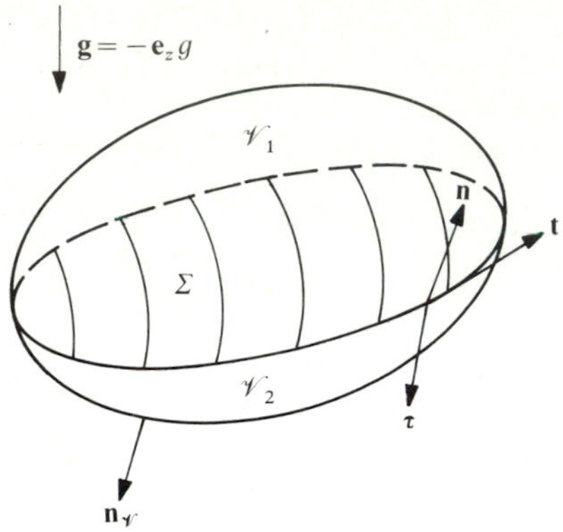

Fig. 96.1: Volumes, surfaces and distinguished directions for two fluids separated by an interface

$\mathscr{V}_1$ and $\mathscr{V}_2$ are the volumes occupied by the two fluids, $\mathscr{V}=\mathscr{V}_1\cup\mathscr{V}_2$,
$\mathbf{n}_i$ is the outward normal to $\partial\mathscr{V}_i$ $(i=1,2)$,
$\Sigma=\partial\mathscr{V}_1\cap\partial\mathscr{V}_2$ is the interface separating the two fluids,
$\mathbf{n}$ is the normal to Σ pointing from $\mathscr{V}_2$ to $\mathscr{V}_1$,
$\partial\mathscr{V}=\partial\mathscr{V}_1\cup\partial\mathscr{V}_2-\Sigma$ is the exterior boundary; parts of this boundary may be made of rigid solids,
$\mathbf{n}_{\mathscr{V}}$ is the outward normal on $\partial\mathscr{V}$,
$\partial\Sigma=\Sigma\cap\partial\mathscr{V}$ is the contact line; l is an arc length on this line,
$\mathbf{t}$ is the tangent vector on $\partial\Sigma$,
$\boldsymbol{\tau}=\mathbf{t}\wedge\mathbf{n}$ is the normal to $\partial\Sigma$ on Σ,
$\mathbf{u}_i$ is the velocity of the fluid in $\mathscr{V}_i$ $(i=1,2)$,
$\mathbf{U}$ is the velocity of a point of the contact line.

where the subscripts on the variables behind the integrals have been suppressed but are understood. Eqs. (96.2) may be written as

$$\frac{d}{dt}[\mathscr{E}+\mathscr{P}]=\int_{\partial\mathscr{V}_1}\mathbf{u}\cdot\mathbf{T}\cdot\mathbf{n}_1+\int_{\partial\mathscr{V}_2}\mathbf{u}\cdot\mathbf{T}\cdot\mathbf{n}_2-\int_{\mathscr{V}}\operatorname{tr}\mathbf{T}\mathbf{D}[\mathbf{u}] \tag{96.3}$$

where

$$\mathscr{V}=\mathscr{V}_1\cup\mathscr{V}_2,$$

$$\mathscr{E}=\int_{\mathscr{V}}\rho\frac{|\mathbf{u}^2|}{2}$$

and

$$\mathscr{P}=\int_{\mathscr{V}}\rho g z\,.$$

The potential $\mathscr{P}$ arises from the relation

$$\int_{\mathscr{V}}\rho\mathbf{g}\cdot\mathbf{u}=\int_{\mathscr{V}}\rho g\frac{dz}{dt}=\frac{d}{dt}\int_{\mathscr{V}}\rho g z\,.$$

One further transformation of Eq. (96.3) is useful in helping one to understand the dynamics of interfaces:

$$\int_{\partial\mathscr{V}_1}\mathbf{u}\cdot\mathbf{T}\cdot\mathbf{n}_1+\int_{\partial\mathscr{V}_2}\mathbf{u}\cdot\mathbf{T}\cdot\mathbf{n}_2=\int_{\partial\mathscr{V}}\mathbf{u}\cdot\mathbf{T}\cdot\mathbf{n}+\int_{\Sigma}[\mathbf{u}_2\cdot\mathbf{T}_2-\mathbf{u}_1\cdot\mathbf{T}_1]\cdot\mathbf{n}\,. \tag{96.4}$$

The last term of (96.4) may be written as

$$\int_{\Sigma}\{\mathbf{u}_2\cdot(\mathbf{T}_2-\mathbf{T}_1)\cdot\mathbf{n}+(\mathbf{u}_2-\mathbf{u}_1)\cdot\mathbf{T}_1\cdot\mathbf{n}\}\,. \tag{96.5}$$

Combining (96.3), (96.4) and (96.5) we find that

$$\frac{d}{dt}[\mathscr{E}+\mathscr{P}]=\int_{\Sigma}[\mathbf{u}_2\cdot(\mathbf{T}_2-\mathbf{T}_1)\cdot\mathbf{n}+(\mathbf{u}_2-\mathbf{u}_1)\cdot\mathbf{T}_1\cdot\mathbf{n}]+\int_{\partial\mathscr{V}}\mathbf{u}\cdot\mathbf{T}\cdot\mathbf{n}-\int_{\mathscr{V}}\operatorname{tr}\mathbf{T}\mathbf{D}[\mathbf{u}]\,. \tag{96.6}$$

The relation

$$(\mathbf{u}_2-\mathbf{u}_1)\cdot\mathbf{T}_1\cdot\mathbf{n}=0 \quad \text{on} \quad \Sigma \tag{96.7}$$

holds in two situations:

(a) $\mathbf{u}_2-\mathbf{u}_1=0$ (no-slip)

and

(b) $\tilde{\mathbf{t}}\cdot\mathbf{T}_1\cdot\mathbf{n}=0$ (no tangential tractions on side 1).

In continuous motions $(\mathbf{u}_2-\mathbf{u}_1)\cdot\mathbf{n}=0$ on Σ; hence $\mathbf{u}_2-\mathbf{u}_1=\alpha\tilde{\mathbf{t}}$ where $\tilde{\mathbf{t}}$ lies in Σ and if (b), then (96.7).

Further reduction of the terms on Σ which produce energy requires constitutive assumptions. We are going to confine our attention to the classical theory of surface tension which is embodied in Eq. (55.10). This may be written as

$$(\mathbf{T}_2-\mathbf{T}_1)\cdot\mathbf{n}=\nabla_{\mathrm{II}}\sigma+2H\sigma\mathbf{n} \tag{96.8}$$

where we have set $\mathbf{T}_e=\mathbf{T}_1$, $\mathbf{T}=\mathbf{T}_2$, $2H$ is the sum of the principal curvatures and σ is the surface tension. Combining Eqs. (96.6), (96.7) and (96.8), we find that

$$\frac{d}{dt}[\mathscr{E}+\mathscr{P}]=\int_{\Sigma(t)}\{\mathbf{u}_2\cdot\nabla_{\mathrm{II}}\sigma+2H\sigma\mathbf{u}_2\cdot\mathbf{n}\}+\int_{\partial\mathscr{V}}\mathbf{u}\cdot\mathbf{T}\cdot\mathbf{n}-\int_{\mathscr{V}}\operatorname{tr}\mathbf{T}\mathbf{D}[\mathbf{u}]\,. \tag{96.9}$$

For incompressible Newtonian fluids, $\mathbf{T}=-p\mathbf{1}+2\mu\mathbf{D}[u]$, $\operatorname{tr}\mathbf{D}[\mathbf{u}]=0$ and the last two terms of (96.9) may be written as

$$-\int_{\partial\mathscr{V}}p(\mathbf{u}\cdot\mathbf{n})+2\mu\int_{\partial\mathscr{V}}\mathbf{u}\cdot\mathbf{D}[\mathbf{u}]\cdot\mathbf{n}-2\mu\int_{\mathscr{V}}\mathbf{D}[\mathbf{u}]:\mathbf{D}[\mathbf{u}]\,. \tag{96.10}$$

The next reduction of the energy equation (96.9) makes use of the identity (55.15) (where, in (55.15) $\hat{S}(t)=\Sigma(t)$, is the fluid interface, $\mathrm{V}=\mathbf{U}$ is the velocity of the interface, $\hat{\mathrm{V}}=\mathbf{u}_2$ is the material velocity and $\sigma=\hat{\phi}$ is the surface tension)

$$\frac{d}{dt}\int_{\Sigma}\sigma\,d\Sigma=\int_{\Sigma}\left[\frac{\partial\sigma}{\partial t}-\mathbf{u}_2\cdot\nabla_{\mathrm{II}}\sigma-2\sigma H\mathbf{u}_2\cdot\mathbf{n}\right]d\Sigma+\oint_{\partial\Sigma}\sigma\boldsymbol{\tau}\cdot\mathbf{U}\,dl\,. \tag{96.11}$$

In the above equation we have set $\mathbf{U}\cdot\mathbf{n}=\mathbf{u}_2\cdot\mathbf{n}$ on Σ; this equation expresses the fact that fluid particles do not cross Σ. Elimination of the mean curvature between (96.9) and (96.11) now gives

$$\frac{d}{dt}\left[\mathscr{E}+\mathscr{P}+\int_\Sigma \sigma\, d\Sigma\right]=\int_\Sigma \frac{D\sigma}{Dt}\, d\Sigma+\oint_{\partial\Sigma}\sigma\boldsymbol{\tau}\cdot\mathbf{U}\,dl+\int_{\partial\mathscr{V}}\mathbf{u}\cdot\mathbf{T}\cdot\mathbf{n}-\int_{\mathscr{V}}\operatorname{tr}\mathbf{TD}[\mathbf{u}] \tag{96.12}$$

where

$$\frac{D\sigma}{Dt}=\frac{\partial\sigma}{\partial t}(\eta,\mu,t) \tag{96.13}$$

is a material derivative and η and μ are material coordinates on Σ (see Exercise 55.3).

An alternative form of (96.11) which follows from the identity (55.16) is of interest:

$$\frac{d}{dt}\int_{\Sigma(t)}\sigma\, d\Sigma=\frac{d}{dt}\int_{S_2(t)}\sigma\, dS_2+\oint_{\partial\Sigma}\sigma\boldsymbol{\tau}\cdot(\mathbf{U}-\mathbf{u}_2)\,dl \tag{96.14}$$

where $S_2(t)$ is the material surface which instaneously coincides with $\Sigma(t)$. Eq. (96.14) shows that a fluid interface for which $\mathbf{U}\neq\mathbf{u}_2$ cannot be a material interface.

At this point the reader may wonder how the velocity on the contact line can be different than the fluid velocity. The answer is that a viscous fluid does not slip relative to a bounding wall but it need not adhere to the wall. On the other hand the contact line, by definition, always "adheres" to the bounding solid but it may slip. It is possible to tear a fluid from a solid in much the same way as one can pull an adhesive tape from a solid surface. It is clear that the normal component of velocity of the adhesive tape is discontinuous, zero where the tape adheres and not zero elsewhere. It is the same for a fluid. Were it not the same, it would never be possible for one fluid to displace another on a solid surface (see Dussan V. and Davis (1974) for a thorough discussion of this type of situation).

Dussan's equation (96.12) governs the evolution of the energy sum $\mathscr{E}+\mathscr{P}+\int_\Sigma\sigma\, d\Sigma$ where $\mathscr{E}$ is the kinetic energy and $\mathscr{P}+\int_\Sigma\sigma\, d\Sigma$ is a free energy.

The last integral in (96.12) is dissipative when the stress power $\operatorname{tr}\mathbf{TD}[\mathbf{u}]$ is positive. The motion may be driven by variations in surface tension through the term $D\sigma/Dt$. The integral

$$\int_{\partial\mathscr{V}}\mathbf{u}\cdot\mathbf{T}\cdot\mathbf{n}$$

gives the power of the working of the traction vector $\mathbf{T}\cdot\mathbf{n}$ on $\partial\mathscr{V}$; this term is not zero when the motion is forced from the boundary.

The integral

$$\oint_{\partial\Sigma}\sigma\boldsymbol{\tau}\cdot\mathbf{U}\,dl\,, \tag{96.15}$$

which gives the power associated with the work of the traction vector $\sigma\boldsymbol{\tau}$ in moving the contact line, is of particular interest. The physical laws which govern the motion of this line are not understood. On some solids, in certain situations, the contact line is stubborn and remains fixed even when there are substantial unsteady motions of the free surface and bulk fluid (see Fig. 95.3 for a good example). This resistance of the contact line to motion is just what one would expect of a viscous fluid following an adherence condition $\mathbf{U}=0$. But resistance to motion is, speaking generally, an unacceptable consequence of the adherence condition, since it contradicts experience with liquid drops running off solid surfaces and other cases in which one fluid replaces another on a solid surface. On the other hand, it is clear that a no-slip condition does not imply a stationary line of contact.

We may resolve $\boldsymbol{\tau}$ into components

$$\tau=\sin\theta\,\mathbf{n}_{\mathscr{V}}-\cos\theta\,\mathbf{n}_w \tag{96.16}$$

where $\mathbf{n}_w=\mathbf{n}_{\mathscr{V}}\wedge\mathbf{t}$, $\mathbf{n}_{\mathscr{V}}\cdot\mathbf{n}=-\cos\theta$ and θ is the contact angle. The work of the contact line is then given by

$$\oint_{\partial\Sigma}\sigma\left[\sin\theta(\mathbf{n}_{\mathscr{V}}\cdot\mathbf{U})-\cos\theta(\mathbf{n}_w\cdot\mathbf{U})\right]dl\,. \tag{96.17}$$

The contact line does no work when the wall is stationary and the fluid adheres ($\mathbf{U}=0$), when the contact line moves parallel to itself ($\mathbf{n}_{\mathscr{V}}\cdot\mathbf{U}=\mathbf{n}_w\cdot\mathbf{U}=0$) or when the contact is perpendicular to the wall ($\cos\theta=0$) and the wall moves parallel to itself ($\mathbf{n}_{\mathscr{V}}\cdot\mathbf{U}=0$). In static problems it is usual to consider problems in which the position of the contact line is prescribed (Dirichlet problems) or problems in which the contact angle is fixed (Neumann problems). The prescribed contact line condition makes sense as an adherence condition ($\mathbf{U}|_{\partial\Sigma}=0$) in the dynamic problem. The dynamic significance of the prescribed contact is unclear. In the static case the contact angles are usually assumed to be given by the constitutive equation of Young and Dupré. This elementary constitutive relation does not seem to explain the experimental observations in all cases. A proper constitutive theory relating the traction $\sigma\boldsymbol{\tau}$ and the velocity $\mathbf{U}$ to the material at the point of triple contact in the dynamic case is not yet known. It is sometimes useful to think in terms of a yield condition associated with critical angles θ_c. The contact line remains fixed whenever $\theta_{c1}<\theta<\theta_{c2}$. When θ is outside this interval, the fluid may lift off or deposit on the solid boundary. In this example of a constitutive assumption the critical angles and the value $\mathbf{U}$ would, of course, depend on the materials at the point of triple contact.

The conditions which replace adherence at a moving contact line are not known. A good theory of stability of fluid surfaces is not possible without a better understanding of the laws which govern the motion of the contact line.

Exercise 96.1: Consider the form of the energy equation (96.12) for a pendent drop moving on a wall $z=0$. Assume that the liquid on the wall does not slip but it may lift off or deposit fluid on Σ at the contact line $\partial\Sigma$ and also assume that the air-liquid interface is a shear-free surface so that

$$\int_{\partial\mathscr{V}}\mathbf{u}\cdot\mathbf{T}\cdot\mathbf{n}=0\,.$$

Suppose further that σ is a constant and that the equation of the surface Σ is

$$f = z - Z(x,y,t) = 0\,. \tag{96.18}$$

On the plane $z=0$, $x = X(l,t)$ and $y = Y(l,t)$ are given parametrically. Show that

$$d\Sigma = \mathbf{n}\cdot(\mathbf{x},_x \wedge \mathbf{x},_y)\,dx\,dy = dx\,dy/(\mathbf{e}_z\cdot\nabla f/|\nabla f|) = [1 + Z_{,x}^2 + Z_{,y}^2]^{1/2}\,dx\,dy\,,$$

$$\mathscr{P} = -\frac{\rho g}{2}\iint_A Z^2(x,y)\,dx\,dy$$

where A is the projection onto $z=0$ of the surface Σ_1,

$$\frac{d[\mathscr{E}+J]}{dt} = -\sigma\int_{\partial\Sigma}\cos\theta\left[\cos\phi\frac{dX}{dt} + \sin\phi\frac{dY}{dt}\right]dl - \int Tr\,\mathbf{T}\cdot\mathbf{D}[\mathbf{u}] \tag{96.19}$$

where $\mathbf{e}_x\cdot\mathbf{n}_w = \cos\phi$, $X(l(t)t)$, $Y(l(t),t)$ are parametric representations of the contact line, $dl = \sqrt{dX^2 + dY^2}$ and $J = \mathscr{P} + \sigma\int_\Sigma d\Sigma$.

Exercise 96.2: Consider the problem posed in Exercise 96.1 when the liquid drop is axisymmetric and the free surface is given by $r = R(z,t)$, $0 \geqslant z \geqslant -H$ where $0 = R(-H,t)$. Show that

$$\frac{d[\mathscr{E}+J]}{dt} = -2\pi\sigma\cos\theta\,R(0,t)\,R,_t(0,t) - \int \mathrm{tr}\,\mathbf{T}\cdot\mathbf{D}[\mathbf{u}] \tag{96.20}$$

and

$$\frac{d}{dt}\int_0^{-H} R^2(z,t)\,dz = 0$$

where

$$J = \pi\rho g\int_0^{-H} zR^2(z,t)\,dz + 2\pi\sigma\int_0^{-H} R(1+R_{,z}^2)^{1/2}\,dz\,. \tag{96.21}$$

Suppose that θ is prescribed and that the stress-power is positive. Show that $\mathscr{E} + J_1$ is decreasing where

$$J_1 = J + \pi\sigma\cos\theta\,R^2(0,t)\,. \tag{96.22}$$

Equation 96.3: Let $\{\xi_\alpha : \alpha = 1,2$ be surface coordinates on $\partial\mathscr{V}$. The position vector $\mathbf{x}(\xi_1(l,t),\ \xi_2(l,t))$ gives the position of the contact line an terms of a parametric representation with arc length parameter l. Show that

$$\begin{aligned}\oint_{\partial\Sigma}\sigma\boldsymbol{\tau}\cdot\mathbf{U}\,dl = &-\frac{d}{dt}\oint_{\partial\Sigma}\sigma\cos\theta(\mathbf{n}_\gamma\wedge\mathbf{t})\,\mathbf{x},_\alpha\xi_\alpha\,dl\\ &+\oint_{\partial\Sigma}\xi_\alpha\frac{\partial}{\partial t}[\sigma\cos\theta(\mathbf{n}_\gamma\wedge\mathbf{t})\cdot\mathbf{x},_\alpha]\,dl\\ &+[\sigma\sigma\cos\theta(\mathbf{n}_\gamma\wedge\mathbf{t})\cdot\mathbf{x},_\alpha\xi_\alpha]_{l=L}\frac{dL}{dt}\end{aligned}$$

where $\partial/\partial t$ in the second integral on right is taken holding l fixed, $\xi_\alpha(0,t) = \xi_\alpha(L(t),t)$, $L(t)$ gives the total length of the contact line and $\mathbf{U} = \mathbf{x},\alpha\dfrac{\partial\xi_\alpha}{\partial t}$.

§ 97. Stability of the Interface between Motionless Fluids When the Contact Line is Fixed

We shall assume that the fluid (which need not be Newtonian) adheres to the walls $\partial\mathscr{V}$ and that these walls are stationary so that $\mathbf{u}|_{\partial\mathscr{V}}=0$, that the contact line does not move, $\mathbf{U}=0$; that surface tension is constant so that $D\sigma/Dt=0$, and that the stress power is positive; $\operatorname{tr}\mathbf{TD}[\mathbf{u}]\geqslant 0$ with equality only when $\mathbf{u}\equiv 0$. Then (96.12) becomes

$$\frac{d}{dt}(\mathscr{E}+J)=-\int_{\mathscr{V}}\operatorname{tr}\mathbf{T}\cdot\mathbf{D}[\mathbf{u}], \qquad J=\mathscr{P}+\int_{\Sigma}\sigma\, d\Sigma\,. \tag{97.1}$$

The first result to be proved from (97.1) is a *theorem about the decrease of* J:

Suppose the free surface suffers a displacement without velocity. Then either $J(t)=J(0)$ *and* $\mathscr{E}(t)=0$ *when* $t>0$, *or* $J(t)$ *decreases at early times. Moreover,* J *must decrease at any instant for which* $\mathscr{E}(t)\neq 0$ *is stationary* $\left(\dfrac{d\mathscr{E}}{dt}=0\right)$.

Proof: If $\mathbf{u}\equiv 0$ at $t=0$, then $\mathscr{E}(t)=0$ or $\mathscr{E}(t)>0$ for $t>0$. If $\mathscr{E}(t)=0$ for $t>0$ then $d\mathscr{E}/dt=0$ and (97.1) shows that J decreases. If $\mathscr{E}(t)>0$ for some t, it must first increase. Then $d\mathscr{E}/dt>0$ and $dJ/dt<0$, proving the first assertion. If $\mathscr{E}(t)$ is stationary, then $d\mathscr{E}/dt=0$ and

$$\frac{dJ}{dt}=-\int_{\mathscr{V}}\operatorname{tr}\mathbf{TD}[\mathbf{u}]<0\,, \tag{97.2}$$

proving the second assertion.

Since there are no energy sources to drive a permanent motion, the kinetic energy of a disturbance should decay. The following theorem about the integrability of $\mathscr{E}(t)$ specifies this decay:
(Dussan V., 1975) *Suppose that* J *is bounded from below and that*

$$\int_{\mathscr{V}}\operatorname{tr}\mathbf{TD}[\mathbf{u}]\geqslant \xi^2\mathscr{E} \tag{97.3}$$

where $\xi^2>0$. *Then* $\mathscr{E}(t)$ *is integrable, that is*

$$\lim_{t\to\infty}\frac{1}{T}\int_t^{t+T}\mathscr{E}(\tau)\,d\tau=0\,. \tag{97.4}$$

We note that J is necessarily bounded from below in every closed container of finite size; the proof of this statement and of the theorem are set as Exercise 97.1.

The decay of the energy of a disturbance does not here imply stability. The decay of $\mathscr{E}(t)$ implies that the rest state is the terminal form of every disturbance. But the rest state need not be unique. The interface separating motionless fluids may have different shapes.

It is usual to follow one of two approaches when investigating the stability of systems containing a free surface. The usual method is to consider the mathematical problem generated by disturbing some basic state. This method, correctly developed, has the best potential for obtaining the stability properties of the basic state. The limitations of this method are associated with the mathematical difficulties arising in the study of very non-linear problems in regions in which the shape of the boundary must also be determined. A second approach, which is sometimes used to study the stability of static states, gives the stable configuration as the one which minimizes a certain free energy (Gibbs; 1961, p. 276). The identification of this energy follows from static considerations involving virtual work. (A convenient reference is Landau and Lifshitz, 1959, p. 230.) These considerations lead to the statement (but no proof) that in a *stable* equilibrium J is a minimum among a suitable class of variational competitors. As we have seen, a partial justification of this static criterion follows from Dussan's energy equation. In the next section we shall work an explicit problem using the static method.

Exercise 97.1 (Dussan V., 1975): Show that $\mathscr{P}$ is bounded from above and below when the two fluids are confined to any container which can be contained between two parallel planes which are perpendicular to gravity.

Show that (97.3) holds when the fluid is Newtonian and $\mathscr{V}$ is a region which can be contained, say, in a sphere of finite radius.

Prove that $\mathscr{E}(t)$ is integrable. (*Hint:* the proof follows along lines laid out in § 69.)

Exercise 97.2: Consider a collection of bubbles of one fluid rising or settling, without collisions, in a second fluid. Suppose that the fluid extends to infinity or is bounded by rigid stationary walls and that the surface tension on each bubble is constant. Show that the total energy

$$\mathscr{E}+\mathscr{P}+\int \sigma d\Sigma$$

of the fluid and bubbles is nonincreasing.

§ 98. Stability of a Column of Liquid Resting on a Column of Air in a Vertical Tube—Static Analysis

A vertical capillary tube of small diameter will support a stable equilibrium configuration in which liquid is over air. In tubes of large diameter the liquid is always on the bottom. The top-heavy configuration can be stable because the column of water is held up by surface tension in the liquid film. The stability properties of the liquid over air configuration may be verified in several simple experiments. The experimental apparatus for the first experiment consists of two tubes whose diameters are larger and smaller than the critical diameter. The length of the tubes is not important. It is necessary to stop the ends of the tubes and it is easiest to use tubes which are short enough to be closed by the thumb and forefinger of one hand. A tube is partially filled with liquid and turned into a vertical position with the liquid on the top. In the small tube the liquid will stay on the top; in the large tube the liquid will fall to the bottom. In a second experiment the tubes are closed on one end and the corner on the interior diameter of the open end is made sharp. The tubes are completely filled with liquid and turned so that the open end points down. Again, the liquid in the large tube runs out but the liquid in the small tube does not.

We shall consider the static analysis which corresponds to the second experiment. We shall imagine the tube is entirely filled with a volume of liquid exactly equal to the volume of the tube. This implies that a flat free surface is possible. Moreover, free surfaces have a strong affinity for sharp edges so that the requirement that the contact line stay fixed at the edge is physically realistic.

To simplify the analysis we shall study a plane two-dimensional problem where the liquid is held in a channel which is closed on the top (see Fig. 98.1). Quantities are assumed not to vary in the third direction Y, perpendicular to the (X,Z) plane.

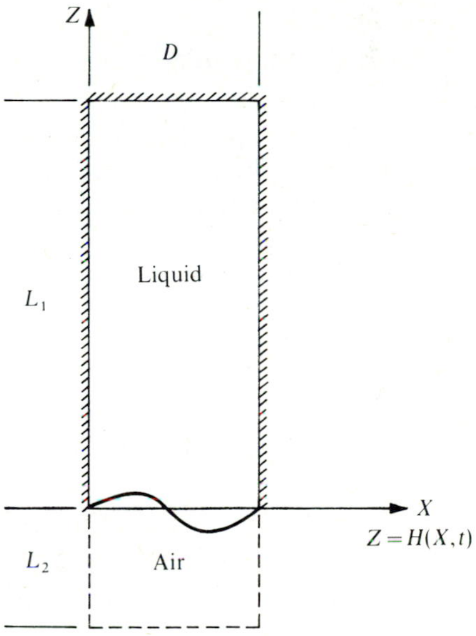

Fig. 98.1: The liquid completely fills the channel; that is, the volume per unit length of liquid is $L_1 D$. The volume of the liquid is a constant and the fluid sticks to the wall and to the sharp edge

We shall restrict our attention to single-valued free surfaces which may be represented by an equation of the form

$$Z=H(X,t).$$

In the static analysis we shall write $Z=H(X)$ and a prime, as in H', always denotes differentiation with respect to X. It is also assumed that the liquid adheres to the channel walls, that it satisfies (97.1), that the surface tension σ is constant and that the stress-power is positive.

The potential energy of the liquid over air configuration in the channel and the control surface shown in Fig. 98.1 is

$$\mathscr{P} = \int_0^D dX\left[\int_{L_2}^H \rho_a g Z\, dZ + \int_H^{L_1} \rho g Z\, dZ\right] = \tfrac{1}{2}(\rho_a - \rho) g \int_0^D H^2(X,t)\, dX + \mathscr{P}_0$$

where ρ is the density of the liquid and $\mathscr{P}_0$ is the potential when $H=0$. In the two-dimensional case being considered here the surface area per unit length is given by an integral of the arc length. It follows that

$$\int_\Sigma \sigma\, d\Sigma = \sigma \int_0^D (1 + H'^2)^{1/2}\, dX\,.$$

Dussan's energy equation (97.1) may be written as

$$\frac{d(\mathscr{E} + J - J_0)}{dt} = -\int_{\mathscr{V}} \operatorname{tr} \mathbf{T}\mathbf{D}[\mathbf{u}] \tag{98.1}$$

where

$$J - J_0 = \sigma\left[\int_0^D (1 + H'^2)^{1/2}\, dX - D\right] + \tfrac{1}{2}(\rho_a - \rho) g \int_0^D H^2\, dX\,. \tag{98.2}$$

We shall first adopt the conventional procedures of thermostatics and assume, consistent with the dynamic results of the last section, that stable equilibrium configurations minimize the free energy J. In the next section we return to the dynamic problem.

The analysis starts from a dimensionless form of the functional (98.2)

$$J[h;G] = \langle (1 + h'^2)^{1/2} \rangle - 1 - \frac{G}{2}\langle h^2 \rangle \tag{98.3}$$

and a linear space Γ of admissible competitors

$$\Gamma = \{h : h(0) = h(1) = \int_0^1 h\, dx = 0,\ h \in C^1[0,1]\}$$

where

$$\langle \circ \rangle = \int_0^1 \circ\, dx\,,$$

$$[x, z, h, J] = [X/D, Z/D, H/D, (J - J_0)/\sigma D]\,,$$

and

$$G = (\rho - \rho_a) D^2 g/\sigma\,.$$

G is the only stability parameter which could arise from a variational analysis of the functional $J[h;G]$ in Γ.

The thermostatic statement of stability is:

Stable free surfaces $h = \tilde{h}$ minimize J in Γ. (98.4)

Minimizing free surfaces $\tilde{h}$ satisfy the relation

$$J[\tilde{h}; G] = \tilde{J} = \min_{h \in \Gamma} J[h; G].$$

We are going to prove the following theorem of static stability:

$$\tilde{J} = J[0, G] = 0, \qquad G \leqslant 4\pi^2; \tag{98.5a}$$

$$\tilde{J} \textit{ does not exist}, \qquad G > 4\pi^2. \tag{98.5b}$$

This criterion implies the stability of a flat interface with water over air when D is smaller than about two centimeters (see Exercise 98.2). The thermostatic stability result is in agreement with dynamic analysis of the same problem for small disturbances. But this result is in disagreement with dynamic analysis for large disturbances (§ 99); there are large disturbances of h which will not decay to $h=0$. This shows that the static criterion (98.4) of stability is incomplete. It is important to know the limitations of the criteria of static analysis. These limitations can be obtained only by dynamic analysis. Such conditions are presently unknown; it appears, however, that the static and dynamic criteria do coincide when the disturbances are small (Huh, 1969).

Proof of (98.5a, b): Let

$$h(x) = F(x) + \gamma k(x) \quad \text{for all} \quad h \in \Gamma$$

where $F \in \Gamma$, $k \in \Gamma$, γ is a real parameter and $J[F; G]$ is stationary in the sense that

$$\frac{d}{d\gamma} J[F + \gamma k; G]\big|_{\gamma=0} = 0 \quad \text{for all} \quad k \in \Gamma. \tag{98.6a}$$

Eq. (98.6a) is a necessary condition for the minimum $\tilde{J}$ of J. It leads, through integrations by parts to Euler's equation

$$-\left\langle k\left[\left(\frac{F'}{\sqrt{1+F'^2}}\right)' + GF\right]\right\rangle = 0 \tag{98.6b}$$

for the functions $F \in \Gamma$ which make J stationary. The function F which makes J stationary will also be a minimizing function $F = h$ for J if

$$\frac{d^2}{d\gamma^2} J[F + \gamma k; G]\big|_{\gamma=0} > 0 \qquad \text{for all} \quad k \in \Gamma \tag{98.7a}$$

that is, if

$$\left\langle \frac{k'^2}{(1+F'^2)^{3/2}} \right\rangle - G\langle k^2 \rangle > 0. \tag{98.7b}$$

To find the stable free surface on the open end of the channel in Fig. 98.1 we must study Eqs. (98.6) and (98.7). From (98.6b) we learn that either $F=0$ for all $G(-\infty<G<\infty)$ or else

$$\left(\frac{F'}{\sqrt{1+F'^2}}\right)'+GF=C_0 \tag{98.8}$$

where C_0 is a constant (see Exercise 98.1). There are solutions of (98.8) only for certain values of G. For all solutions of (98.8) with $F\in\Gamma$, we note that

$$\left\langle\frac{F'^2}{\sqrt{1+F'^2}}\right\rangle=G\langle F^2\rangle\,. \tag{98.9}$$

From (98.7b) we learn that the stationary solution $F=0$ is also a minimum when

$$\langle k'^2\rangle-G\langle k^2\rangle\geqslant(4\pi^2-G)\langle k^2\rangle>0 \tag{98.10}$$

since

$$\min_\Gamma\langle k'^2\rangle/\langle k^2\rangle=4\pi^2$$

when $k=c_1\sin 2\pi x$. It follows from (98.10) that the stationary solution $F\equiv 0$ is minimizing when

$$G<4\pi^2\,.$$

The stationary solutions $F\neq 0$ of (98.8) with $G>0$ are never minimizing because there is a $k\in\Gamma$, namely, $k=F$, which makes the functional on the left of (98.7b) negative. To show this we consider (98.7b) with $k=F$ and use (98.9) to show that

$$\begin{aligned}\left\langle\frac{F'^2}{(1+F'^2)}\,\frac{1}{\sqrt{1+F'^2}}-GF^2\right\rangle&=[1+F'^2(\bar{x})]^{-1}\left\langle\frac{F'^2}{\sqrt{1+F'^2}}\right\rangle-G\langle F^2\rangle\\&=G\left[\frac{1}{1+F'^2(\bar{x})}-1\right]\langle F^2\rangle<0\end{aligned} \tag{98.11}$$

where $0<\bar{x}<1$ is a mean-value such that $F'^2(\bar{x})>0$. This completes the proof of (98.5 a, b).

At this stage the result of the analysis of the static criterion is complete. We close with a brief discussion of the solution $F\in\Gamma$ of (98.8). All such solutions have an energy integral

$$-(1+\Phi'^2)^{-1/2}+G\Phi^2/2=C_2=-1+G\Phi_n^2/2 \tag{98.12}$$

where

$$\Phi=F-C_0/G\,, \tag{98.13}$$

and $\Phi_n = \Phi(x_n) = \pm C_3/G$ where $\Phi'(x_n) = F'(x_n) = 0$ $(n = 1, 2, \ldots, N)$ and N is the number of interior extreme values of $F(x)$ and C_2 and C_3 are constants. From (98.12) we find that

$$(-1)^n \frac{u}{\sqrt{1-u^2}} d\Phi = dx \quad \text{for} \quad x_n \leqslant x \leqslant x_{n+1} \tag{98.14}$$

where $x_0 = 0$, $x_{N+1} = 1$,

$$u = \frac{G}{2}[\Phi^2 - \Phi_n^2] + 1 = (1 + \Phi'^2)^{-1/2}$$

and

$$u(x_n) = 1 \quad \text{for} \quad (1 \leqslant n \leqslant N).$$

We now multiply (98.14) by Φ and integrate over the interval $[x_n, x_{n+1}]$ where $(1 \leqslant n \leqslant N-1)$. Since $\Phi d\Phi = du/G$ we find that $(0 \leqslant n \leqslant N)$

$$(-1)^{n+1} \int_{\sqrt{1-u^2(x_n)}}^{\sqrt{1-u^2(x_{n+1})}} d\sqrt{1-u^2} = G \int_{x_n}^{x_{n+1}} \Phi dx = G \int_{x_n}^{x_{n+1}} F dx - C_0(x_{n+1} - x_n).$$

It follows from the equations under (98.14) and

$$\Phi(0) = \Phi(1) = -C_0/G, \quad u(0) = u(1) = \frac{1}{2G}(C_0^2 - C_3^2) + 1, \quad \langle F \rangle = 0$$

that

$$\begin{aligned} 0 &= \textstyle\sum_{n=0}^N \{(-1)^{n+1} \int_{\sqrt{1-u^2(x_n)}}^{\sqrt{1-u^2(x_{n+1})}} d\sqrt{1-u^2} + C_0(x_{n+1} - x_n)\} \\ &= \sqrt{1-u^2(0)} + (-1)^{N+1}\sqrt{1-u^2(1)} + C_0 . \end{aligned}$$

Hence

$$\left(1 - \left[\frac{1}{2G}(C_0^2 - C_3^2) + 1\right]^2\right)^{1/2} \{1 + (-1)^{N+1}\} = -C_0 . \tag{98.15}$$

If N is even, $C_0 = 0$. Solutions of (98.8) with $N = 1$ are impossible. When $N = 1$ there is only one extremum of $F(x)$ for $0 < x < 1$; this is not consistent with $\langle F \rangle = 0$.

A second consequence of (98.14) is that

$$x_{n+1} - x_n = (-1)^n \int_{\Phi(x_n)}^{\Phi(x_{n+1})} u \, d\Phi/(1-u^2)^{1/2} \tag{98.16}$$

for $0 \leqslant n \leqslant N$. Since $\Phi(x_n) = \pm C_3/G$ for $n = 1, 2, \ldots, N$, and if $\Phi(x_n) = C_3/G$ then $\Phi(x_{n+1}) = -C_3/G$, we find that the intervals $x_{n+1} - x_n$ are of equal size where $1 \leqslant n \leqslant N$ and $x_1 = 1 - x_N$. Adding the intervals given by (98.16) we get

$$1 = 2 \int_{-C_0/G}^{|C_3|/G} u \, d\Phi/(1-u^2)^{1/2} + (N-1) \int_{-|C_3|/G}^{|C_3|/G} u \, d\Phi/(1-u^2)^{1/2} . \tag{98.17}$$

When N is even, $C_0=0$ and (98.17) is a functional relation between the eigenvalue G and the amplitude C_3 which can be obtained in the form $G(C_3)$. If $N>1$ is odd, (98.15) determines a function $C_0(G, C_3)$ which when combined with (98.17) again leads to a functional relation between G and C_3. The function $C_0(G(C_3), C_3)$ is not generally zero. The shape of the functions Φ and F are the same when N is given. The value C_0/G is the constant which must be added to $\Phi(x)$ to insure that (98.14) holds when $N>1$ is odd.

Exercise 98.1: Suppose that

$$\int_{\mathscr{V}} k(\mathbf{x})\, G(\mathbf{x})\, d\mathscr{V} = 0$$

for some function $G(\cdot) \in C^0(\mathscr{V})$ and for all $k(\cdot) \in C^0(\mathscr{V})$ such that

$$\int_{\mathscr{V}} k(\mathbf{x})\, d\mathscr{V} = 0\,.$$

Show that

$$G(\mathbf{x}) = \text{const}\,. \tag{98.18}$$

Exercise 98.2: Consider the stability of heavy liquid over air when the tube has an arbitrary cross section A but is otherwise as in Fig. 98.1. Show that

$$J[h, G] = \iint_A [1 + (\partial_x h)^2 + (\partial_y h)^2]^{1/2}\, dx\, dy - \tfrac{1}{2} G \iint_A h^2\, dx\, dy \tag{98.19}$$

where $Z = h(x, y)$ is the equation of the free surface. Suppose A is the cross section of a circular tube of dimensionless radius 1, where d is the tube radius. Show that the flat free surface is stable in the static sense whenever

$$G < (3.83171)^2 = \min_{k \in \Gamma} \frac{\langle k'^2 \rangle}{\langle k^2 \rangle} = \lambda_0 \tag{98.20}$$

where, here, the angle brackets denote integration over A

$$\Gamma = \{k(r, \theta) : k(1, \theta) = 0,\, k(0, \theta) < \infty,\, \iint_A k\, r\, dr\, d\theta = 0\}$$

and the minimum is attained when

$$k(r, \theta) = (2/\pi)^{1/2} J_1(\sqrt{\lambda_0}\, r) \cos\theta\,.$$

For water, $\sigma \simeq 70$ dynes/cm and the stability criterion (98.20) reduces to

$$d \leqslant 3.83171 \sqrt{\frac{\sigma}{g(\rho - \rho_a)}} \simeq 1 \text{ cm}\,.$$

Hence, water over air is stable in tubes whose diameter is smaller than about 2 centimeters.

Exercise 98.3 (Huh, 1969; Pitts, 1973): Formulate the static stability problem for the axisymmetric pendent drop which hangs from the ceiling. Why should the volume of the drop be held fixed when perturbing the minimizing solutions? Introduce a Lagrange multiplier λ and explain why a criterion for the static stability of the pendent drop is determined from minimizing the functional

$$J_1 + \lambda \pi \int_0^{-H} R^2(z)\, dz$$

where J_1 is defined by (96.22). Show that functional is stationary when

$$\frac{1}{R_{,z}}=0 \quad \text{at} \quad z=-H,$$

$$\frac{R_{,z}}{(1+R_{,z}^2)^{1/2}}=\cos\theta \quad \text{at} \quad z=0,$$

$$R(\rho z+\lambda)+\sigma(1+R_{,z}^2)^{1/2}=\sigma\left[\frac{R R_{,z}}{(1+R_{,z}^2)^{1/2}}\right]_{,z}$$

for $0>z>-H$.

§ 99. Stability of a Column of Liquid Resting on a Column of Air in a Vertical Tube—Dynamic Analysis

In the last section we showed that the functional $J[h;G]$ has no minimum value for $h\in\Gamma$ when $G>4\pi^2$. Now we are going to define a new functional from J which is bounded from below in Γ. The existence of this bounded from below functional in Γ coupled with the energy equation (97.1) leads to two dynamic stability theorems.

Suppose that we divide the functions in Γ into sets parameterized by their amplitude in $L^2[0,1]$:

$$\langle h^2\rangle=\varepsilon^2\equiv\delta\geqslant 0.$$

We then introduce the scaled function

$$h=\varepsilon\phi; \qquad \langle\phi^2\rangle=1$$

and define the functional

$$\mathbb{J}[\phi;\varepsilon^2,G]\equiv J[\varepsilon\phi;G]=\langle\sqrt{1+\varepsilon^2\phi'^2}\rangle-1-\frac{G\varepsilon^2}{2}\langle\phi^2\rangle. \tag{99.1}$$

The functional $\mathbb{J}[\phi;\varepsilon^2,G]$ *is bounded from below among functions* $\phi\in\tilde{\Gamma}$ where

$$\tilde{\Gamma}=[\phi:\phi\in\Gamma,\langle\phi^2\rangle=1].$$

Unlike $\Gamma,\tilde{\Gamma}$ is not a linear space.

Proof:

$$1=\langle\phi^2\rangle^{1/2}=\langle[\textstyle\int_0^x\phi'\,dx]^2\rangle^{1/2}\leqslant\langle\langle\sqrt{\phi'^2}\rangle\langle\sqrt{\phi'^2}\rangle\rangle^{1/2}=\langle\sqrt{\phi'^2}\rangle.$$

Hence,

$$\mathbb{J}\geqslant\varepsilon\langle\sqrt{\phi'^2}\rangle-1-\frac{G}{2}\varepsilon^2\geqslant\varepsilon-1-\frac{G}{2}\varepsilon^2. \tag{99.2}$$

We next introduce the homogeneous functional

$$\mathbb{J}[\phi;\varepsilon^2,G]\equiv\bar{\mathbb{J}}[\phi/\langle\phi^2\rangle^{1/2};\varepsilon^2,G]=\Lambda[\phi;\varepsilon^2]-1-G\varepsilon^2/2 \tag{99.3}$$

where

$$\Lambda[\phi;\varepsilon^2]=\langle(1+\varepsilon^2\phi'^2/\langle\phi^2\rangle)^{1/2}\rangle.$$

We define

$$\bar{j}(\varepsilon^2,G)=\min_{\phi\in\tilde{\Gamma}}\bar{\mathbb{J}}[\phi;\varepsilon^2,G] \tag{99.4a}$$

and show that

$$\bar{j}(\varepsilon^2,G)=\min_{\phi\in\Gamma}\mathbb{J}[\phi;\varepsilon^2,G]=1-G\varepsilon^2/2+\min_{\phi\in\Gamma}\Lambda[\phi;\varepsilon^2]. \tag{99.4b}$$

The last equality of (99.4b) follows directly from (99.3). To prove the first equality, we note that

$$\min_{\tilde{\Gamma}}\bar{\mathbb{J}}[\phi;\varepsilon^2,G]=\min_{\tilde{\Gamma}}\bar{\mathbb{J}}[\phi/\langle\phi^2\rangle^{1/2};\varepsilon^2,G]\geqslant\min_{\Gamma}\mathbb{J}[\phi;\varepsilon^2,G]$$

where the equality follows from the fact that $\langle\phi^2\rangle=1$ when $\phi\in\tilde{\Gamma}$ and the inequality follows from (99.3) and the fact that $\tilde{\Gamma}\subset\Gamma$. Let the constant a be chosen so that the minimizing element $\tilde{\phi}$ for $\mathbb{J}$ in Γ is related to $\tilde{\tilde{\phi}}\in\tilde{\Gamma}$ by $\tilde{\phi}=a\tilde{\tilde{\phi}}$. Then, continuing the previous inequality, we have

$$\begin{aligned}\min_{\Gamma}\mathbb{J}[\phi;\varepsilon^2,G]&=\mathbb{J}[\tilde{\phi};\varepsilon^2,G]=\mathbb{J}[\tilde{\tilde{\phi}};\varepsilon^2,G]\geqslant\min_{\Gamma}\mathbb{J}[\phi;\varepsilon^2,G]\\&=\min_{\tilde{\Gamma}}\bar{\mathbb{J}}\left[\frac{\phi}{\langle\phi^2\rangle^{1/2}},\varepsilon^2,G\right]=\min_{\Gamma}\bar{\mathbb{J}}[\phi;\varepsilon^2,G].\end{aligned}$$

This proves (99.4). In the first problem of (99.4a) we seek the minimum of a functional on a manifold; in the last problem (99.4b) we have eliminated the manifold by using it to define a homogeneous functional of degree zero having the same minimum.

The Euler equation for (99.4) is

$$[\phi'/\sqrt{1+\varepsilon^2\phi'^2/\langle\phi^2\rangle}]'+\lambda\phi=C_0, \tag{99.5a}$$

where $\lambda(\varepsilon^2)$ is an eigenvalue. Since $\langle\phi\rangle=0$ we find, multiplying (99.5) by ϕ and integrating, that

$$\lambda(\varepsilon^2)=\bar{\lambda}[\phi;\varepsilon^2]=\langle\phi'^2/\sqrt{1+\varepsilon^2\phi'^2/\langle\phi^2\rangle}\rangle/\langle\phi^2\rangle \tag{99.5b}$$

and $\phi\in\Gamma$. This problem is almost identical to the problem (98.8) for the stationary values $F\in\Gamma$ of $J[h;G]$. To relate these two problems we introduce the function

$$j(\tilde{\delta})=\max_{\delta=\varepsilon^2\geqslant 0}\bar{j}(\delta,G)=\bar{j}(\tilde{\delta},G). \tag{99.6a}$$

If $\tilde{\delta}$ is an interior value $0<\tilde{\delta}<M<\infty$ we find, using (99.4), that

$$\frac{d\bar{j}}{d\delta}(\delta, G)=\frac{\partial \Lambda}{\partial \delta}[\tilde{\phi};\delta]-\frac{G}{2}=\tfrac{1}{2}\{\bar{\lambda}[\tilde{\phi};\delta]-G\} \tag{99.6 b}$$

must vanish when $\delta=\tilde{\delta}$. (The last equality of (99.6 b) follows from the rule for differentiating stationary functionals with respect to a parameter (see Exercise B 4.6).) Comparing (99.5) and (99.6) with (98.8) and (98.9), we find that

$$j(\varepsilon^2)=J[F;G]$$

when

$$\langle F^2\rangle=\varepsilon^2\,.$$

It follows that the function $\tilde{\phi}=F/\varepsilon$, which makes $J[\varepsilon\tilde{\phi};G]$ stationary when $\langle F^2\rangle=\varepsilon^2$, actually minimizes the functional $\mathbb{J}[\phi;\varepsilon^2,G]$ when (ε^2,G) satisfy (99.6 b). Of course $F=\varepsilon\tilde{\phi}$ cannot minimize $J[F;G]$ when the amplitude of F is unrestricted.

The next preliminary to the statement and proof of Dussan's two dynamic stability theorems requires that we describe the curves $\bar{j}(\varepsilon^2, G)$. We have already obtained the lower bound (99.2)

$$\bar{j}(\varepsilon^2, G)\geqslant\varepsilon-1-\frac{G}{2}\varepsilon^2\,.$$

For an upper bound we use Schwarz's inequality and $\langle\phi'^2\rangle\geqslant 4\pi^2\langle\phi^2\rangle$ to find that

$$\begin{aligned}\bar{j}(\varepsilon^2, G)&\leqslant\min_{\phi\in\Gamma}(1+\varepsilon^2\langle\phi'^2\rangle/\langle\phi^2\rangle)^{1/2}-1-G\varepsilon^2/2\\&=\sqrt{1+4\pi^2\varepsilon^2}-1-G\varepsilon^2/2\,.\end{aligned} \tag{99.7}$$

The exact values of $\bar{j}(\varepsilon^2, G)$ could be obtained by the method of quadrature used in § 98. It is also possible to express $\bar{j}(\varepsilon^2, G)$ for small values of ε^2 as a perturbation series in powers of $\varepsilon^2\equiv\delta$. We insert the representation

$$\begin{bmatrix}\tilde{\phi}(x,\delta)\\ \lambda(\delta)\end{bmatrix}=\sum_{n=0}^{\infty}\begin{bmatrix}\phi_n(x)\\ \lambda_n\end{bmatrix}\delta^n \tag{99.8}$$

into (99.5) and equate the coefficients of separate powers of δ to zero. This gives rise to ordinary differential equations whose solutions are given below

$$\begin{bmatrix}\phi_0(x)\\ \lambda_0\end{bmatrix}=\begin{bmatrix}\sqrt{2}\sin 2\pi x\\ 4\pi^2\end{bmatrix}$$

$$\begin{bmatrix}\phi_1(x)\\ \lambda_1\end{bmatrix}=\begin{bmatrix}(3/2^{9/2})4\pi^2\sin 6\pi x\\ -(3/4)16\pi^4\end{bmatrix}$$

and

$$\begin{bmatrix}\phi_2(x)\\ \lambda_2\end{bmatrix}=$$

$$\begin{bmatrix}-(3^2/2^{21/2})16\pi^4\sin 2\pi x-(3^3/2^{19/2})16\pi^4\sin 6\pi x+(5^2/2^{19/2})16\pi^4\sin 10\pi x\\ (93/2^6)64\pi^6\end{bmatrix}.$$

The computation of these values and the justification for expanding the solution $\phi(x,\delta)$ which is antisymmetric around $x=1/2$ and has a single interior zero is left as an Exercise (99.2).

Given the series

$$\lambda=4\pi^2[1-(3/4)4\pi^2\delta+(93/2^6)16\pi^4\delta^2+\ldots],$$

we use the first of Eq. (99.6 b) to find that

$$\bar{j}(\delta,G)=\tfrac{1}{2}\{[4\pi^2-G]\delta-(3/8)16\pi^4\delta^2+(93/2^6)64\pi^6\delta^3/3+\ldots\} \tag{99.9}$$

and

$$\max_\delta \bar{j}(\delta,G)=\bar{j}(\tilde{\delta},\lambda(\tilde{\delta}))=j(\tilde{\delta}).$$

In Fig. 99.1 we have sketched the graph of $\bar{j}(\delta, G)$ in the (δ, J) plane with G as a parameter. The following properties of $\bar{j}(\delta, G)$ follow from Eqs. (99.2), (99.7) and (99.9)

(i) $\lim_{\delta\to\infty}\bar{j}(\delta,G)\to -G\delta/2$.

(ii) $\bar{j}(\delta,G)$ has an interior maximum $j(\tilde{\delta})>0$, $0<\tilde{\delta}<\infty$ whenever $0<G<4\pi^2$.

(iii) $\bar{j}(\delta,G)$ has a maximum values $j(\tilde{\delta})$ whenever $G>0$. The only possible maximum value when δ is small and $G\geqslant 4\pi^2$ is $\bar{j}(0,G)=0$.

(iv) $\bar{j}(\delta,G)$ has no maximum value when δ is small and $G<0$.

To examine the consequences of Dussan's energy equation we define

$$\hat{J}(t,G)=J[\sqrt{\delta(t)}\,\phi(x,t),G] \tag{99.10a}$$

where $\delta(t)=\sqrt{\langle h^2(x,t)\rangle}$ and $\phi(x,t)=h(x,t)/\sqrt{\delta(t)}$. Since $\phi\in\tilde{\Gamma}$ we have

$$\bar{j}(\delta(t),G)\leqslant\hat{J}(t,G) \tag{99.10b}$$

Now we recall that Dussan's energy equation for the liquid filling a tube can be written as an inequality $d(\mathscr{E}+\hat{J})/dt\leqslant 0$, where $\hat{J}$ is defined by (99.10a). Hence

$$\hat{J}(t,G)\leqslant\mathscr{E}(0)+\hat{J}(0,G). \tag{99.11}$$

Dussan observed that $\hat{J}(t,G)$ could not fall below the function $\bar{j}(\delta(t),G)$ and for certain values of $\delta(0)$, $\mathscr{E}(0)$ and $\hat{J}(0,G)$ it was impossible to have $\delta(t)\to 0$ along any solution of the energy inequality. These observations lead to a *dynamic instability-stability theorem.*

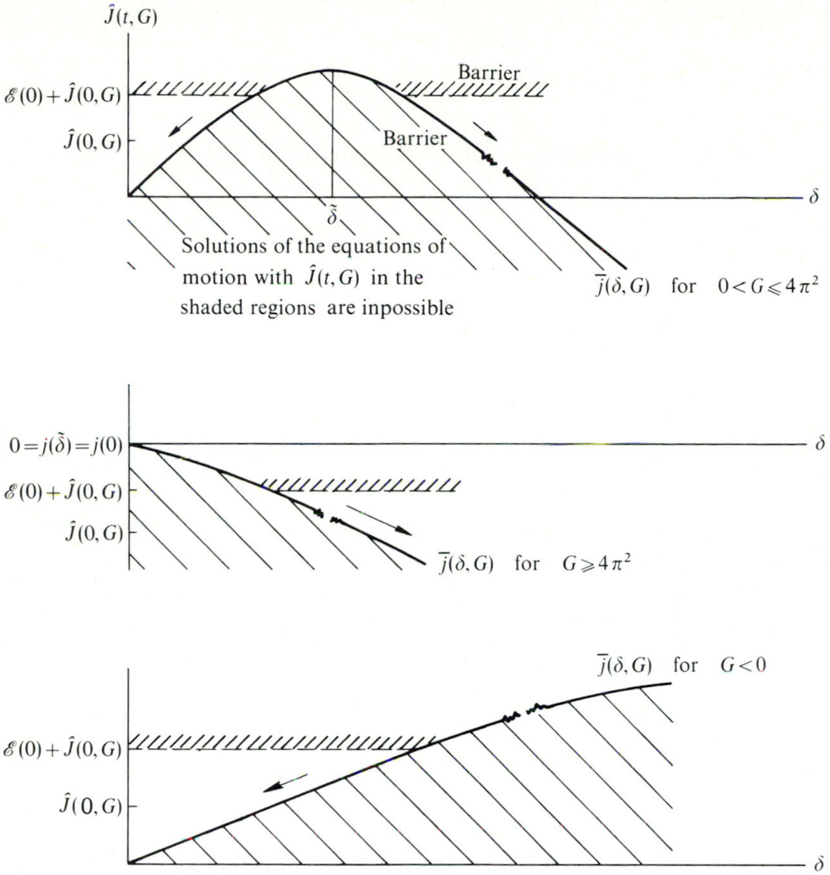

Fig. 99.1: The form the minimizing function $\bar{j}(\delta, G)$ for different values of G. The existence of a maximum value $j(\tilde{\delta})$ forms an energy barrier which prevents certain disturbances from reaching the flat free interface $\delta=0$, $\hat{J}=J=0$

Suppose that $G>0$,

$$\mathscr{E}(0)+\hat{J}(0, G)<\bar{j}(\tilde{\delta}, G) \tag{99.12}$$

and

$$\delta(t)>\tilde{\delta} \quad \text{when} \quad t=0. \tag{99.13}$$

Then the free energy cannot get over the energy barrier,

$$\hat{J}(0,G)+\mathscr{E}(0)\geqslant\hat{J}(t,G)\geqslant\bar{j}(\delta(t),G) \quad \text{and} \quad \delta(t)>\tilde{\delta}, \quad \text{for all} \quad t>0,$$

and the flat free surface (which has $\delta=0$, $J[0,G]=0$) *cannot be attained.*

Suppose that (99.12) *holds,* $G<4\pi^2$ *and*

$$\delta(t)<\tilde{\delta} \quad \text{when} \quad t=0\,. \tag{99.14}$$

Then the kinetic energy is integrable and the free energy

$$\hat{J}(t, G)\to J[0, G]=0$$

of the flat free surface is the smallest value which is attainable.

Proof: The proof of this theorem is nearly obvious from Fig. 99.1. Suppose $0\leqslant G$ and $\delta(0)>\tilde{\delta}$. Then from (99.10), (99.11) and (99.12) we note that

$$\bar{j}(\delta(t), G)\leqslant \hat{J}(t, G)\leqslant \hat{J}(0, G)+\mathscr{E}(0) \tag{99.15}$$

and $\delta(t)>\tilde{\delta}\geqslant 0$ in this region. Since $\delta(t)\neq 0$, the flat free surface is unstable to initial conditions (99.12) and (99.13).

When (99.12) holds with $G<4\pi^2$, the free energy is again bounded from above and below by (99.15). If (99.14) holds, then the minimum attainable value of $\bar{j}$ is on the left of the energy barrier and

$$\min_{\delta}\bar{j}(\delta, G)=\bar{j}(0, G)=0\,.$$

Since J is bounded from below, the theorem about the integrability of $\mathscr{E}$ (§ 97) applies.

The only stable configuration in the upside down tube is the one with a flat free surface and $G\leqslant 4\pi^2$. When $0<G<4\pi^2$, the flat free surface is conditionally stable. Sufficiently large disturbances satisfying (99.12) and (99.13) persist; they can never decay to the flat free surface because the equations of motion do not allow these disturbances to cross the energy barrier. Such disturbances are outside the domain of attraction of the basic state.

The barrier criterion (99.11) leaves open the possibility that very large stable deformations of the free surface can be realized by creating a suitable initial velocity field whose initial energy $\mathscr{E}(0)$ is above the threshold value. The distributions of pressures associated with such a surface might then draw a large deformation, which would otherwise be unstable, back into the equilibrium configuration with a flat free surface.

Exercise 99.1: Show that solutions $\phi\neq 0$ bifurcate from (99.5) at every eigenvalue $\lambda=n^2\pi^2(n>1)$ of the problem which follows from (99.5) when it is linearized around $\phi=0$. Show how these bifurcating branches may be constructed by quadrature (see § 98). Show that the smallest value $\bar{j}(\delta, G)$ corresponds to the smallest value of $\bar{\lambda}[\phi, \delta]$. Show that $n=2$ makes $\bar{\lambda}$ smallest when δ is small. Compute the bifurcating branches given under (99.8).

Exercise 99.2 (Dussan V., 1975): Derive the expressions for the perturbation coefficients given under (99.8).

Notes for Chapter XIV

The material in this Chapter is inspired by the excellent work of E.B. Dussan V. Unattributed theorems and exercises, the static analysis of § 98, and the method of analysis used in § 99 are due to Joseph. The basic energy identity and the energy instability theorem were discovered by Dussan V.

A beautiful experiment which corresponds closely to the stability problem sketched in Fig. 98.1 can be seen in the film "Flow instabilities" by E. Mollo Christensen[1]. This stability problem has a certain formal resemblance to the problem of stability of a flat interface on the whole x, y plane which was treated by Taylor (1950) with surface tension and viscosity neglected, and by Bellman and Pennington (1954) with surface tension and viscosity included. These authors use the linearized method of "normal modes". In the problem treated by Taylor and by Bellman and Pennington there is never stability, even linear stability, since there are always wave numbers for which the flat free surface is unstable. An interesting bifurcation analysis for Taylor's problem has been given by Pimbley (1972). An analysis of the static stability of fluid surfaces under more general circumstances has been given by Gillette and Dyson (1974). The interested reader can find more references to the problem of interfacial stability in the paper of Dussan V (1975) and the thesis of Huh (1969).

[1] Produced by the National Council for Fluid Mechanics Films. Encyclopedia Brittanica Educational Corp.

References

(The section in the text where the given reference is cited appears as a cross-reference in the square bracket which follows the citations listed below.)

Aris, R.: Vectors, Tensors and the Basic Equations of Fluid Mechanics. Englewood Cliffs, N.J.: Prentice-Hall 1962. [55]

Auchmuty, J.F.: Bounding flows for turbulent convection. Arch. Rational Mech. Anal. **51**, 219 (1973). [83, 84]

Ayyaswamy, P.S.: On the stability of plane parallel flow between differentially heated, tilted planes. J. Appl. Mech. **41**, 554 (1974). [63]

Barcilon, V., Pedlosky, J.: Linear theory of rotating stratified fluid motions. J. Fluid Mech. **29**, 1 (1967). [64]

Batchelor, G.K.: Heat transfer by free convection across a closed cavity between vertical boundaries at different temperatures. Quart. Appl. Math. **12**, 209 (1954). [64]

Bauer, L., Keller, H.B., Reiss, E.L.: Multiple eigenvalues lead to secondary bifurcation. SIAM Review. **17**, 101 (1975). [Add to X]

Beavers, G.S., Joseph, D.D.: Tall Taylor cells in polyacrilamide. Phys. Fluids. **17**, 650 (1974). [95]

Beavers, G.S., Joseph, D.D.: The rotating rod viscometer. J. Fluid Mech. **69**, 475 (1975). [95]

Beavers, G.S., Joseph, D.D.: Novel Weissenberg effects. (forthcoming) [95]

Beavers, G.S., Sparrow, E.M.: Non-Darcy flow through fibrous porous media. J. Appl. Mech. **36**, 711 (1969). [70]

Beck, J.L.: Convection in a box of porous material saturated with fluid. Phys. Fluids **15**, 1377 (1972). [70]

Bellman, R., Pennington, R.H.: Effects of surface tension and viscosity on Taylor instability. Quart. Appl. Math. **12**, 151 (1954). [Notes for XIV]

Bénard, H.: Les tourbillons cellulaires dans une nappe liquide. Revue générale des Science pures et appliquées **11**, 1261 and 1309 (1900). [60]

Bénard, H.: Les tourbillons cellulaires une nappe liquide transportant de la chaleur par convection en régime permanent. Ann. Chim. Phys. **23**, 62 (1901). [60, 73]

Berg, J.C., Acrivos, A., Boudart, M.: Evaporation convection, Adv. Chem. Ergrg. **6**, 61 (1966). [60]

Bhattacharyya, S.P., Jain, P.C.: Hydrodynamic stability in the presence of a magnetic field. Arch. Rational Mech. Anal. **44**, 281 (1972). [Add. for IX]

Bird, R.E., Stewart, W.E., Lightfoot, E.: Transport Phenomena. New York: Wiley 1960. [55]

Block, M.J.: Surface tension as the cause of Bénard cells and surface deformation in a liquid film. Nature **178**, 650 (1956). [60]

Boussinesq, J.: Théorie analytique de la chaleur Vol. 2. Paris: Gauthier-Villars 1903. [54]

Buretta, R.J.: Thermal convection in a fluid-filled porous layer with uniform heat sources. Ph. D. Thesis: Dept. of Aero. Eng., University of Minn. 1972. [92]

Buretta, R.J., Berman, A.S.: Convective heat transfer in a liquid saturated porous layer. J. App. Mech. **43**, 249 (1976). [92]

Busse. F.H.: Dissertation, Munich (1962). See also: The stability of finite amplitude cellular convection and its relation to an extremum principle. J. Fluid Mech. **30**, 625 (1967). [Int. to X, 76, 77, 81, 82]

Busse, F. H.: Bounds on the transport of mass and momentum by turbulent flow between parallel plates. Jour. App. Math. Phys. (ZAMP), **20**, 1 (1969A). [Int. to XII]

Busse, F. H.: On Howard's upper bound for heat transport by turbulent convection. J. Fluid Mech. **37**, 457 (1969B). [Int. to XII)

Busse, F. H.: Stability regions of cellular fluid flow: IUTAM Symposium on Instability of Continuous Systems. Ed. H. Leipholz. Berlin-Heidelberg-New York: Springer 1971. [Int. to XI]

Busse, F. H.: Patterns of convection in spherical shells J. Fluid Mech. **72**, 67 (1975). [74]

Busse, F. H., Joseph, D. D.: Bounds for heat transport in a porous layer, J. Fluid Mech. **54**, 521 (1972). [Int. to XII, 85, 88]

Carmi, S., Lalas, D. P.: Universal stability of hydromagnetic flows. J. Fluid Mech. **43**, 711 (1970). [Add. for IX]

Chan, S.: Infinite Prandtl number turbulent convection. Studies in Appl. Math. **50**, 1: 13 (1971). [90, 91]

Chandrasekhar, S.: Hydrodynamic and Hydromagnetic Stability, Oxford University Press 1961 [59, 60, 62, 74, 77, Add. for IX]

Charlson, G. S., Sani, R. L.: Thermoconvective instability in a bounded cylindrical layer. Int. J. Heat Mass Transfer, **13**, 1479 (1970). [62]

Charlson, G. S., Sani, R. L.: Finite amplitude axisymmetric thermoconvective flows in a bounded cylindrical layer of fluid. J. Fluid Mech. **71**, 209 (1975). [62, 73]

Clever, R. M., Busse, F. H.: Transition to time-dependent convection. J. Fluid Mech. **65**, 625 (1974). [81]

Coleman, B. D.: Kinematical concepts with applications in the mechanics and thermodynamics of incompressible viscoelastic fluids. Arch. Rational Mech. Anal. **9**, 273 (1962). [93, 94]

Coleman, B. D., Markovitz, H.: Normal stress effects in second-order fluids. J. Appl. Phys. **35**, 1 (1964). [94]

Coleman, B. D., Mizel, V. J.: Breakdown of laminar shearing flows for second-order fluids in channels of critical width. Z. Angew. Math. Mech. **46**, 445 (1966). [93, 94]

Coleman, B. D., Noll, W.: Helical flow of general fluids. J. Appl. Physics **30**, 1508 (1959). [93, 95]

Coleman, B. D., Noll, W.: An approximation theorem for functionals, with applications in continuum mechanics. Arch. Rational Mech. Anal. **6**, 355 (1960). [93, 94]

Coleman, B. D., Noll, W.: Foundation of linear viscoelasticity. Rev. Mod. Phys. **33**, 239 (1961). [93, 94, 95]

Coleman, B. D., Duffin, R. J., Mizel, V. J.: Instability, uniqueness, and nonexistence theorems for the equation $U_t = U_{xx} - U_{xtx}$ on a strip. Arch. Rational Mech. Anal. **19**, 100 (1965). [94]

Coleman, B. D., Markovitz, H., Noll, W.: Viscometric Flows of Non-Newtonian Fluids. (Springer Tracts in Natural Philosophy, Vol. 5). Berlin-Heidelberg-New York 1966. [93]

Coles, D.: Transition in circular Couette flow. J. Fluid Mech. **21**, 385 (1965). [95]

Combarnous, M., LeFur, B.: Transfert de chaleur par convection naturelle dans une couche poreuse horizontale. Comptes Rendus **269** B, 1009 (1969). [92]

Craik, A. D.: A note on the static stability of an elastico-viscous fluid. J. Fluid Mech. **33**, 33 (1968). [94]

Darcy, H. P. G.: Les fontaines publiques de la ville de Dijon, Paris 1856. [70]

Davis, S. H., von Kerczek, D.: A reformulation of energy stability theory. Arch. Rational Mech. Anal. **52**, 112 (1973). [58]

Davis, S. H.: Convection in a box: linear theory. **30**, 465 (1967). [62]

Davis, S. H.: On the principle of exchange of stabilities. Proc. Roy. Soc. A **310**, 341 (1969A). [75]

Davis, S. H.: Buoyancy-surface tension instability by the method of energy. J. Fluid Mech. **39**, 347 (1969B). [60]

Davis, S. H., Segel, L. A.: Effects of surface curvature and property variation in cellular convection. Phys. Fluids **11**, 470 (1968). [Int. to X, 74]

DiPrima, R. C., Habetler, G. J.: A completeness theorem for non-selfadjoint eigenvalue problems in hydrodynamic stability. Arch. Rat. Mech. Anal. **34**, 218 (1969). [70]

Dudis, J. J., Davis, S. H.: Energy stability of the buoyancy boundary layer. J. Fluid Mech. **47**, 381 (1971A). [64]

Dudis, J. J., Davis, S. H.: Energy stability of the Ekman boundary layer. J. Fluid Mech. **47**, 405 (1971B). [64]

Dunn, J. E., Fosdick, R. L.: Thermodynamics, stability and boundedness of fluids of complexity 2 and fluids of second grade. Arch. Rational Mech. Anal. **56**, 191 (1974). [94]

Dussan V., E. B., Davis, S. H.: On the motion of a fluid-fluid interface along a solid surface. J. Fluid Mech. **65**, 71 (1974). [96]
Dussan V., E. B.: Hydrodynamic stability and instability of fluid systems with interfaces. Arch. Rational Mech. Anal. **57**, 363 (1975). [Int. to XIV, 97].
Eckhaus, W.: Studies in Non-linear Stability Theory. Berlin-Heidelberg-New York: Springer 1965. [Int. to XI]
Elder, J. W.: Steady free convection in a porous medium heated from below. J. Fluid Mech. **27**, 29 (1967). [92]
Fife, P.: The Bénard problem for general fluid dynamical equations and remarks on the Boussinesq equations. Indiana Univ. Math. J. **20**, 303 (1970). [54]
Fife, P., Joseph, D. D.: Existence of convective solutions of the generalized Bénard problem which are analytic in their norm. Arch. Rat. Mech. Anal. **33**, 116 (1969). [76]
Forchheimer, P. H.: Wasserbewegung durch Böden. Z. ver. dtsch. Ing. **45**, 1782 (1901). [70]
Galdi, G. P.: Su un criterio di stabilità non lineaire in presenza di perturbazioni iniziali ed al contorno e sua applicazione a due classi di moti magnetoidrodinamici. Boll. U.M.I. **8**, 336, (1973). [Add. to IX]
Gershuni, G. Z., Zhukhovitskii, E. M.: Stability of plane parallel convective motion with respect to spatial perturbations. Prik. Mat. Meh. **33**, 855 (1969). [63]
Gibbs, J. W.: The Scientific Papers of J. Willard Gibbs. New York: Dover Publications 1961. [97]
Giesekus, H.: On instabilities in Poiseuille and Couette flow of viscoelastic fluids in Progress in Heat and Mass Transfer, V, Eds.: W. R. Schowalter, A. V. Luikov, W. J. Minkowycz and N.H. Afgan. Pergamon 1972. [95]
Gill, A. E.: The boundary-layer regime for convection in a rectangular cavity. J. Fluid Mech. **26**, 515 (1966). [64]
Gill, A. E., Davey, A.: Instabilities of a buoyancy-driven system. J. Fluid Mech. **35**, 775 (1969). [64]
Gillette, R. D., Dyson, D. C.: Stability of static configurations with applications to the theory of capillarity. Arch. Rational Mech. Anal. **53**, 150 (1974). [Notes for XIV]
Graham, A.: Shear patterns in an unstable layer of air. Phil. Trans. Roy. Soc. A**232**, 285 (1933). [63, 77, 79]
Green, A. E., Rivlin, R. S.: The mechanics of non-linear materials with memory, Part I, Arch. Rational Mech. Anal. **1**, 1 (1957). [94]
Gumerman, R. J., Homsy, G. M.: Convective instabilities in concurrent two phase flow: Part I. Linear stability, Part II. Global stability. Amer. Inst. Ch. Eng. J. (to appear). [60, 63]
Gumerman, R. J., Homsy, G. M.: The stability of uniformly accelerated flows with application to convection driven by surface tension. J. Fluid Mech. **68**, 191 (1975). [60, 63]
Gupta, V. P.: Upper bound on heat transfer across a porous fluid layer heated from below. Ph. D. Thesis. Dept. of Aero. Eng., Univ. of Minn. 1972. [91]
Gupta, V. P., Joseph, D. D.: Bounds for heat transport in a porous layer. J. Fluid Mech. **57**, 491 (1973). [Int. to XII, 86, 88, 90, 91, 92]
Hales, A. L.: Convection currents in geysers. Mon. Not. Roy. Astron. Soc., Geophys. Suppl. **4**, 122 (1937). [62]
Hancock, H.: Elliptic Integrals. Dover 1958. [86]
Hart, J. E.: Stability of the flow in a differentially heated inclined box. J. Fluid Mech. **47**, 547 (1971). [63]
Hoare, R. A.: Problems of heat transfer in Lake Vanda, a density stratified Antarctic lake. Nature **210**, 787 (1966). [65]
Homsy, G. M.: Global stability of time-dependent flows: impulsively heated or cooled fluid layers. J. Fluid Mech. **60**, 129 (1973). [75]
Homsy, G. M.: Global stability of time-dependent flows, Part 2. Modulated fluid layers. J. Fluid Mech. **62**, 387 (1974). [75]
Horton, C. W., Rogers, Jr., F. T.: Convection currents in a porous medium. J. Appl. Phys. **16**, 367 (1945). [78]
Howard, L. N.: Heat transport by turbulent convection. J. Fluid Mech. **17**, 405 (1963). [62, Int. to XII, 83, 84, 85, 87]
Howard, L. N.: Bounds on flow quantities. Ann. Rev. of Fluid Mech. **4**, 473 (1972). [87]
Huh, Chun: Capillary hydrodynamics—Interfacial instability and the solid-liquid-fluid contact line, Ph. D. Thesis, Department of Chemical Eng., University of Minn. 1969. [98, Notes for XIV]

Irmay, S.: On the theoretical derivation of Darcy and Forchheimer formulas. Trans. Am. Geophys. Un. **39**, 702 (1958). [70]

Jakob, M.: Heat Transfer, Vol. 1. New York: John Wiley and Sons 1949. [54]

Jeffreys, H.: The instability of a compressible fluid heated below. Proc. Camb. Phil. Soc. **26**, 170 (1930). [54]

Joseph, D. D.: On the stability of the Boussinesq equations. Arch. Rational Mech. Anal. **20**, 59 (1965). [62]

Joseph, D. D.: Nonlinear stability of Boussinesq equations by the method of energy. Arch. Rational Mech. Anal. **22**, 163 (1966). [62, 63]

Joseph, D. D.: Global stability of the conduction-diffusion solution. Arch. Rational Mech. Anal. **36**, 285 (1970). [69]

Joseph, D. D.: Stability of convection in containers of arbitrary shape. J. Fluid Mech. **47**, 257 (1971). [77]

Joseph, D. D.: Energy stability of hydromagnetic flow in: Proceedings of a Conference on Mathematical Topics in Stability Theory. Dept. of Math., Wash. State Univ., 1972. [Add. for IX]

Joseph, D. D.: Perturbations of the domain: a Lagrangian derivation of the higher order theory of infinitesimal water waves. Arch. Rational Mech. Anal. **51**, 295 (1973). [94]

Joseph, D. D.: Repeated supercritical branching of solutions arising in the variational theory of turbulence. Arch. Rational Mech. Anal. **53**, 101 (1974A). [Int. to XII, 68, 88, 89, 91]

Joseph, D. D.: Slow motion and viscometric motion, stability and bifurcation of the rest state of a simple fluid. Arch. Rational Mech. Anal. **56**, 99 (1974C). [Int. to XIII]

Joseph, D. D., Beavers, G. S., Fosdick, R. L.: The free surface on a liquid between cylinders rotating at different speeds. Part II. Arch. Rational Mech. Anal. **49**, 381 (1973).

Joseph, D. D., Beavers, G. S.: The free surface on a simple fluid between cylinders undergoing torsional oscillations. Arch. Rational Mech. Anal. (forthcoming) [Int. to XIII, 95]

Joseph, D. D., Carmi, S.: Subcritical convective instability. Part 2. Spherical Shells. J. Fluid Mech. **26**, 769 (1966). [61, 74]

Joseph, D. D., Sattinger, D. H.: Bifurcating time periodic solutions and their stability. Arch. Rational Mech. Anal. **45**, 79 (1972). [92]

Joseph, D. D., Shir, C. C.: Subcritical convective instability. Part 1. Fluid layers. J. Fluid Mech. **26**, 753 (1966). [75]

Joseph, D. D., Sturges, L.: The free surface on a liquid in a trench heated from its side. J. Fluid Mech. **69**, 565 (1975). [94]

Kato, T.: Perturbation theory for linear operators (Die Grundlehren der math. Wissenschaften, Bd. 132). Berlin-Heidelberg-New York: Springer 1966 [Add. to X]

Kirchgässner, K., Kielhöfer, H.: Stability and bifurcation in fluid mechanics. Rocky Mountain J. Math. **3**, (2) 275 (1972). [77]

Kirchgässner, K., Sorger, P.: Stability analysis of branching solutions of the Navier-Stokes equations. Proc. 12th Intl. Cong. Appl. Mech., Stanford, p. 257. Berlin-Heidelberg-New York: Springer 1969. [80]

Kogleman, S., DiPrima, R. C.: Stability of spatially periodic supercritical flows in hydrodynamics. Phys. Fluid **13**, 1 (1970). [Int. to XI]

Koschmieder, E. L.: Bénard convection: Advances in Chemical Physics (Ed.: I. Prigogine and S.A. Rice) **26**, 177. New York: Wiley 1974. [62, 73]

Koschmieder, E. L.: On the wavelength of convective motions. J. Fluid Mech. **35**, 527 (1969). [82]

Krishnamurti, R.: Finite amplitude convection with changing mean temperature. Part I. Theory. J. Fluid Mech. **33**, 445 (1968A). [Int. to X, 75, 76]

Krishnamurti, R.: Finite amplitude convection with changing mean temperature. Part II. An experimental test of the theory. **33**, 457 (1968B). [Int. to X, 75]

Krishnamurti, R.: On the transition to turbulent convection. Part I. The transition from two-to three-dimensional flow. J. Fluid Mech. **42**, 295 (1970A). [92]

Krishnamurti, R.: On the transition to turbulent convection. Part 2. Transition to time-dependent flow. J. Fluid Mech. **42**, 309 (1970B). [92]

Kulacki, F. A., Goldstein, R. J.: Thermal convection in a horizontal fluid layer with uniform volumetric energy sources. J. Fluid Mech. **55**, 271 (1972). [75]

Kulacki, F. A., Goldstein, R. J.: Hydrodynamic instability in fluid layer with volumetric energy sources. Appl. Sci. Res. **31**, 81 (1975). [75]

Lalas, D.P., Carmi, S.: Nonlinear thermal convection in conducting fluids. Phys. Fluids **15**, 2182 (1972). [Add. for IX]

Landau, L. D., Lifshitz, E. M.: Fluid Mechanics. Oxford: Pergamon Press 1959. [97]

Langlois, W.E., Rivlin, R.S.: Slow steady-state flow of visco-elastic fluids through non-circular tubes. Rendiconti di Matematica, **22**, 169 (1963). [94]

Lapwood, E.R.: Convection of a fluid in a porous medium. Proc. Camb. Phil. Soc. **44**, 508 (1948). [78]

Liang, S.F., Vidal, A., Acrivos, A.: Buoyancy-driven convection in cylindrical geometries. J. Fluid Mech. **36**, 239 (1969). [73, 76, 77)

Lorenz, L.: Über das Leitungsvermögen der Metalle für Wärme und Elektrizität. Annalen der Physik und Chemie **13**, 581 (1881). [54]

Malkus, W.V.R.: Discrete transitions in turbulent convection. Proc. Roy. Soc. A **225**, 185 (1954). [92]

Markovitz, H.: Normal stress effect in polyisobutylene solutions, II. Classification and application of rheological theories. Trans. Soc. Rheol. **1**, 37 (1957). [93]

Markovitz, H., Coleman, B.D.: Incompressible second-order fluids in: Advances in Applied Mechanics, VIII, Ed.: G. Kuerte. New York: Academic Press 1964. [93]

McLeod, J.B., Sattinger, D.H.: Loss of stability and bifurcation at a double eigenvalue. J. Functional Anal. **14**, 62 (1973). [Add. A to X]

Mihaljan, J.M.: A rigorous exposition of the Boussinesq approximation applicable to a thin layer of fluid. Astrophys. J. **136**, 1126 (1962). [54]

Miller, R.K.: Asymptotic stability properties of linear Volterra integrodifferential equations. J. Diff. Eqs. **10**, 485 (1971). [94]

Mollo-Christensen, E.: "Flow Instabilities", a film produced by Education Dev. Center Nat. Com. Fluid Mechs. Films. Chicago: Encyclopedia Britannica Ed. Corp. 1969. [Notes for XIV]

Mott, T., Joseph, D.D.: Stability of parallel flow between concentric cylinders. Phys. Fluids. **11**, 2065 (1968). [100]

Newell, A.C., Whitehead, J.A.: Finite bandwidth, finite amplitude convection. J. Fluid Mech. **38**, 279 (1969). [Int. to XI, 82]

Nield, D.A.: Surface tension and buoyancy effects in cellular convection. J. Fluid Mech. **19**, 341 (1964). [60]

Nield, D.A.: The thermohaline Rayleigh-Jeffreys problem. J. Fluid Mech. **29**, 545 (1967). [67]

Nield, D.A.: Onset of thermohaline convection in a porous medium. Water Resources Res. **4**, 553 (1968). [78]

Nield, D.A.: Thermal convection between horizontal planes with mean temperature gradient inclined to vertical and mean horizontal velocity. (forthcoming) (1975). [63]

Noll, W.: A mathematical theory of the mechanical behavior of continuous media. Arch. Rational Mech. Anal. **2**, 197 (1958). [93]

Oberbeck, A.: Über die Wärmeleitung der Flüssigkeiten bei der Berücksichtigung der Strömungen infolge von Temperaturdifferenzen. Annalen der Physik und Chemie **7**, 271 (1879). [54].

Oberbeck, A.: Über die Bewegungserscheinungen der Atmosphäre. Sitz. Ber. K. Preuss. Akad. Miss., 383 and 1129 (1888). Translated by C. Abel in Smiths Misc. Coll., 1891. [54]

Palm, E.: On the tendency towards hexagonal cells in steady convection. J. Fluid Mech. **8**, 183 (1960). [Int. to X, 77]

Palm, E.: Nonlinear thermal convection. Ann. Rev. Fluid Mech. **7**, 39 (1975). [Int. to X, 74, 77]

Palm, E., Øiann, H.: Contribution to the theory of cellular thermal convection. J. Fluid Mech. **19**, 353 (1964). [77]

Palm, E., Weber, J.E., Kvernvold, O.: On steady convection in a porous medium. J. Fluid Mech. **54**, 153 (1972). [86, 92]

Patil, R., Rudraiah, U.: Instability of hydromagnetic thermoconvective flow through porous medium. J. Appl. Mech. **40**, 879 (1973). [Add. to IX]

Pearson, J.R.A.: On convection cells induced by surface tension. J. Fluid Mech. **4**, 489 (1958). [60]

Pearson, J.R.A.: Instability in Non-Newtonian Flow. Ann. Rev. Fluid Mech. **8**, Palo Alto: Annual Reviews Inc., 1976. [95]

Pellew, A., Southwell, R.V.: On maintained convective motion in a fluid heated from below. Proc. Roy. Soc. A **176**, 312 (1940). [62]

Petrie, C.J.S., Denn, M.M.: Instabilities in polymer processing. AIChE J., **22**, 209 (1976). [95]

Pimbley, G. H., Jr.: An Analysis of the Rayleigh-Taylor problem of superposed fluids. The inviscid incompressible irrotational case; steady-state solutions. Los Alamos Science Laboratory LA4839, UC32 and 34, Aug. 1972. [Notes to XIV]

Pipkin, A. C.: Small finite deformations of viscoelastic solids. Rev. Mod. Phys. **36**, 1034 (1964). [93, 94]

Pipkin, A. C.: Small displacements superposed on viscometric flow. Trans. Soc. Rheology. **12**, 397 (1968). [93]

Pipkin, A. C., Owen, D. R.: Nearly viscometric flows. Phys. Fluids. **10**, 836 (1967). [93]

Pipkin, A. C., Tanner, R. I.: A survey of theory and experiment in viscometric flows of viscoelastic liquids in Chapter VI of Mechanics Today, Vol. I, 1972. [93, 95]

Pitts, E.: The stability of pendent liquid drops. Part 1. Drops formed in a narrow gap. J. Fluid Mech. **59**, 753 (1973). [98]

Pitts, E.: The stability of pendent liquid drops. Part 2. Axial symmetry J. Fluid Mech. **63**, 487 (1974). [98]

Prandtl, L.: Essentials of Fluids Dynamics. London: Blackie 1952. [64]

Rabinowitz, P. H.: Existence and nonuniqueness of rectangular solutions of the Bénard problem. Arch. Rational Mech. Anal. **29**, 32 (1968). [62, 77]

Rabinowitz, P. H.: A priori bounds for some bifurcation problems in fluid dynamics. Arch. Rational Mech. Anal. **49**, 270 (1973). [71]

Rayleigh, Lord: On convection currents in a horizontal layer of fluid, when the higher temperature is on the under side. Phil. Mag. **32**, 529 (1916). Collected papers, **6**, 432. [54, 60]

Reid, W. H., Harris, D. L.: Some further results on the Bénard problem. Phys. Fluids **1**, 102 (1958). [62]

Rionero, S.: Sulla stabilità asintotica in media in magnetoidinamica. Ann. Mat. Pura Appl. **LXXVI**, 75 (1967). [Add. to IX]

Rionero, S.: Metodi variazionali per la stabilità asintotica in media in magnetoidrodinamica. Ann. Mat. Pura Appl. **LXXVIII**, 339 (1968A). [Add. to IX]

Rionero, S.: Sulla stabilità magnetoidrodinamica non lineaire asintotica in media con vari tipi di condizioni al contorno. Ricerche di Mat. **XVII**, 64 (1968B). [Add. to IX]

Rionero, S.: Sulla stabilità asintotica in media nella dinamica dei miscugli fluidi. Boll. U.M.I., **4**, 364 (1971A). [58]

Rionero, S.: Sulla stabilità magnetofluidodinamica non lineare asintotica in media in presenza di effetto Hall. Ricerche di Mat. **XX**, 285 (1971B). [Add. to IX]

Rivlin, R. S.: Solution of some problems in the exact theory of visco-elasticity. J. Rational Mech. Anal. **5**, 179 (1956A). [93]

Rivlin, R. S.: Further remarks on the stress-deformation relations for isotropic materials. J. Rational Mech. Anal. **4**, 681 (1956B). [94]

Rivlin, R. S., Ericksen, J. L.: Stress-deformation relations for isotropic materials. J. Rational Mech. Anal. **4**, 323 (1955). [94]

Rivlin, R. S., Sawyers, K. N.: Nonlinear continuum mechanics of viscoelastic fluids. Ann. Rev. of Fluid Mech. **3**, 117 (1971). [93]

Sani, R. L.: Ph. D. Thesis. Minneapolis: Dept. of Chem. Eng., Univ. of Minn. 1963. [67]

Sani, R. L.: On the non-existence of subcritical instabilities in fluid layers heated from below. J. Fluid Mech. **20**, 315 (1964). [62]

Sani, R. L.: On the finite amplitude roll-cell disturbances in a fluid layer subjected to heat and mass transfer. A. I. Ch. E. J. **11**, 971 (1965). [69]

Scanlon, J. W., Segel, L. A.: Finite amplitude cellular convection induced by surface tension. J. Fluid Mech. **30**, 149 (1967). [Int. to X]

Schechter, R. S., Velarde, M. G., Platten, J. K.: The two-component Bénard problem: Advances in Chemical Physics (Ed.: I. Prigogine and S. A. Rice) **26**, 265. New York: Wiley 1974. [65]

Schlüter, A., Lortz, D., Busse, F. H.: On the stability of steady finite amplitude cellular convection. J. Fluid Mech. **23**, 129 (1965). [Int. to XI, 75, 77]

Schmidt, R. J., Saunders, O. A.: On the motion of a fluid when heated from below. Proc. Roy. Soc. A **165**, 216 (1938). [92]

Schneider, K. J.: Investigation of the influence of free thermal convection on heat transfer through granular material. 11th Int. Cong. of Refrigeration (Munich), Paper, II-4, Oxford: Pergamon Press 1963.

Schowalter, W. R.: Mechanics of Non-Newtonian Fluids. Pergamon Press (forthcoming). [93]

Schwab, T.H.: Dynamic spatial patterns generated by transport and transformation. Ph. D. Thesis. Dept. Chem. Eng., University of Minn. 1974.

Schwiderski, E.W., Schwab, H.J.: Convection experiments with electrolytically heated fluid layers. J. Fluid Mech. **48**, 703 (1971). [75]

Schwiderski, E.W., (with Appendix by M. Jarnagin): Bifurcation of convection in internally heated fluid layers. Phys. Fluids **15**, 1882 (1972). [Int. to X, 77]

Scriven, L.E., Sternling, C.V.: On cellular convection driven by surface-tension gradients: effects of mean surface tension and surface viscosity. J. Fluid Mech. **19**, 321 (1964). [60]

Segel, L.A.: Nonlinear hydrodynamic stability theory and its application to thermal convection and curved flows in Non-Equilibrium Thermodynamics: Variational Techniques and Stability. (Ed.: R.J. Donnelly, I. Prigogine and R. Herman). University of Chicago Press 1966. [72, 82]

Segel, L.A., Stuart, J.T.: On the question of the preferred mode in cellular thermal convection. J. Fluid Mech. **13**, 289 (1962). [Int. to X, 77]

Shir, C.C., Joseph, D.D.: Convective instability in a temperature and concentration field. Arch. Rational Mech. Anal. **30**, 38 (1968). [58, 67]

Shirtcliffe, T.G.L.: Lake Bonney, Antarctica: Cause of the elevated temperatures. J. Geophys. Res. **69**, 5257 (1964). [65]

Slemrod, M.: A hereditary partial differential equation with applications in the theory of simple fluids. Arch. Rational Mech. Anal. (forthcoming). [93, 94, 95]

Smith, K.A.: On convective instability induced by surface-tension gradients. J. Fluid Mech. **24**, 401 (1966). [60]

Sokolov, M., Tanner, R.I.: Convective stability of a general viscoelastic fluid heated from below. Phys. Fluids **15**, 534 (1972). [94]

Sorokin, V.S.: Variational method in the theory of convection. (In Russian) Prikl. Mat. Mekh. **17**, 39 (1953). [62]

Sorokin, V.S.: Stationary motions in a fluid heated from below. (In Russian) Prikl. Mat. Mekh. **18**, 197 (1954). [62]

Sparrow, E.M., Goldstein, R.J., Jonsson, V.K.: Thermal instability in a horizontal fluid layer: effect of boundary conditions and non-linear temperature profile. J. Fluid Mech. **18**, 513 (1964). [75]

Spencer, A.J.M., Rivlin, R.S.: Further results in the theory of matrix polynomials. Arch. Rational Mech. Anal. **4**, 214 (1959/60). [94]

Spiegel, E.A., Veronis, G.: On the Boussinesq approximation for a compressible fluid. Astrophys. J. **131**, 422 (1960). [54]

Stern, M.E.: The "salt fountain" and thermohaline convection. Tellus **12**, 172 (1960). [65]

Sternling, C.V., Scriven, L.E.: Interfacial turbulence: hydrodynamic instability and the Marangoni effect. A.I.Ch.E. Jour., **5**, 514 (1959). [60]

Stommel, H., Aarons, A.B., Blanchard, D.: An oceanographical curiosity: the perpetual salt fountain. Deep Sea Res. **3**, 152 (1956). [65]

Stork, K., Müller, U.: Convection in boxes: experiments. J. Fluid Mech. **54**, 599 (1972). [62]

Stork, K., Müller, U.: Convection in boxes. An experimental investigation in vertical cylinders and annuli. J. Fluid Mech. **71**, 231 (1975). [62, 73]

Straus, J.M.: Large amplitude convection in porous media. J. Fluid Mech. **64**, 51 (1974). [81]

Stuart, J.T.: On the cellular patterns in thermal convection. J. Fluid Mech. **18**, 481 (1964). [77]

Swallow, J.C., Crease, J.: Hot salty water at the bottom of the Red Sea. Nature **205**, 165 (1965). [65]

Tabor, H.: Large-area solar collections for power production. Solar Energy **7**, 189 (1963). [65]

Tabor, H., Matz, R.: Solar pond project. Solar Energy **9**, 177 (1965). [65]

Taylor, G.I.: The instability of liquid surfaces when accelerated in a direction perpendicular to their planes. I. Proc. Roy. Soc. A **201**, 192 (1950). [Notes to XIV]

Ting, T.W.: Certain non-steady flows of second order fluids. Arch. Rational Mech. Anal. **14**, 1 (1963). [94]

Tippelskirch, H.: Über Konvektionszellen, insbesondere im flüssigen Schwefel. Beitr. Phys. Atmos. **29**, 37 (1956). [77]

Tritton, D.J., Zarraga, M.N.: Convection in horizontal layers with internal heat generation. Experiments. J. Fluid Mech. **30**, 21 (1967). [75]

Truesdell, C.: The mechanical foundations of elasticity and fluid dynamics. J. Rational Mech. Anal. **1**, 125 (1952). [94]

Truesdell, C.: Rational fluid mechanics 1687. Editors introduction to Vol. II, 12 of Euler's Opera Omnia. Orell Füssli, Zürich, 1954. [59]

Truesdell, C.: The meaning of viscometry in fluid dynamics. Ann. Rev. of Fluid Mech. **6**, 111, Palo Alto: Annual Reviews Inc., 1974. [93]

Truesdell, C., Noll, W.: The non-linear field theories of mechanics. Handbuch der Physik III/3. Berlin-Heidelberg-New York: Springer 1965. [93, 94, 95]

Turner, J.S.: Buoyancy Effects in Fluids. Cambridge University Press 1973. [65]

Ukhovskii, M.R., Yudovich, V.I.: On the equations of steady-state convection. J. Appl. Math. Mech. **27**, 432 (1963). [62]

Veronis, G.: Penetrative convection. Astrophys. J. **137**, 641 (1963). [54, Int. to X, 70]

Veronis, G.: On finite amplitude instability in thermohaline convection. J. Marine Res. **23**, 1 (1964). [67, 69]

Veronis, G.: Effect of a stabilizing gradient of solute on thermal convection. J. Fluid Mech. **34**, 315 (1968). [69]

Vertgeim, B.A.: On the condition for the formation of convection in a binary mixture. (In Russian) Prikl. Mat. Mekh. **19**, 745 (1955). [54]

Ward, J.C.: Turbulent flow in porous media. J. Hydraul. Div. Proc. A.S.C.E. **90** (HY5), 1 (1964). [70]

Weatherburn, C.E.: Differential Geometry of Three Dimensions. Cambridge University Press 1927 [55]

Weinberger, H.: The physics of the solar pond. Solar Energy **VIII**, No. 2, 45 (1964). [65]

Westbrook, D.R.: The stability of convective flow in a porous medium. Phys. Fluids **12**, 1547 (1969). [78]

Yih, C.-S.: Spectral theory of Taylor vortices, Part I: Structure of unstable modes. Arch. Rational Mech. Anal. **46**, 218 (1972A). [75]

Yih, C.-S.: Spectral theory of Taylor vortices, Part II: Proof of non-oscillation. Arch. Rational Mech. Anal. **47**, 288 (1972B). [75]

Yudovich, V.I.: On the origin of convection. J. Appl. Math. Mech. **30**, 1193 (1966). [73, 77]

Subject Index

Springer Tracts in Natural Philosophy

Vol. 1 Gundersen: Linearized Analysis of One-Dimensional Magnetohydrodynamic Flows.
With 10 figures. X, 119 pages. 1964.

Vol. 2 Walter: Differential- und Integral-Ungleichungen und ihre Anwendung bei Abschätzungs- und Eindeutigkeitsproblemen
Mit 18 Abbildungen. XIV, 269 Seiten. 1964.

Vol. 3 Gaier: Konstruktive Methoden der konformen Abbildung
Mit 20 Abbildungen und 28 Tabellen. XIV, 294 Seiten. 1964.

Vol. 4 Meinardus: Approximation von Funktionen und ihre numerische Behandlung
Mit 21 Abbildungen. VIII, 180 Seiten, 1964.

Vol. 5 Coleman, Markovitz, Noll: Viscometric Flows of Non-Newtonian Fluids. Theory and Experiment
With 37 figures. XII, 130 pages. 1966.

Vol. 6 Eckhaus: Studies in Non-Linear Stability Theory
With 12 figures. VIII, 117 pages. 1965.

Vol. 7 Leimanis: The General Problem of the Motion of Coupled Rigid Bodies About a Fixed Point
With 66 figures. XVI, 337 pages. 1965.

Vol. 8 Roseau: Vibrations non linéaires et théorie de la stabilité
Avec 7 figures. XII, 254 pages. 1966.

Vol. 9 Brown: Magnetoelastic Interactions
With 13 figures. VIII, 155 pages. 1966.

Vol. 10 Bunge: Foundations of Physics
With 5 figures. XII, 312 pages. 1967.

Vol. 11 Lavrentiev: Some Improperly Posed Problems of Mathematical Physics
With 1 figure. VIII, 72 pages. 1967.

Vol. 12 Kronmüller: Nachwirkung in Ferromagnetika
Mit 92 Abbildungen. XIV, 329 Seiten. 1968.

Vol. 13 Meinardus: Approximation of Functions: Theory and Numerical Methods
With 21 figures. VIII, 198 pages. 1967.

Vol. 14 Bell: The Physics of Large Deformation of Crystalline Solids
With 166 figures. X, 253 pages. 1968.

Vol. 15 Buchholz: The Confluent Hypergeometric Function with Special Emphasis on its Applications
XVIII, 238 pages. 1969.

Vol. 16 Slepian: Mathematical Foundations of Network Analysis
XI, 195 pages. 1968.

Vol. 17 Gavalas: Nonlinear Differental Equations of Chemically Reacting Systems
With 10 figures. IX, 107 pages. 1968.